Aggregation in Large-Scale Optimization

Applied Optimization
Volume 83

Series Editors:

Panos M. Pardalos
University of Florida, U.S.A.

Donald W. Hearn
University of Florida, U.S.A.

Aggregation in Large-Scale Optimization

by

Igor Litvinchev

and

Vladimir Tsurkov

Computing Center,
Russian Academy of Sciences,
Moscow, Russia

SPRINGER-SCIENCE+BUSINESS MEDIA, B.V.

Electronic Services <http://www.wkap.nl>

Library of Congress Cataloging-in-Publication

Litvinchev, Igor / Tsurkov, Vladimir
Aggregation in Large-Scale Optimization
ISBN 978-1-4613-4812-2 ISBN 978-1-4419-9154-6 (eBook)
DOI 10.1007/978-1-4419-9154-6

Printed on acid-free paper.

Contents

Preface

When analyzing systems with a large number of parameters, the dimension of the original system may present insurmountable difficulties for the analysis. It may then be convenient to reformulate the original system in terms of substantially fewer aggregated variables, or macrovariables. In other words, an original system with an n-dimensional vector of states is reformulated as a system with a vector of dimension much less than n. The aggregated variables are either readily defined and processed, or the aggregated system may be considered as an approximate model for the original system. In the latter case, the operation of the original system can be exhaustively analyzed within the framework of the aggregated model, and one faces the problems of defining the rules for introducing macrovariables, specifying loss of information and accuracy, recovering original variables from aggregates, etc.

We consider also in detail the so-called iterative aggregation approach. It constructs an iterative process, at every step of which a macroproblem is solved that is simpler than the original problem because of its lower dimension. Aggregation weights are then updated, and the procedure passes to the next step. Macrovariables are commonly used in coordinating problems of hierarchical optimization.

In this book we deal with large-scale extremal problems. A fundamental issue in the use of optimization models is the tradeoff between the level of detail to employ and the ease of analyzing and solving the problem. Special approaches such as aggregation-disaggregation and decomposition have been developed to facilitate solution and data manipulation for different levels of decision makers. Aggregation-disaggregation techniques provide a set of tools to cope with large optimization problems by: a) combining data, b) using an auxiliary (aggregated) problem which is reduced in size and/or complexity relative to the original problem, and c) analyzing the error due to the fact that a simpler problem than the original one has been solved. In iterative aggregation all or some of these steps are realized repeatedly.

The first publications on aggregation appeared about 50 years ago. The book contains 3 chapters and focuses on results obtained in this area for the last one and a half decades in the Complex Systems Department of the Computing Center of Russian Academy of Sciences.

Chapter 1 addresses the error arising from to the use of the aggregated problem instead of the original detailed one. Applying a specific transformation, called disaggregation, an optimal solution of the aggregated problem yields a feasible (disaggregated) solution to the original problem. The aggregation error is measured by the difference between the original optimal objective value and the value for the disaggregated solution. To calculate the aggregation error exactly requires solution of both the original and the aggregated problems. The bounds on the aggregation error are derived without solving the original problem. Two types of the bounds are considered - a priori and a posteriori. A priori bounds are calculated after aggregation, but before optimization of the aggregated problem, whereas a posteriori error bounds require an optimal solution (primal and/or dual) of the aggregated problem. In this chapter the bounds for the aggregation error are studied for linear, convex, and some special-structured and integer programming problems. Illustrative examples and numerical indications are provided to compare the bounds and estimate the correspondence to the actual error.

Chapter 2 is devoted to iterative aggregation. The main question studied here is how to adjust the aggregated problem to reduce the aggregation error. Iterative aggregation procedures are realized in a two-level solution scheme. The upper level solves the aggregated problem for the fixed values of certain aggregation parameters, characterizing the combination of the original data, while the lower level updates these parameters. For various special-structured problems iterative aggregation results in decomposition algorithms, splitting either upper or lower level problems, or both. To decompose the problems special types of aggregation/disaggregation are considered. The resemblance and the difference between this approach and classical decomposition schemes is analyzed. The convergence of iterative aggregation schemes to the original optimal solution is studied. Iterative aggregation techniques are justified and applied to mathematical programming and optimal control problems, including the problems arising in mathematical physics. In the later case both spatial and time aggregation of controls is considered. Illustrative and numerical examples are provided to demonstrate the effectiveness of iterative aggregation techniques.

In Chapter 3 parametric optimization problems are studied. Here the effect of changes in problem data (parameters) on the optimal objective needs to be analyzed and the parameters resulting in better optimal objective value need to be identified. Thus the computation burden is associated not only with optimization itself, but also with searching the parameters resulting in a preferable optimization problem. The question is how to construct a parametric aggregated problem having the following property. If a change of parameters results in an improvement of the aggregated problem, then it improves the original problem as well, and if the aggregated problem has no improving changes of parameters, then the original problem also has none. Such an aggregation is said to be consistent with respect to parameter variations. The approach of this chapter is as follows. Suppose a few parameters are changed. The data associated with the unchanged parameters are aggregated in a special way to form the consistent aggregated problem. The latter problem is used to find the changes of parameters resulting in an improvement of the original non-aggregated problem. In this way the parametric analysis of the original problem is, in a certain sense, approximated by the analysis of the parametric aggregated problem. Consistent aggregated problems are constructed for the linear and convex programming, optimal control and integer programming. Examples demonstrating the usefulness of consistent aggregation are provided.

We would like to thank our former colleagues from Computing Center of Russian Academy of Sciences, Victor Ven and Yury Krysov, for fruitful cooperation that gave rise to many results presented in this book. We kindly acknowledge that without the input from their side this book could not have been written. We greatly enjoyed stimulating discussions with many colleagues from all over the world in the past years. We mention Ada Alvarez and Oscar Chacon (both from the State University of Nuevo Leon, Mexico), Kurt Jörnsten (Norwegian School of Economics and Business Administration, Bergen), Rainer Leisten (Gerhard Mercator University of Duisburg, Germany), Socorro Rangel and Geraldo Silva (both from Sao Paulo State University, Brasil), David Rogers (University of Cincinnati, USA), Sverre Storøy (University of Bergen, Norway), and Jørgen Tind (University of Copenhagen, Denmark). We are also grateful to John Martindale and his assistants (Kluwer Academic Publishers) for their kind practical help, and to Panos Pardalos, the managing editor of the series "Applied Optimization" for his deep interest in modern optimization methods and his constant encouragement.

The reader is assumed to be acquainted with mathematical programming, optimal control, and with mathematical physics. References to additional literature are given in comments to the chapters. The material of each chapter can be considered independently.

In the book, two numbers are used to identify each formulas and theorems. The first number denotes the number of the section in this chapter, the second is the number of the formula or theorem. This should be specially noted when a reference to a theorem or a formula from another chapter is made.

This book continues and extends the monograph *Tsurkov V.I., Large-Scale Optimization — Problems and Methods, Dordrecht, Boston, London: Kluwer Acad. Publ., 2001.*

Chapter 1

Aggregated Problem and Bounds for Aggregation

Aggregation-disaggregation techniques provide a number of methods for handling large optimization problems by: a) combining data, b) using auxiliary (aggregated) problems which are reduced in size and/or complexity relative to the original problem, and c) analyzing the error caused by the replacement of the original problem by a simpler one. In this chapter, we consider realizations of these steps for various classes of optimization problems. The main attention is focused on estimating the aggregation error, which does not require solving the original problem.

In Section 1 we study aggregation error bounds for variable aggregated linear programs, while Section 3 is concerned with the case of convex programming. In Sections 2 and 5 the aggregation bounds are specified and further strengthened for the generalized transportation problem and the generalized assignment problem, respectively. Section 4 is devoted to an aggregation approach in integer programs.

§1. Linear Programming Aggregation

The approach of this section is as follows. The aggregated problem for a linear programming problem is formed by partitioning the variables of the original program and replacing the columns and objective coefficients in each group by their weighted averages. Applying a special transformation, called disaggregation, to an optimal solution of the aggregated problem yields a feasible (disaggregated) solution of the original problem. The aggregation error is measured by the difference between the original optimal objective value and the optimal value for the disaggregated solution. The exact calculation of this aggregation error requires solving both the original and aggregated problems, but bounds for the aggregation error can be derived without solving the original problem.

1.1. Definitions and Preliminary Results

Consider the original linear programming problem in the form

$$z^* = \max cx,$$
$$\text{s.t. } Ax \leq b, \tag{1.1}$$
$$x \geq 0,$$

where $x = \{x_j\}$ is an n-vector of decision variables, $c = \{c_j\}$ is an n-vector, $b = \{b_j\}$ is an m-vector, $A = [a_{ij}]$ is an $m \times n$ matrix having columns a_j. It is assumed that (1.1) has a finite optimal solution.

We restrict ourselves to column aggregation, i.e. we consider only the case of aggregating variables in (1.1). Let σ be a partition of the column indices $\{1, ..., n\}$ into a set of clusters S_k, $k = 1, ..., K$, such that $S_k \cap S_p = \varnothing$, $\cup_{k=1}^{K} S_k = \{1, ..., n\}$. Let $|S_k| = n_k$, $\sum\limits_k n_k = n$. Define A^k to be the submatrix of A consisting of those columns whose indices are in S_k. Define the subvectors c^k and x^k similarly. Consider a nonnegative n_k-vector $g^k = \{g_j^k\}$, fulfilling the following normalizing condition

$$g \in G = \left\{ g \mid \sum_{j \in S_k} g_j^k = 1,\ g_j^k \geq 0,\ k = 1, ..., K,\ j \in S_k \right\},$$

and set

$$\overline{A}^k = A^k g^k,\ \overline{c}^k = c^k g^k,\ k = 1, ..., K,$$

such that $\overline{A}^k$ is a column m-vector equal to the linear combination of the columns $a_j, j \in S_k$ with the coefficients g_j^k, and $\overline{c}^k$ is a scalar defined similarly. Define the matrix $\overline{A} = [\overline{A}^1, ..., \overline{A}^K] \equiv [\overline{a}_{ik}]$, and the vector $\overline{c} = \{\overline{c}^1, ..., \overline{c}^K\}$. Then the problem

$$\overline{z} = \max \overline{c}X,$$
$$\text{s.t. } \overline{A}X \leq b, \tag{1.2}$$
$$X \geq 0$$

defines the (column or variable) *aggregated problem* corresponding to (1.1). Here X is a K-vector of aggregated variables. For a given original problem, then, an aggregated problem is determined by the pair (σ, g). It is assumed that (σ, g) has been chosen so that (1.2) is feasible. We denote its optimal primal and dual solutions as $(\overline{X}, \overline{u})$. The vectors g^k are called *weighting vectors* or the *weights of aggregation.*

Remark 1.1. Consider a $n \times K$ matrix $g = blockdiag\{g^k, k = 1, ...K\}$ satisfying the normalizing condition $g \in G$. The kth diagonal block of the matrix g is then a column n_k-vector g^k with nonnegative components whose sum equals 1. We refer to g as the *weighting matrix.* It is easy to verify that the aggregated problem (1.2) can be obtained by substituting the linear transformation $x = gX$ into the original problem (1.1), and hence (1.2) can be written in the form

$$\begin{aligned} \overline{z} &= \max cgX, \\ \text{s.t. } AgX &\leq b, \\ X &\geq 0. \end{aligned} \tag{1.3}$$

Formally, the condition $x \geq 0$ in (1.1) should take the form $gX \geq 0$ in the aggregated problem (1.3). But since $g \in G$, then every vector g^k has at least one positive component and hence the condition $gX \geq 0$ is reduced to $X \geq 0$. Moreover, the conditions $g \in G$ and $x = gX$ yield $X_k = \sum_{l \in S_k} x_l^k$ and hence the aggregated variable is equal to the sum of all components in the cluster. In general, we can use an arbitrary weighting matrix or other normalizations. The choice $g \in G$ simplifies the aggregated problem; in addition, under such a choice, our formulations for linear programming aggregation are in a one-to-one correspondence with those of Zipkin [33].

The simplest way to deduce a feasible solution of the original problem (1.1) from an aggregated solution of (1.2) is to use the *fixed-weight disaggregation*: $\overline{x}_j = g_j^k \overline{X}_k$, $j \in S_k$, $k = 1, ..., K$. The (fixed-weight) *disaggregated solution* $\overline{x} = \{\overline{x}_j\}$ is feasible to (1.1) since

$$b \geq \overline{A}\overline{X} \equiv \sum_k \overline{A}^k \overline{X}_k \equiv \sum_k A^k g^k \overline{X}_k \equiv A\overline{x}.$$

Moreover, $c\overline{x} = \overline{z} \leq z^*$.

An improved disaggregated solution can be obtained by performing the *optimal disaggregation* of $\overline{X}$, i.e. by solving the subproblems

$$\begin{aligned} z_k(\overline{X}_k) &= \max c^k x^k, \\ \text{s.t. } A^k x^k &\leq \overline{A}^k \overline{X}_k, \\ x^k = \{x_j^k,\ j \in S_k\} &\geq 0 \end{aligned} \tag{1.4}$$

for each $k = 1, ..., K$. The solution of these subproblems $\widetilde{x} = \{\widetilde{x}^k\}$ yields a feasible solution to the original problem, since $A\widetilde{x} \equiv \sum_k A^k \widetilde{x}^k \leq \sum_k \overline{A}^k \overline{X}_k \equiv \equiv \overline{AX} \leq b$. Moreover, the optimal disaggregation $\widetilde{x}$ is at least as good as the fixed-weight disaggregation $\overline{x}$, since every fixed-weight disaggregated solution (restricted to the subproblem k) defines a feasible solution to (1.4). Hence the subproblems (1.4) are always feasible and $c\widetilde{x} \geq c\overline{x} = \overline{z}$.

For any partition σ there exists an *optimal weighting* vector g^* which yields $\overline{z} = z^*$. Unfortunately, construction of g^* requires an optimal solution to (1.1), x^*. Let

$$X_k^* = \sum_{j \in S_k} x_j^*,$$

$$g_j^{k*} = \begin{cases} \text{arbitrary weighting vector, if } X_k^* = 0, \\ x_j^* / X_k^*, \text{ if } X_k^* > 0. \end{cases}$$

Then g^{k*} is optimal in the sense that X^* is feasible for the aggregated problem (1.2) for $g = g^*$, and $\overline{c}X^* = cx^* = z^*$.

In view of the fact that optimal weights exist for any given partition of the variables, methods of iteratively improving a given set of weights have been proposed. Some of them will be considered later in Chapter 2. Most of them use dual information from the aggregated problem. The following proposition characterizes the optimal weights from this point of view.

Proposition 1.1 [29]. *The optimal solution to the aggregated problem constructed with the use of optimal weights is nondegenerate if and only if the optimal solution to the original problem is nondegenerate and its basic variables are all aggregated into separate variables.*

Proof. Let x^* be a nondegenerate optimal solution of (1.1), and suppose, without loss of generality, that the m first components are basic. The vector b may then be expressed as

$$x_1^* a_1 + x_2^* a_2 + ... + x_m^* a_m = b. \tag{1.5}$$

Assume further that we have a partition σ where the first q columns of A, $q \geq 2$, are aggregated to be the first column of $\overline{A}$ and the remaining $m - q$ columns in (1.5) are each in separate groups, i.e.

$$S_1 = \{1, ..., q\}, \ n_1 = q,$$

$$S_k = \{q + k - 1\}, \ n_k = 1, \ k = 2, ..., m - q + 1.$$

Using optimal weights the aggregated version of (1.5) becomes:

$$X_1^*\overline{A}^1 + X_2^*\overline{A}^2 + ... + X_{m-q+1}^*\overline{A}^{m-q+1} = b, \tag{1.6}$$

where

$$\overline{A}^1 = \frac{1}{\sum_{j\in S_1} x_j^*}(x_1^*a_1 + ... + x_q^*a_q)$$

and

$$\overline{A}^k = a_{q+k-1},\ k = 2, ..., m-q+1.$$

Since all the columns a_j's in (1.5) are linearly independent, so are all the columns $\overline{A}^l$'s in (1.6). But since the number of $\overline{A}^l$'s in (1.6) is $m-q+1$, and since $m-q+1 < m$ for $q \geq 2$, we see that b lies in the linear hull of a set of less than m linear independent columns of $\overline{A}$, which means that X^* is a degenerate solution (see [22]).

If x^* is degenerate, we see immediately that X^* will also be degenerated even if all of the basic variables of x^* are aggregated into separate groups.

On the other hand, if x^* is nondegenerate and all of the basic variables are aggregated into separate variables, (1.5) and (1.6) will be identical.

Thus we may expect that in the vicinity of the optimal weights the dual information may be nonunique depending on the partition of the variables. This follows immediately from Proposition 1.1:

Corollary 1.1. *When two or more optimal basic variables of the original problem (1.1) are aggregated into the same variable of the aggregated problem (1.2) using optimal weights, the optimal solution of the aggregated problem is degenerate.*

An appropriate measure for the loss in accuracy caused by aggregation is the difference between the actual optimal value of the objective function and the optimal objective value for the aggregated problem. Of course, exactly calculate this difference requires solving both the original and aggregated problems. However, bounds for the loss of accuracy can be found without solving the original problem. Usually, two types of error bounds are considered, a priori and a posteriori. *A priori error bounds* are calculated after the aggregation but before the optimization of the aggregated problem, whereas calculating *a posteriori error bounds* requires knowing an optimal solution (primal and/or dual) of the aggregated problem.

1.2. A Posteriori Bounds

The objective function value of the aggregated problem (1.2), as well as the sum of the optimal objective function values of all optimally disaggregated subproblems (1.4), provide lower bounds on the actual optimal objective of the original problem:

$$\overline{z} \leq \sum_{k=1}^{K} z_k(\overline{X}_k) \leq z^*.$$

This is so because the fixed-weight disaggregated solution, as well as the optimal disaggregated solution, are feasible solutions to the original problem.

To obtain upper bounds for z^*, we make the following assumption about an (unknown)optimal solution x^* to (1.1).

Assumption 1.1. Let $\sigma' = \{S'_k, k = 1, ..., K'\}$ be a partition of $\{1, ..., n\}$, not necessary the same as aggregation partition σ. Suppose that d_j, $j = 1, ...n$ are known positive numbers, p_k, $k = 1, ..., K'$ are known nonnegative numbers, and there is an (unknown) optimal solution x^* such that

$$\sum_{j \in S'_k} d_j x_j^* \leq p_k, \quad k = 1, ..., K'. \tag{1.7}$$

Proposition 1.2 [33]. *If Assumption 1.1 holds for some x^* optimal for (1.1), then for any m-vector $u \geq 0$* we have

$$z^* \leq ub + \sum_{k=1}^{K'} \max_{j \in S'_k} \left[\frac{c_j - ua_j}{d_j}\right]^+ p_k, \tag{1.8}$$

where $[\alpha]^+ = \max\{0, \alpha\}$.

Proof. For any $u \geq 0$ we have

$$z^* = cx^* \leq cx^* + u(b - Ax^*) = ub + (c - uA)x^*.$$

The value $(c - uA)x^*$ can be bounded by an estimation of the original optimal solution x^*. Then for arbitrary $u \geq 0$ it follows

$$(c - uA)x^* = \sum_{k=1}^{K'} \sum_{j \in S'_k} (c_j - ua_j)x_j^* \leq$$

$$\leq \sum_{k=1}^{K'} \max_{j \in S'_k} \left[\frac{c_j - ua_j}{d_j}\right] \sum_{j \in S'_k} d_j x_j^* \leq \sum_{k=1}^{K'} \max_{j \in S'_k} \left[\frac{c_j - ua_j}{d_j}\right]^+ p_k.$$

Combining the above inequalities yields the desired upper bound for z^*.

In the sequel, we suppose the partition σ' and constraints (1.7) to be unchanged during the bound computation process and consider the approaches dealing with the generation and evaluation of the duals u.

Zipkin [33] utilized the optimal dual solution $u = \overline{u}$ of the aggregated problem (1.2). By the duality theorem we have $\overline{u}b = c\overline{x} \equiv \overline{z}$ and hence by (1.8)

$$z^* \leq \overline{z} + \sum_{k=1}^{K'} \max_{j \in S'_k} \left[\frac{c_j - \overline{u}a_j}{d_j} \right]^+ p_k. \tag{1.9}$$

Mendelssohn [21] looked for the best (i.e. bound-minimizing) dual solution on the ray starting at the origin and passing through $\overline{u} : u = \rho\overline{u}$, where ρ is a non-negative stepsize. Substituting this u in (1.8) we get

$$z^* \leq \min_{\rho \geq 0} \left\{ \rho\overline{z} + \sum_{k=1}^{K'} \max_{j \in S'_k} \left[\frac{c_j - \rho\overline{u}a_j}{d_j} \right]^+ p_k \right\}. \tag{1.10}$$

Since the expression in the figure brackets in (1.10) is convex and piecewise linear in ρ, only the points of the non-differentiability of this expression should be analyzed to find the minimum in (1.10), i.e. the points of the form

$$\frac{c_j - \rho\overline{u}a_j}{d_j} = 0, \quad j = 1, ..., n, \tag{1.11}$$

$$\frac{c_j - \rho\overline{u}a_j}{d_j} = \frac{c_{j'} - \rho\overline{u}a_j}{d_{j'}}, \tag{1.12}$$

$j, j' = 1, ..., n$ and j, j' belong to the same cluster S'_k, $k = 1, ..., K'$.

Mendelssohn evaluates only (1.11). Knolmayer [11] points out that (1.12) has to be evaluated too in order to find the minimum in (1.10). In practice, some kind of line search procedure, e.g., a Fibonacci search, may be the most efficient approach to finding the optimal value of ρ and exhaustive determination of the candidate values for ρ is not worthwhile.

Shetty and Taylor [28] also start with $\overline{u}$ but they choose the vector of capacities b as a search direction. More precisely, to ensure the nonnegativity of the duals for some positive stepsize ρ they utilized the modified search direction

$$\tilde{b}_i = \begin{cases} 0 \text{ if } \overline{u}_i = 0 \text{ and } b_i < 0, \\ b_i \text{ otherwise} \end{cases}$$

and considered the duals in the form: $u = \overline{u} + \rho\tilde{b}$, $\rho \geq 0$. They argue that this choice tends to move $\overline{u}$ into the direction of the feasible set for the original dual problem. Note that if u is feasible to the original dual, then the second term in (1.8) is zero since $c_j - ua_j \leq 0$. The choice of Shetty and Taylor results in the bound

$$z^* \leq \overline{z} + \min_{\rho \geq 0, \overline{U} + \rho\tilde{b} \geq 0} \left\{ \sum_{k=1}^{K'} \rho b \tilde{b} + \max_{j \in S'_k} \left[\frac{c_j - (\overline{u} + \rho\tilde{b})a_j}{d_j} \right]^+ p_k \right\}. \tag{1.13}$$

Because of convexity and piecewise linearity in ρ of the expression in (1.13), the determination of the optimal stepsize can be handled similar to the previous bound.

Remark 1.2. The expression (1.8) is convex in $u \geq 0$, Therefore, the upper bound can be minimized with the use of standard convex programming methods. In contrast to the approaches considered above, these methods give dual solutions aimed only at minimizing the upper bound, the other characteristics of the problem being disregarded. Some (sub)gradient approaches to minimizing (1.8) with respect to u are outlined in [14]. These approach may give a slow rate of convergence if the number of constraints in the original problem is large; nevertheless, they might be used to further improve the dual solution generated by one of the other approaches.

1.3. A Priori Bounds

In fixed-weight aggregation we always have $z^* - \overline{z} \geq 0$. This difference can be estimated from above with the use of (1.9). To obtain an a priori bound, the values $c_j - \overline{u}a_j$ need to be estimated without optimizing the aggregated problem. Suppose that $\{s_j\}$ are numbers such that $s_j \geq c_j - \overline{u}a_j$, $j = 1, ..., n$, and (1.7) holds. Defining $s^k = \max_{j \in S'_k}\{s_j/d_j\}$, $k = 1, ..., K'$ and using (1.9) we obtain

$$z^* - \overline{z} \leq \varepsilon^a \equiv \sum_{k=1}^{K'} [s^k]^+ p_k. \tag{1.14}$$

The problem is to specify $\{s_j\}$ without knowledge of $\overline{u}$. If it is met, (1.14) provides an a priori bound on the accuracy loss.

For each $j \in S_k$, $k = 1, ..., K$, consider the following problem:

$$s_j \equiv c_j - \min_u ua_j, \tag{1.15}$$

$$\text{s.t. } u\overline{A}^k \geq \overline{c}^k,$$

$$u \geq 0.$$

If this problem is finite for all j, then $\{s_j\}$ yields a valid a priori bound, since $\overline{u}$ is feasible for the dual of the aggregated problem (1.2), and hence satisfies the constraints of (1.15). Note that (1.15) has only one linear constraint and can be solved easily.

General conditions for the problem (1.15) to be finite are stated in [33] using Farkas lemma. A particular case when (1.15) is finite for all j is $A \geq 0$ and $c \geq 0$. In this case,

$$s_j = c_j - \overline{c}^k \min_i \{a_{ij}/\bar{a}_{ik} \mid \bar{a}_{ik} > 0\}, \quad j \in S_k, \; k = 1, ..., K$$

and (1.14) becomes

$$\varepsilon^a = \sum_{k=1}^{K'} \left[\max_{i,j} \{(c_j - (\overline{c}^k a_{ij}/\bar{a}_{ik}))/d_j \mid j \in S'_k, \; \bar{a}_{ik} > 0 \right]^+ p_k. \tag{1.16}$$

Remark 1.3. (a) Suppose that $\sigma' = \sigma$. If (c_j, a_j), $j \in S_k$, are identical, then the kth term in (1.16) is zero. This result is intuitive, because, if identical variables are aggregated, there is no loss in accuracy. The same result holds if (c_j, a_j), $j \in S_k$ differ only by scalar multipliers. (b) If a_j, $j \in S_k$, are identical, then s^k, the kth term in (1.16), is the maximal weighted positive deviation of the c_j's, $j \in S_k$, from their weighted mean. And if c_j, $j \in S_k$, are identical, s^k can be interpreted as a measure of weighted maximal negative deviation of the $\{a_{ij}\}$ from their weighted means. (c) These observations suggest an interpretation of s^k as a measure of dispersion among $\{(c_j, a_j), \; j \in S_k\}$. The value ε_a will be smaller if the columns of each subset are less dissimilar in this sense. This conclusion supports the intuitive notion that one would rather aggregate entities which are "similar".

Problem (1.15) is a special case of the following general problem:

$$s_j \equiv c_j - \min_u u a_j,$$

subject to constraints which constitute a relaxation of the constraints of the dual to the aggregated problem (1.2), and yield a finite s_j. The following tactics suggest the range of possible variants of (1.15). For example, we can choose k arbitrarily in (1.15), i.e., not k for which $j \in S_k$. Moreover, to obtain a relaxation, we can use a restriction $(u\overline{A} - \overline{c})w \geq 0$, where w is any fixed nonnegative K-vector. Suppose $\overline{u}_i \leq q_i$, $i = 1, ..., m$, a priori. Then appending $u_i \leq q_i$, $i = 1, ..., m$ to (1.15) or its generalization we obtain the problem which is always finite and easy to solve.

1.4. Some Generalizations

In this section we demonstrate that many results of Sections 1.2 and 1.3 can be further strengthening by applying the Lagrange duality in a certain aggregation-disaggregation context.

To derive the upper bound (1.8) for z^* we have assumed that certain knapsack-type constraints of the form (1.7) are known explicitly (As- sumption 1.1). In this section we only make the following general assumption about an optimal solution to (1.1).

Assumption 1.2. A set $W \subseteq \mathbf{R}^n$ is known, such that $x^* \in W$.

We will refer to W as the *localization of* x^*, or simply the *localization.* The set W can be defined by manipulating the constraints of the original problem or by querying a decision maker. We distinguish between an *a priori localization*, , when the set W is defined *before* an aggregated problem has been solved, and *a posteriori localization*, , when W is defined or corrected/sharpened *after* an aggregated problem has been optimized.

To simplify further considerations, we assume that W is compact (closed and bounded) and contains the condition $x \geq 0$, unless otherwise specified.

1.4.1. A Posteriori Bounds. The following result can be treated as a generalization of Proposition 1.2.

Proposition 1.3. *If the assumption 1.2 holds for some x^* optimal to (1.1), then for any fixed m-vector $u \geq 0$ we have*

$$z^* \leq UB(u, W) \equiv \sum_{i=1}^{m} u_i b_i + \max_{x \in W} \left\{ \sum_j \left(c_j - \sum_{i=1}^{m} u_i a_{ij} \right) x_j \right\}. \tag{1.17}$$

Proof. By the definition of the localization we can attach the condition $x \in W$ to the constraints of the original problem (1.1) without changing its optimal solution, such that

$$z^* = \max cx,$$
$$\text{s.t. } Ax \leq b,$$
$$x \in W.$$

Let $L(x, u) = cx - u(Ax - b)$ be the Lagrangian function associated with the constraints $Ax \leq b$. Since the set W is compact, by the duality theorem (see, e.g., [12]) we have

$$z^* = \min_{u \geq 0} \max_{x \in W} L(x, u) \leq \max_{x \in W} L(x, u) \equiv UB(u, W)$$

for any $u \geq 0$.

Corollary 1.2. *Let some original constraints be included in the definition of W, such that*

$$W = \{x \mid \sum_j a_{ij}x_j \le b_i,\ i \in I \subseteq \{1,...,m\} \equiv M,\ x \in \widetilde{W}\},$$

where $\widetilde{W}$ is another localization of x^*. Then for any fixed $u \ge 0$ we have

$$\max_{x\in W}\{cx - \sum_{i\in M} u_i(\sum_j a_{ij}x_j - b_i)\} \equiv UB(u, W) \ge$$
$$\ge \max_{x\in W}\{cx - \sum_{i\in M\setminus I} u_i(\sum_j a_{ij}x_j - b_i)\} \ge z^*.$$

Proof. The last inequality is derived similar to the proof of Proposition 1.3 if we define the Lagrangian in the form

$$L(x,u) = cx - \sum_{i\in M\setminus I} u_i(\sum_j a_{ij}x_j - b_i),$$

including only those constraints which are not used in the definition of W. The first inequality holds since

$$\sum_{i\in M} u_i(\sum_j a_{ij}x_j - b_i) = \sum_{i\in I} u_i(\sum_j a_{ij}x_j - b_i) + \sum_{i\in M\setminus I} u_i(\sum_j a_{ij}x_j - b_i)$$

and $\sum_{i\in I} u_i(\sum_j a_{ij}x_j - b_i) \le 0$ for any $u \ge 0$ and $x \in W$.

The corollary shows that if a localization W is defined by some original constraints, then we should not include these constraints in the objective of the problem to compute the upper bound.

We will denote by $u^W \in \mathbf{R}^{m-|I|}$ a nonnegative vector containing the components u_i, $i \in M\setminus I$ of the nonnegative vector $u \in \mathbf{R}^m$. In other words, to form u^W we should delete from u all the components associated with the original constraints used to define W. We use the notation u^W to highlight that the dimension and the components of u^W depend on the localization W formed by some of the original constraints. Defining

$$UB(u^W, W) \equiv \sum_{i\in M\setminus I} u_i b_i + \max_{x\in W}\left\{\sum_j \left(c_j - \sum_{i\in M\setminus I} u_i a_{ij}\right) x_j\right\} \tag{1.18}$$

we have by Corollary 1.2 that $UB(u, W) \ge UB(u^W, W) \ge z^*$.

A critical question in upper bound computations is how to construct a proper localization W. From (1.17), (1.18) it follows that the sharper the set W, the stronger the upper bound. That is, for example, defining W by the original constraints, we should retain as many constraints as possible. Alternatively, localization W should be "simple" enough to guarantee that the computational burden associated with calculation of an upper bound is not too expensive. Below we consider some choices of W which result in analytical computation of the bound.

To compute the bounds by (1.17) or (1.18) we should maximize a linear objective under the condition $x \in W$. Note, that the form of this problem is independent on the way we have obtained the aggregated problem. Denote by δ_j the coefficients of the maximization problem in the right-hand side of (1.17), (1.18). The values δ_j depend on u and are different for the different types of bounds. But for the time being we are not interested in these differences and use the notation δ_j to highlight that these maximization problems have linear objectives for any fixed $u \geq 0$.

Let W be as follows

$$W = \left\{ x \mid \sum_j d_j x_j \leq p,\ 0 \leq x_j \leq r_j,\ j = 1, ..., n \right\},\ r_j > 0. \tag{1.19}$$

A linear constraint and upper bounds for variables could be taken either directly from the original problem or (and) obtained by constraint manipulations. Note, that the set W is convex, closed and bounded for any values of d_j, p. A localization W in the form (1.19) can be easily obtained, if the original problem has knapsack-type constraints $\sum\limits_j a_{ij} x_j \leq b_i,\ a_{ij}, b_i > 0$. We can put, for example, $r_j = \min_i \{b_i / a_{ij}\}$ and take one original constraint or a linear combination of the original constraints.

To compute an upper bound using the localization (1.19) we should solve the problem

$$\max \sum_j \delta_j x_j, \tag{1.20}$$

$$\text{s.t.} \ \sum_j d_j x_j \leq p,\ 0 \leq x_j \leq r_j,\ j = 1, ..., n.$$

For simple consideration we restrict ourselves only to the knapsack-like case with $d_j, p > 0$. The general case is treated similarly.

An optimal solution $\widehat{x}$ of (1.20) is easily obtained by the Kuhn-Tucker conditions. The form of an optimal solution reflects the following well-known "knapsack" idea: the objects in the knapsack should be placed beginning from the most "valuable" one until either the upper bound r_j is reached or the knapsack is completely full.

Let the variables be arranged such that $\delta_1/d_1 \geq \delta_1/d_1 \geq ....$ An optimal solution $\widehat{x}$ of (1.20) is such that

$$\delta_j \widehat{x}_j = [\delta_j]^+ \min \left\{ r_j, \frac{p - \sum_{l=1}^{j-1} d_l \widehat{x}_l}{d_j} \right\}, \quad j = 1, ..., n, \tag{1.21}$$

where the term $[\delta_j]^+$ guarantees that $\widehat{x}_j = 0$ for $\delta_j < 0$. If $\delta_j = 0$, then $\widehat{x}_j$ is nonunique, such that

$$0 \leq \widehat{x}_j \leq \min \left\{ r_j, \frac{p - \sum_{l=1}^{j-1} d_l \widehat{x}_l}{d_j} \right\}.$$

Assume now that instead of one linear constraint in (1.20) we have a number of independent linear constraints. Namely, let $\sigma' = \{S'_k,\ k = 1, ..., K'\}$ be a partition of $\{1, ..., n\}$, not necessary the same as aggregation partition σ, and W is as follows:

$$W_b = \left\{ x \mid \sum_{j \in S'_k} d_j x_j \leq p_k,\ k = 1, ..., K',\ 0 \leq x_j \leq r_j,\ j = 1, ..., n \right\}.$$

Then the problem (1.20) decomposes into K' independent subproblems of the same type and an optimal solution of each subproblem is computed similar to the previous case. The overall optimal objective is the sum of the subproblem's optimal objectives.

An important example, where we can easily construct decomposable localization, is the problem with generalized transportation constraints, which will be discussed in Section 2 of this Chapter. Various types of decomposable localizations can be obtained in special-structured problems, and play an important role in iterative decomposition-aggregation techniques studied in Chapter 2.

Remark 1.4. Let W_Z be the localization defined by the knapsack constraints (1.7) and additional restriction $x \geq 0$. Then it is not hard to verify that the bound (1.8) coincides with the upper bound $UB(u, W_Z)$. Since $W_b \subseteq W_Z$, we may conclude that $UB(u, W_b)$ is at least as good as Zipkin bound (1.8). Note that Zipkin [33] mentioned the possibility of deriving upper bounds by weak duality, but this approach was not used any further.

Assume that the bound $UB(u, W)$ has been calculated for a fixed $u \geq 0$, say $u = \overline{u}$. Then it is natural to look for another value of u to diminish the upper bound. In other words, we would like to solve (approximately) the problem $\min_{u \geq 0} UB(u, W)$.

Denote by $\theta(u)$ the objective-value function of the maximization problem in (1.17):

$$\theta(u) \equiv \max_{x \in W} \left\{ \sum_j \left(c_j - \sum_{i=1}^m u_i a_{ij} \right) x_j \right\}.$$

Assume that W is convex and the maximization problem has a unique optimal solution $\widehat{x}$ for $u = \overline{u}$. Then by the marginal value theorem (see, e.g.[15]) $\theta(u)$ is differentiable at $u = \overline{u}$, such that

$$\nabla_{u_i} \theta(\overline{u}) = -\sum_j a_{ij} \widehat{x}_j.$$

Respectively,

$$\nabla_{u_i} UB(\overline{u}, W) = b_i - \sum_j a_{ij} \widehat{x}_j, \quad i = 1, 2, ..., m. \tag{1.22}$$

Now take one step of the projected gradient technique for the problem $\min_{u \geq 0} UB(u, W)$, starting from $u = \overline{u}$. This gives

$$u(\rho) = \Pi_+(\overline{u} - \rho \nabla_u UB(\overline{u}, W)),$$

where $\rho \geq 0$ is a stepsize and $\Pi_+(\tau) = \{[\tau_i]^+, i = 1, 2, ..., m\}$. The stepsize $\widetilde{\rho}$, associated with the steepest descent, could be defined by the one-dimensional search: $\widetilde{\rho} = \arg\min\{UB(u(\rho), W) \mid \rho \geq 0\}$. Note that $UB(u(\rho), W)$ is a convex piecewise-linear function of ρ and the standard techniques, such as Fibonacci or golden-cut search , could be applied.

The choice of $\widetilde{u} = u(\widetilde{\rho})$ guarantees a strict decrease of the upper bound, that is either $UB(\widetilde{u}, W) < UB(\overline{u}, W)$ or $UB(\overline{u}, W) = \min_{u \geq 0} UB(u, W)$.

Note that neither Mendelssohn´s, nor Shetty-Taylor´s approaches do not guarantee this property. Generally speaking, the search directions used in these two approaches are not the descend directions for the upper bound (see the example in [17]).

The proposed approach of decreasing the upper bound can be realized iteratively. We briefly state the basic steps of the algorithm:

$S0$. (Initialization). Set the iteration counter $t := 0$. Determine an initial value of u^0, e.g. $u^0 = \overline{u}$.

$S1$. For $u = u^t$ solve (1.21) and find its optimal solution $\widehat{x}^t$ (it is assumed that $\widehat{x}^t$ is unique). Compute the upper bound $UB(u^t, W)$ by (1.17):

$$UB(u^t, W) = \sum_{i=1}^{m} u_i^t b_i + \sum_j \left(c_j - \sum_{i=1}^{m} u_i^t a_{ij} \right) \widehat{x}_j^t.$$

$S2$. For $\widehat{x} = \widehat{x}^t$ compute the components of the gradient in (1.22):

$$\nabla_{u_i} UB(u^t, W) = b_i - \sum_j a_{ij} \widehat{x}_j, \quad i = 1, 2, ..., m.$$

$S3$. For $u^t(\rho) = \Pi_+(u^t - \rho \nabla_u UB(u^t, W))$ compute

$$\rho_t = \arg\min\{UB(u^t(\rho), W) \mid \rho \geq 0\}.$$

Increase the iteration counter $t := t + 1$ and go to $S1$ with $u^{t+1} := u^t(\rho_t)$.

To terminate the algorithm we can either fix the maximal number of iterations or use the stopping rules associated with the first-order methods [25]. For example, we can stop when $[UB(u^t, W) - UB(u^{t+1}, W)]/UB(u^t, W) \leq \gamma$, where $\gamma > 0$ is a certain small threshold.

If the optimal solution $\widehat{x}$ to (1.21) is nonunique, then by the marginal value theorem $\theta(u)$ is directionally differentiable and subgradient techniques (see, e.g. [25]) can be applied to decrease the upper bound. The subgradient approach, generally speaking, does not guarantee a monotone decrease of the upper bound from iteration to iteration.

Along with linear localization we consider also an ellipsoidal one, such that

$$W_e = \left\{ x \mid \sum_j q_j^2 x_j^2 \leq R^2, \; x \geq 0 \right\}, \quad q_j > 0, \; R > 0.$$

Such a localization is constructed, for example, in interior point methods, or can be derived based on a feasible solution to the original problem (see, e.g.,

[19, 20] and the references therein). We do not discuss here how to obtain the ellipsoidal localization. Instead, we show how to solve the problem

$$UB(\cdot, W_e) = \max\{\sum_j \delta_j x_j \mid \sum_j q_j^2 x_j^2 \leq R^2,\ x \geq 0\}$$

arising in upper bound computations. Denote by $\widehat{x}$ an optimal solution of this problem and let J_+, J_- and J_0 be the index sets of positive, negative and zero components of the vector $\delta = \{\delta_j\}$. Obviously, $\widehat{x}_j = 0$ for $j \in J_-$.

Let $J_+ \neq \varnothing$. The Kuhn-Tucker conditions give $\delta_j - 2\lambda q_j^2 \widehat{x}_j \leq 0$, $\widehat{x}_j(\delta_j - 2\lambda q_j^2 \widehat{x}_j) = 0$, $j \in J_+ \cup J_0$, where $\lambda \geq 0$ is the optimal Lagrange multiplier. By algebraic manipulations we obtain

$$\widehat{x}_j = \delta_j / 2\lambda q_j^2,\ j \in J_+ \cup J_0,\ \lambda = \left[\sum_{j \in J_+} (\delta_j / q_j)^2\right]^{1/2} (2R)^{-1},$$

$$\sum_j \delta_j \widehat{x}_j = R \left[\sum_{j \in J_+} (\delta_j / q_j)^2\right]^{1/2}.$$

If $J_+ = \varnothing$, then the optimal solution is nonunique. The set of all optimal solutions is defined by the conditions $\sum\limits_{j \in J_0} q_j^2 x_j^2 \leq R^2$, $x_j \geq 0$, $j \in J_0$, and $x_j = 0$, $j \in J_-$, such that $\sum\limits_j \delta_j \widehat{x}_j = 0$.

For the bound $UB(\overline{u}, W_e)$ we have $\overline{z} \leq z^* \leq \overline{z} + \sum\limits_j \delta_j \widehat{x}_j$, where $\delta_j = c_j - \sum\limits_i \overline{u}_i a_{ij}$. If $\sum\limits_j \delta_j \widehat{x}_j = 0$, then $z^* = \overline{z}$ and hence $\overline{x}$ is optimal to the original problem. Thus the problem associated with the calculation of $UB(\overline{u}, W_e)$ has a unique optimal solution, except for the case $z^* = \overline{z}$.

It is interesting to note that based on the ellipsoidal constraint we can derive a linear (cubic) localization

$$W_c = \{x \mid 0 \leq x_j \leq R/q_j,\ j = 1, ..., n\}$$

for which we have $\max\{\sum\limits_j \delta_j x_j \mid x \in W_c\} = R \sum\limits_{j \in J_+} (\delta_j / q_j)$. Since

$$R \left[\sum_{j \in J_+} (\delta_j / q_j)^2\right]^{1/2} < R \sum_{j \in J_+} (\delta_j / q_j)$$

we may conclude that $UB(u, W_e) < UB(u, W_c)$. Hence it is better to use the ellipsoidal constraint directly rather than utilize the linear localization obtained from the same ellipsoid.

1.4.2. A Priori Bounds. Now suppose W is an a priori localization. By (1.17) we have for $u = \overline{u}$:

$$z^* - \overline{z} = \max_{x \in W} \sum_j (c_j - \sum_i \overline{u}_i a_{ij}) x_{ij}.$$

To obtain an a priori bound by this expression we need to estimate the values $c_j - \sum_i \overline{u}_i a_{ij}$ without optimization of the aggregated problem. If the numbers s_j are such that $s_j \geq c_j - \sum_i \overline{u}_i a_{ij}$, $j = 1, ..., n$, then

$$z^* - \overline{z} \leq \max_{x \in W} \sum_j s_j x_j \equiv \varepsilon^a(W) \tag{1.23}$$

provides the a priori bound. Similar to Section 1.3 we may put

$$s_j \equiv c_j - \min_{u \in \overline{U}} \sum_i u_i a_{ij},$$

where $\overline{U} \subseteq \mathbf{R}^m$ is such that $\overline{u} \in \overline{U}$ and it is assumed that the minimization problem has a finite solution.

In particular, if $A \geq 0, c \geq 0$, we may define $\overline{U}$ by a single constraint of the dual aggregated problem:

$$\overline{U}_k = \{u \mid \sum_i u_i \overline{a}_{ik} \geq \overline{c}_k, \ u \geq 0\}$$

and then set

$$s_j = c_j - \max_k \min_{u \in \overline{U}_k} \sum_i u_i a_{ij}.$$

In this case s_j can be calculated explicitly:

$$s_j = c_j - \Delta_j, \quad \Delta_j = \max_k \overline{c}_k \min_i \left\{ \frac{a_{ij}}{\overline{a}_{ik}} \mid \overline{a}_{ik} > 0 \right\}, \ j = 1, ..., n. \tag{1.24}$$

Suppose s_j's were estimated and the a priori bound $\varepsilon^a(W)$ was calculated for some localization W. If W was defined by some, but not all constraints of the original problem, there still is a room for strengthening the a priori bound $\varepsilon^a(W)$.

Let

$$W = \{x \mid \sum_j a_{ij}x_j \le b_i, \ i \in I \subset M \equiv \{1, ..., m\}, \ x \in \widetilde{W}\},$$

where $\widetilde{W}$ is a convex localization , and the original linear constraints were not used explicitly in the definition of $\widetilde{W}$.

Denote by W_{all} the localization defined by all constraints of the original problem. Then by the weak Lagrangian duality we have for $u \ge 0$:

$$\varepsilon^a(W_{all} \cap \widetilde{W}) \equiv \max_{x \in W_{all} \cap \widetilde{W}} \sum_j s_j x_j \le$$

$$\le \max_W \left\{ \sum_j s_j x_j - \sum_{i \in M \setminus I} u_i \left(\sum_j a_{ij} x_j - b_i \right) \right\} \equiv \varepsilon^a(u^W, W), \qquad (1.25)$$

where, as in Section 1.4.1, $u^W \in \mathbf{R}^{m-|I|}$. Obviously, $\varepsilon^a(W)$ corresponds to $u^W = 0 : \varepsilon^a(W) = \varepsilon^a(0, W)$.

Suppose $\varepsilon^a(W)$ was calculated and let $\widehat{x}$ be a unique optimal solution of the respective optimization problem. Then by the marginal-value theorem [15] $\varepsilon^a(u^W, W)$ is differentiable at $\varepsilon^a(W)$ with

$$\nabla_{u_i} \varepsilon^a(0, W) = -(\sum_j a_{ij}\widehat{x}_j - b_i), \ i \in M \setminus I.$$

Now the projected gradient scheme can be applied for searching the duals to improve $\varepsilon^a(W)$. The first step $(\varepsilon^a(W))$ gives

$$u^W(\rho) = \rho \Pi_+(-\nabla_u \varepsilon^a(0, W)),$$

where $\rho \ge 0$ is defined by a one-dimensional search. Applying this technique iteratively as in Section 1.4.1 we can, in principle, decrease $\varepsilon^a(W)$ down to $\varepsilon^a(W_{all} \cap \widetilde{W})$.

Note, that the inequality (1.25) still holds if we penalize by the Lagrangian only one restriction which has not been included in the definition of W. For example, we may choose the restriction which is most violated by $\widehat{x}$.

Along with absolute deviation $z^* - \overline{z}$, the aggregation error can be characterized by a relative deviation, say, $(z^* - \overline{z})/\overline{z}$. To obtain the bound for the relative error consider the localization W_r defined by summing up all original constraints with nonnegative coefficients $\overline{u}_i$:

$$W_r = \{x \mid \sum_j x_j \sum_i \overline{u}_i a_{ij} \le \sum_i \overline{u}_i b_i \equiv \overline{z}, \ x \ge 0\}.$$

By (1.17) we have

$$z^* - \overline{z} \leq \max_{x \in W_r} \sum_j (c_j - \sum_i \overline{u}_i a_{ij}) x_j.$$

Suppose $A > 0, c \geq 0$. Solving the maximization problem we obtain

$$\frac{z^* - \overline{z}}{\overline{z}} \leq \max_j \left(\frac{c_j}{\sum\limits_i \overline{u}_i a_{ij}} \right) - 1. \tag{1.26}$$

Note, that unless $z^* = \overline{z}$, the optimal solution $\overline{u}$ of the aggregated dual problem is infeasible to the original dual, and hence $c_j > \sum\limits_i \overline{u}_i a_{ij}$ at least for one j.

To get an a priori bound for the relative error we should estimate $\sum\limits_i \overline{u}_i a_{ij}$. This can be done, for example, similar to (1.24):

$$\sum_i \overline{u}_i a_{ij} \geq \Delta_j \equiv \max_k \overline{c}_k \min_i \{a_{ij}/\overline{a}_{ik}\}$$

and (1.26) becomes

$$\frac{z^* - \overline{z}}{\overline{z}} \leq \left[\max_k \min_{i,j} \left(\frac{\overline{c}_k}{\overline{a}_{ik}} \right) / \left(\frac{c_j}{a_{ij}} \right) \right]^{-1} - 1 \equiv \varepsilon^a_{rel}. \tag{1.27}$$

Remark 1.5. If the relative deviation is characterized by $(z^* - \overline{z})/z^*$, then by (1.27) and straightforward algebraic manipulations we obtain $(z^* - \overline{z})/z^* \leq$ $\leq \varepsilon^a_{rel}/(1 + \varepsilon^a_{rel})$.

1.5. Illustrative Example and Numerical Indications

Consider the following problem [33, 13]:

$$\begin{aligned} z^* = \max \quad & 2.5x_1 + 3x_2 + 4x_3 + 5x_4, \\ \text{s.t.} \quad & 4x_1 + 5x_2 + 7x_3 + 10x_4 \leq 54, \\ & x_1 + 2x_2 + x_3 + 2x_4 \leq 10, \\ & x_j \geq 0. \end{aligned}$$

The optimal solution is $x_1^* = 16/3$, $x_3^* = 14/3$, $x_2^* = x_4^* = 0$, $z^* = 32$. The optimal dual solution is $u^* = (0.5, 0.5)$.

Suppose that this problem is aggregated such that $K = 2$, $S_1 = \{1, 2\}$, $S_2 = \{3, 4\}$, and $g^1 = (0.5, 0.5)$, $g^2 = (0.75, 0.25)$. The aggregated problem is the following:

$$\begin{aligned} \overline{z} = \max\ & 2.75X_1 + 4.25X_2, \\ \text{s.t.}\quad & 4.5X_1 + 7.75X_2 \le 54, \\ & 1.5X_1 + 1.25X_2 \le 10, \\ & X_k \ge 0. \end{aligned}$$

The optimal solution is $\overline{X} = (5/3, 6)$, $\overline{u} = \frac{1}{96}(47, 35)$, $\overline{z} = 30\frac{1}{12} \simeq 30.083$. The true loss in accuracy is thus $z^* - \overline{z} = 1\frac{11}{12} \simeq 1.97$ and $(z^* - \overline{z})/\overline{z} = \frac{23}{361} \simeq 0.064$.

The optimal disaggregation results in two subproblems:

$$\begin{aligned} z_1(\overline{X}_1) = \max\ & 2.5x_1 + 3x_2, \\ \text{s.t.}\quad & 4x_1 + 5x_2 \le 4.5\overline{X}_1, \\ & x_1 + 2x_2 \le 1.5\overline{X}_1, \\ & x_j \ge 0, \end{aligned}$$

$$\begin{aligned} z_2(\overline{X}_2) = \max\ & 4x_3 + 5x_4, \\ \text{s.t.}\quad & 7x_3 + 10x_4 \le 7.75\overline{X}_2, \\ & x_3 + 2x_4 \le 1.25\overline{X}_2, \\ & x_j \ge 0. \end{aligned}$$

The optimal solution of the first subproblem is $x^1 = (1.875, 0)$, $z_1(\overline{X}_1) = 4.6875$. For the second problem we have $x^2 = (6\frac{9}{14}, 0)$, $z_2(\overline{X}_2) = 26\frac{4}{7}$. Thus the optimal disaggregated solution gives the lower bound $z^* \ge z_1(\overline{X}_1) + z_2(\overline{X}_2) \simeq 31.26$.

Suppose neither problem has been solved. An a priori bound on the relative accuracy loss $(z^* - \overline{z})/\overline{z}$ may be computed by (1.27) as follows. We have

$$\Delta_1 = \max\{2.75 \min\{\frac{4}{4.5}, \frac{1}{1.5}\},\ 4.25 \min\{\frac{4}{7.75}, \frac{1}{1.25}\}\} = \frac{4.25 \cdot 4}{7.75} \simeq 2.19.$$

Similarly $\Delta_2 = \frac{2.75 \cdot 5}{4.5} \simeq 3.056$, $\Delta_3 = \frac{4.25}{1.25} = 3.4$, $\Delta_4 = \frac{4.25 \cdot 10}{7.75} \simeq 5.48$. Then $\max_j\{c_j/\Delta_j\} = c_3/\Delta_3 \simeq 1.176$ and by (1.26), (1.27) we obtain $(z^* - \overline{z})/\overline{z} \le \varepsilon^a_{rel} = 0.176$.

To calculate the a priori bound $\varepsilon^a(W)$ in (1.23) for an a priori localization W we need to estimate s_j. By (1.24) we have $s_j = c_j - \Delta_j$ and hence $s_1 \simeq \simeq 0.306$, $s_2 \simeq -0.56$, $s_3 = 0.6$, $s_4 \simeq -0.48$. Define the a priori localization by the second original constraint

$$W_2 = \{x \geq 0 \mid x_1 + 2x_2 + x_3 + 2x_4 \leq 10\}.$$

Then (1.23) yields $z^* - \overline{z} \leq \varepsilon^a(W_2) \equiv \max\limits_{x \in W_2} \sum\limits_j s_j x_j = 10 s_3 = 6$. The unique optimal solution of the later problem is $\widehat{x} = (0, 0, 10, 0)$.

Let

$$W_0 = \{x \geq 0 \mid x_1 \leq 10,\ x_2 \leq 5,\ x_3 \leq \frac{54}{7},\ x_4 \leq 5\},$$

where the restrictions $x_j \leq r_j$ in W_0 have been obtained from the original problem by $r_j = \min_i\{b_i/a_{ij}\}$. For the a priori localization $W_0 \cap W_2$ we have $z^* - \overline{z} \leq \varepsilon^a(W_0 \cap W_2) = \frac{16}{7} s_1 + \frac{54}{7} s_2 \simeq 5.328$, which is an improvement over $\varepsilon^a(W_2)$.

In the definition of the localization W_2 the second original constraint was not used, and it is violated by $\widehat{x}$. Hence we can try to improve $\varepsilon^a(W_2)$ by searching the dual u_2. By (1.25) we have

$$\varepsilon^a(u_1, W_2) = \max_{x \in W_2}\{\sum_j (s_j - u_1 a_{1j}) x_j + 54 u_1\}.$$

Note that $\nabla_{u_1} \varepsilon^a(0, W_2) = b_1 - \sum\limits_j a_{1j} \widehat{x}_j = 16 > 0$ and hence the bound $\varepsilon^a(W_2) = \varepsilon^a(0, W_2)$ can be strictly decreased by changing u_1 from 0 to some positive value.

To calculate the best (i.e., bound minimizing) dual $u_1 \geq 0$ we can either use a one-dimensional search, or analyze the points of non-differentiability of $\varepsilon^a(u_1, W_2)$. Applying the later approach we obtain from $s_3 - u_1 a_{13} = 0$ the best $\widehat{u}_1 = 0.6/7$. Respectively, $\varepsilon^a(\widehat{u}_1, W_2) = 54 \widehat{u}_1 = 4.629$ which is an improvement over $\varepsilon^a(W_2) = 6$. It is not hard to verify that $\varepsilon^a(\widehat{u}_1, W_2) = = \varepsilon^a(W_{all})$, where the a priori localization W_{all} is defined by original constraints.

The bound $\varepsilon^a(W_0 \cap W_2)$ can also be improved by searching the dual u_1. Doing the same as before, we obtain $\varepsilon^a(\widehat{u}_1, W_0 \cap W_2) = 4.629 < \varepsilon^a(W_0 \cap W_2)$.

Now suppose the aggregated problem has been solved and $\overline{u}$ is known. Define σ' such that $K' = 1$, $S'_1 = \{1, ..., 4\}$. The coefficients d_j, p_1 are defined by the second constraint of the original problem such that $d_1 = d_3 = 1$, $d_2 = = d_4 = 2$, $p_1 = 10$. A vector with components $c_j - \overline{u} a_j$ is $\frac{1}{96}(17, -17, 20, -60)$. Thus by (1.9) we obtain Zipkin bound $z^* \leq \overline{z} + 2\frac{1}{12} \simeq 32.167$.

To compute Mendelssohn-Knolmayer bound (1.10) we need to check the values of the stepsize ρ obtained from (1.11),(1.12). After some algebraic manipulations we get from

$$\frac{c_1 - \rho\overline{u}a_1}{d_1} = \frac{c_3 - \rho\overline{u}a_3}{d_3}$$

the stepsize $\rho = 1\frac{1}{47}$ which by (1.10) gives $z^* \leq 1\frac{1}{47}\overline{z}+10(c_3-\rho\overline{u}a_3) = 32$. This bound is sharp and can not be improved further. The dual vector associated with Mendelssohn-Knolmayer bound is $u = \rho\overline{u} = (0.5, \frac{35}{94}) \simeq (0.5, 0.372)$ and differs from u^*.

In Shetty-Taylor approach the duals are considered in the form $u = \overline{U}+\rho b$, $\rho \geq 0$ since both $\overline{U}$ and b are positive. Analyzing the points of the non-differentiability in (1.13) similar to (1.10) we conclude that the best (i.e. bound-minimizing) stepsize ρ is obtained from the condition

$$\frac{c_1 - \overline{u}a_1 - \rho b a_1}{d_1} = \frac{c_3 - \overline{u}a_3 - \rho b a_3}{d_3}$$

such that $\rho = \frac{1}{5184}$. By (1.13) this gives $z^* \leq \overline{z}+\rho bb+10(c_1-\overline{u}a_1-\rho ba_1) = 32$. The dual vector associated with Shetty-Taylor bound is $u \simeq (0.5, 0.3665)$ and differs from u^*, as well as from the duals obtained by Mendelssohn-Knolmayer approach.

By (1.18) we have $UB(\overline{u}^{W_2}, W_2) = UB(\overline{u}_1, W_2)$. For $\delta = \{c_j - \overline{u_1}a_{1j}\} = \frac{1}{96}(52, 53, 55, 10)$ we obtain $UB(\overline{u}_1, W_2) = \overline{u}_1 b_1 + \max\{\sum_j \delta_j x_j \mid x \in W_2\} = 32\frac{1}{6} \simeq 32.167$. Thus $z^* \leq UB(\overline{u}_1, W_2) \simeq 32.167$, the same as for Zipkin bound calculated earlier. By Comments 1.3, Zipkin bound coincides with $UB(\overline{u}, W_2)$. Hence for the example $UB(\overline{u}_1, W_2) = UB(\overline{u}, W_2)$.

To compute the bound $UB(\overline{u}_1, W_0 \cap W_2)$ we should solve the problem $\max\{\sum_j \delta_j x_j \mid x \in W_0 \cap W_2\}$ for the same δ as before. The unique optimal solution to this problem is $\widehat{x} = \frac{1}{7}(16, 0, 54, 0)$. By (1.18) this yields $UB(\overline{u}_1, W_0 \cap W_2) = \overline{u}_1 b_1 + \sum_j \delta_j \widehat{x}_j = 32\frac{2}{11} \simeq 32.095$, which is an improvement over Zipkin bound.

Now try to decrease $UB(\overline{u}_1, W_0 \cap W_2)$ by searching the dual u_1. By (1.22) we have $\nabla_{u_1} UB(\overline{u}_1, W_0 \cap W_2) = b_1 - \sum_j a_{1j}\widehat{x}_j = -\frac{64}{7}$ and hence $u_1(\rho) = \overline{u}_1 + \rho\frac{64}{7}$, $\rho \geq 0$. The best $\widetilde{\rho}$, minimizing the bound, is $\widetilde{\rho} = \frac{7}{6144}$ and $u_1(\widetilde{\rho}) = 0.5$. This $\widetilde{\rho}$ can be defined from the condition $c_1 - u_1(\rho)a_{11} = c_3 - u_1(\rho)a_{13}$. Respective upper bound is $z^* \leq UB(u_1(\widetilde{\rho}), W_0 \cap W_2) = 0.5b_1 + 5 = 32$.

We briefly discuss now the results of a numerical study realized in [17] to compare the bounds computation approaches for the a posteriori bounds. The continuous multidimensional knapsack problem was used as a benchmark:

$$\max\{\sum_j c_j x_j \mid \sum_j a_{ij} x_{ij} \le b_i,\ i = 1, ..., m,\ 0 \le x_j \le 1,\ j = 1, ..., n\},$$
$$c_j,\ a_{ij},\ b_i > 0. \qquad (1.28)$$

The problems were considered for $n = 100, 250, 500$ and $m = 5, 10, 30$. From each problem instance two problems were randomly generated to form 18 different problems of the type (1.28). In all these problems the variables are arranged into $K = 5,10,15$ clusters which gives 54 aggregated problems. Each cluster has $\lfloor n/K \rfloor$ variables, sorted by their variable index, except for the last cluster which contains all the rest variables. Here $\lfloor n/K \rfloor$ denotes the integer part of n/K. The weights are determined to be equal for all variables in the cluster.

For each combination of the problem size $m \times n$ and cluster numbers K were considered $m+1$ localizations for the bounds of Zipkin, Mendelssohn, and Shetty-Taylor: $W_0 = \{x \mid 0 \le x_j \le 1,\ j = 1, ..., n\}$ and $W_i = \{x \mid \sum_j a_{ij} x_j \le b_i,\ x_j \ge 0\}$, $i = 1, ..., m$. For the bounds derived in Section 1.4 $m+1$ localizations of the form W_0 and $W_i \cap W_0$ were considered. Hence for each bound computation approach 864 bounds have been calculated within the test series.

The dimension of the original problem allows for the exact solution of (1.28). To eliminate the influences of the lower bound on the upper bound evaluation and vice versa, the bounds have been characterized by the values $(z^* - \overline{z})/z^*$ and $(UB - z^*)/z^*$, respectively. These indicators, averaged over all combinations of m, n, K and localizations, can be seen from the first column of the Table 1.1. The rest columns demonstrate the averaged values of the indicators for the fixed m or K. Bound IV in Table 1.1 corresponds to the bound UB in (1.18) calculated for $u = \overline{u}$. Bound V is the improved bound IV, obtained by changing duals by the projected gradient approach. The main computational burden in the approaches II, III, and IV is associated with the line search. To make the computational cost of these approaches comparable we have realized in the approach V only one iteration of the improvement procedure.

It can be seen from Table 1.1 that adding the upper bounds for the variables to the localization (Bound IV) improves Zipkin Bound I significantly.

Bounds (in %)	Average		$m=5$		$m=10$		$m=30$		$K=10$		$K=15$		$K=20$	
	M	SD	M	SD	M	SD	M	SD	M	SD	M	SD	M	SD
Lower bound	17.71	1.53	18.69	1.37	17.07	0.6	16.87	1.57	19.4	1.55	17.73	1.33	16.26	1.95
Upper bounds:														
Bound I	1987.87	1356.45	3210.12	1046.34	1965.81	1411.23	1640.85	1541.14	2177.23	1251.64	1897.93	1316.56	1888.46	1488.65
(Zipkin)														
Bound II	71.55	39.16	91.48	61.98	81.32	41.07	62.22	23.78	80.89	39.32	74.12	39.04	59.64	35.97
(Mendelssohn)														
Bound III	42.27	20.18	60.36	32.84	47.16	18.27	35.17	8.96	45.82	19.22	42.82	20.54	38.18	20.02
(Shetty-Taylor)														
Bound IV	10.11	3.68	6.57	3.51	9.84	3.21	11.26	3.18	11.54	3.53	10.39	4.04	8.40	2.63
Bound V	4.76	3.25	0.69	1.05	2.64	1.95	6.75	2.35	5.40	3.41	5.02	3.42	3.85	2.72

Table 1.1. Bounds in % (m =Number of restrictions, K =Number of clusters, M =Mean, SD =Standard deviation).

Moreover, the bound IV is superior to the bounds of Mendelssohn (II) and Shetty-Taylor (III) within the test examples.

The same tendency takes place for the particular examples within the test series. For every fixed m, n, K, the localization W_i for the bounds I, II, III and the localizations $W_i \cap W_0$ for the bounds IV, V we have I > II > II > IV > V. If we use the localization W_0 for all approaches, then I $\geq$ IV and I > II > II > V. We note also that for every fixed m, n, K the worst (among all localizations) Bound V was smaller then the best (among all localizations) of the Bounds I, II, III.

In [14] an extended numerical study of the aggregation bounds was realized for the problems (1.28), but without upper bounds for the variables. Zipkin-type localization (1.7) was used for $d_j = 1$, $p_k = 1.5 \max_{k=1,...,K'} \sum_{j \in S_k} x_j^*$, $k = 1, ..., K'$ providing x^* is known. The improvement of Zipkin bound (1.9) by Mendelssohn approach (1.10) was rather moderate. However the improvement by Shetty-Taylor procedure (1.13) was enormous within the test examples. The change of the search direction from rays through the origin and the dual solution under consideration (Mendelssohn) to the direction given by the resource vector b (Shetty-Taylor) diminishes the bound significantly. The subgradient approach [14] to improve Zipkin bound provides the best results among dual-searching procedures.

It has to be pointed out that the results of the numerical tests presented in [14, 17] have to be interpreted very carefully, and can not be treated as general results for the comparison of bounds behavior. Further numerical studies, in particular, regarding special structured problems, is a topic for future research. For example, for the generalized transportation problem the numerical results reported in [18] show that Mendelssohn approach results in stronger bounds than Shetty-Taylor procedure.

§2. The Generalized Transportation Problem

Many applications of aggregation/disaggregation theory in optimization involve linear network flow problems and, in particular, the classical transportation problem (TP). For this problem both a priori and a posteriori error bounds were developed and investigated (see [24, 26, 35, 36], and the references therein). In network flow problems the nodes traditionally correspond to constraints and arcs to variables. For most aggregation schemes, aggregation of nodes (constraints) implies or necessitates a corresponding aggregation of arcs (variables). Usually, only variables and constraints of the same type (e.g., sources or destinations) are aggregated.

The TP minimizes the cost of transporting supplies from the set of sources to the set of destinations, and it is assumed that the transportation flow is conserved on every arc between the source and the destination. In the generalized transportation problem (GTP) the amount of flow that leaves the source can differ from the amount of flow that arrives in the destination. A certain multiplier is associated with each arc to represent a "gain" ("loss") of a unit flow on the way between the source and the destination. There are two common interpretations of the arc multipliers [1]. In the first approach the multipliers are treated as modifying the amount of flow of some particular item. This way it is possible to model situations involving physical or administrative transformations. For example, evaporation, seepage, or monetary growth due to the interest rates. In the second interpretation, the multipliers are used to transform one type of item into another (manufacturing, machine loading and scheduling, currency exchanges, etc.). These and other applications of generalized network problems are discussed in [1].

In this section we use the specifics of the GTP to derive a priori and a posteriori error bounds for the aggregation of destinations. The results of a numerical experiment are reported to study the correspondence between the actual error and the error bounds.

2.1. The Aggregated Problem

The original GTP before aggregation is formulated in the following form [5]:

$$z^* = \min \sum_{i \in S} \sum_{j \in T} c_{ij} x_{ij}, \tag{2.1}$$

$$\sum_{j \in T} d_{ij} x_{ij} \leq a_i, \; i \in S,$$

$$\sum_{i \in S} x_{ij} = b_j, \; j \in T,$$

$$x_{ij} \geq 0 \qquad \forall i, j.$$

We will refer to S and T as the set of all sources and all destinations (customers), respectively. It is assumed that all the coefficients in (2.1) are positive, and d_{ij} represents the transformation of the flow on the way from the source i to the destination j.

The role of "sources" and "destinations" in (2.1) can be reversed [5] by a simple transformation of variables: $y_{ij} = d_{ij}x_{ij}$. The resulting problem becomes

$$z^* = \min \sum_{i\in S} \sum_{j\in T} (c_{ij}/d_{ij}) y_{ij},$$

$$\sum_{j\in T} y_{ij} \le a_i, \ i \in S,$$

$$\sum_{i\in S} (1/d_{ij}) y_{ij} = b_j, \ j \in T,$$

$$y_{ij} \ge 0 \qquad \forall i, j.$$

Bearing in mind this correspondence between the sources and the destinations we restrict ourselves to the aggregation of the destinations in (2.1). The aggregation of the sources can be treated similarly.

The aggregation is defined by a partition of the set T of the destinations into a set T^A of clusters $T_k, k = 1, ..., K$ such that $T_k \cap T_p = \varnothing$, $k \ne p$ and $\cup_{k=1}^{K} T_k = T$. For each k consider the non-negative weights g_{ij} fulfilling the following normalizing conditions

$$\sum_{j\in T_k} g_{ij} = 1, \ g_{ij} \ge 0 \qquad \forall i, j. \tag{2.2}$$

As it was mentioned in Remark 1.1, the aggregated problem can be obtained by fixing the normalized weights g_{ij} and substituting the linear transformation

$$x_{ij} = g_{ij} X_{ik}, \ j \in T_k \tag{2.3}$$

into (2.1). Thus formally the aggregated problem is as follows:

$$\overline{z} = \min \sum_{i\in S} \sum_{k\in T^A} X_{ik} \sum_{j\in T_k} c_{ij} g_{ij}, \tag{2.4}$$

$$\sum_{k\in T^A} X_{ik} \sum_{j\in T_k} d_{ij} g_{ij} \le a_i, \ i \in S, \tag{2.5}$$

$$\sum_{i\in S} g_{ij} X_{ik} = b_j, \ j \in T_k, \ k \in T^A, \tag{2.6}$$

$$X_{ik} \ge 0 \qquad \forall i, k. \tag{2.7}$$

To write down problem (2.4)–(2.7) we have used the observation that clustering is simply a rearrangement of indexes and hence $\sum_{j\in T} = \sum_{k\in T^A}\sum_{j\in T_k}$.

Note that from (2.2), (2.3) it follows that for any normalized weights we have $X_{ik} = \sum_{j\in T_k} x_{ij}$ and thus X_{ik} can be treated as the supply from the source i to the aggregated customer T_k. However, problem (2.4)–(2.7) cannot in general be considered as some GTP associated with S sources and T^A destinations. First of all because the number of restrictions (2.6) is $|T|$, the same as in (2.1). This problem can be solved as follows.

Let the weights be chosen as $\bar{g}_{ij} = b_j / \sum_{j\in T_k} b_j$, $j \in T_k$. Then for each $j \in T_k$ condition (2.6) turns out to be $\sum_{i\in S} X_{ik} = \sum_{j\in T_k} b_j$ and problem (2.4)–(2.7) becomes:

$$\bar{z} = \min \sum_{i\in S}\sum_{k\in T^A} \bar{c}_{ik} X_{ik}, \tag{2.8}$$

$$\sum_{k\in T^A} \bar{d}_{ik} X_{ik} \le a_i, \; i \in S,$$

$$\sum_{i\in S} X_{ik} = \bar{b}_k \text{ (repeated } |T_k| \text{ times)}, \; k \in T^A,$$

$$X_{ik} \ge 0 \qquad \forall i, k,$$

where $\bar{c}_{ik} = \sum_{j\in T_k} c_{ij}\bar{g}_{ij}$, $\bar{d}_{ik} = \sum_{j\in T_k} d_{ij}\bar{g}_{ij}$, $\bar{b}_k = \sum_{j\in T_k} b_j$. It is assumed that (2.8) has an optimal solution.

Obviously, the redundant constraints in problem (2.8) can be relaxed without changing its optimal solution to form the problem which is traditionally referred to as the aggregated GTP [5]:

$$\bar{z} = \min \sum_{i\in S}\sum_{k\in T^A} \bar{c}_{ik} X_{ik}, \tag{2.9}$$

$$\sum_{k\in T^A} \bar{d}_{ik} X_{ik} \le a_i, \; i \in S,$$

$$\sum_{i\in S} X_{ik} = \bar{b}_k, \; k \in T^A,$$

$$X_{ik} \ge 0 \qquad \forall i, k.$$

The primal optimal solutions $\overline{X}_{ik}$ of (2.8) and (2.9) are the same, and the fixed-weight disaggregated solution $\bar{x}_{ij} = \bar{g}_{ij}\overline{X}_{ik}$, $j \in T_k$ is feasible to the original problem (2.1). The following proposition shows the correspondence between the optimal dual solutions to problems (2.8), (2.9).

Proposition 2.1. *Let* $\{\overline{u}_i, \overline{v}_k\}$ *be an optimal dual solution to (2.9). Denote*

$$D(\overline{v}) = \{v_j,\ j \in T \mid \sum_{j \in T_k} v_j b_j = \overline{v}_k \overline{b}_k,\ k = 1, ..., K\}.$$

Then any pair $\overline{u}, \widehat{v}$ *with* $\widehat{v} \in D(\overline{v})$ *be an optimal dual solution to (2.8) and*

$$\overline{z} = -\sum_i a_i \overline{u}_i + \sum_k \sum_{j \in T_k} \widehat{v}_j b_j$$

for any $\widehat{v} \in D(\overline{v})$.

Proof. Consider the dual to (2.8) or just the same, the dual to (2.4)–(2.7) for $g_{ij} = \overline{g}_{ij}$:

$$-\sum_i a_i u_i + \sum_k \sum_{j \in T_k} v_j b_j \to \max, \tag{2.10}$$

$$\overline{c}_{ik} + \overline{d}_{ik} u_i - \sum_{j \in T_k} v_j \overline{g}_{ij} \geq 0,\ \forall i, k,$$

$$u_i \geq 0,\ i \in S.$$

The dual to (2.9) is as follows:

$$-\sum_i a_i u_i + \sum_k v_k \sum_{j \in T_k} b_j \to \max, \tag{2.11}$$

$$\overline{c}_{ik} + \overline{d}_{ik} u_i - v_k \geq 0,\ \forall i, k,$$

$$u_i \geq 0,\ i \in S.$$

These two dual problems should have equal optimal values since the primal problems (2.8) and (2.9) have the same optimal solutions. Let $\{\overline{u}_i, \overline{v}_k\}$ be an optimal solution to (2.11). Construct $\{\widehat{u}_i, \widehat{v}_j\}$ as follows: $\widehat{u}_i = \overline{u}_i$ and $\widehat{v}_j$ is such that $\sum_{j \in T_k} \widehat{v}_j \overline{g}_{ij} = \overline{v}_k$. Comparing the restrictions of (2.10) and (2.11) we see that this $\{\widehat{u}_i, \widehat{v}_j\}$ is feasible to (2.10). Moreover, since $\overline{g}_{ij} = b_j / \sum_{j \in T_k} b_j$ it is not hard to verify that the objective value of (2.10) for the constructed duals is equal to the optimal objective of (2.11). Hence this $\{\widehat{u}_i, \widehat{v}_j\}$ is optimal to (2.10).

Note, that the nonuniqueness of the optimal duals in (2.8) follows from the way the aggregation was performed, no matter whether the problem (2.9) has a unique optimal dual solution or not. The nonuniqueness will be used further to derive a priori bounds for the GTP.

2.2. The Error Bounds

To obtain the bounds for aggregation error we can use the results established in the section 1.4, bearing in mind that the original problem (2.1) is a minimization one. Let W be a localization for (2.1). Then by Proposition 1.3 we have

$$\overline{z} - z^* \leq UB(u, v, W) \equiv \overline{z} - (-\sum_{i \in S} u_i a_i + \sum_{j \in T} v_j b_j)$$

$$+ \max_{x \in W} \sum_{i \in S} \sum_{j \in T} (-u_i d_{ij} + v_j - c_{ij}) x_{ij} \tag{2.12}$$

for any fixed $u \geq 0$ and v. Note that here we use the notation $UB(u, v, W)$ for the upper bound to $\overline{z} - z^*$.

Suppose that W is an a priori localization. The critical question in a priori bound calculations is the choice of multipliers $u \geq 0$, v to estimate the right-hand side of (2.12) without knowledge of $\overline{z}$. A possible choice is to look for multipliers, giving

$$\overline{z} = -\sum_{i \in S} u_i a_i + \sum_{j \in T} v_j b_j, \tag{2.13}$$

and then to estimate the coefficients of the maximization problem in (2.12).

Let us define the multipliers $\widetilde{u}, \widetilde{v}$ as follows:

$$\widetilde{u}_i = \overline{u}_i,\ i \in S_k,\ \ \widetilde{v}_j = \overline{v}_k,\ j \in T_k,\ k = 1, 2..., K. \tag{2.14}$$

By the definition of $\overline{b}_k$ it follows that $\widetilde{v} \in D(\overline{v})$. Moreover, by Proposition 2.1, the equation (2.13) is fulfilled for $\widetilde{u}, \widetilde{v}$ defined in (2.14), and hence by (2.12)

$$\overline{z} - z^* \leq \max_{x \in W} \sum_{i \in S} \sum_{k} \sum_{j \in T_k} (-\overline{u}_i d_{ij} + \overline{v}_k - c_{ij}) x_{ij}$$

since $\sum_{j \in T} = \sum_k \sum_{j \in T_k}$.

Consider the classical TP, which is a special case of the GTP (2.1) with $d_{ij} = 1$ for all i, j. By restrictions of the dual problem (2.11) and, since $\overline{d}_{ik} = 1$,we have $-\overline{u}_i + \overline{v}_k \leq \overline{c}_{ik}$. Then it follows that

$$\overline{z} - z^* \leq \max_{x \in W} \sum_{i \in S} \sum_{k} \sum_{j \in T_k} (\overline{c}_{ik} - c_{ij}) x_{ij} \tag{2.15}$$

gives an a priori bound for the classical TP. However, if $d_{ij} \neq 1$, this consideration does not work and the choice of the multipliers in the form (2.14) does not allow to derive an a priori bound for the GTP. To cope with this problem, we will use the nonuniqueness of the multipliers fulfilling (2.13), as stated in Proposition 2.1.

Remark 2.1. As follows from the proof of Proposition 2.1, to fulfill the equality (2.13) it is sufficient to choose the multipliers $\{u_i, v_j\}$ in the form $u_i = \overline{u}_i$, $v_j : \sum\limits_{j \in T} v_j b_j = \sum\limits_k \overline{v}_k \overline{b}_k$. These multipliers will not form in general a feasible solution to the dual problem (2.10).

Proposition 2.2. *Let $d_{ij} = p_i t_j$, where $p_i > 0$, $t_j > 0$. Then there exists $\widehat{v} \in D(\overline{v})$ such that for the pair $\overline{u}, \widehat{v}$ we have*

$$-\overline{u}_i d_{ij} + \widehat{v}_j - c_{ij} \leq (\overline{c}_{ik} d_{ij} - c_{ij} \overline{d}_{ik})/\overline{d}_{ik}, \qquad j \in T_k.$$

Proof. From restrictions of the dual problem (2.11) we have $\overline{u}_i \geq (\overline{v}_k - \overline{c}_{ik})/\overline{d}_{ik}$ and hence

$$-\overline{u}_i d_{ij} + \widehat{v}_j - c_{ij} \leq [(\overline{c}_{ik} d_{ij} - c_{ij} \overline{d}_{ik}) - -(\overline{v}_k d_{ij} - \widehat{v}_j \overline{d}_{ik})]/\overline{d}_{ik}, \ j \in T_k.$$

Now choose $\widehat{v}_j$ to fulfill conditions $\overline{v}_k d_{ij} - \widehat{v}_j \overline{d}_{ik} = 0$ for all $j \in T_k$. Resolving the latter system with respect to $\widehat{v}_j$ for $d_{ij} = p_i t_j$ and $\overline{d}_{ik} = (\sum\limits_{j \in T_k} b_j d_{ij}) \overline{b}_k^{-1}$ we obtain

$$\widehat{v}_j = \overline{v}_k t_j \frac{\overline{b}_k}{\sum\limits_{j \in T_k} b_j t_j}, \ j \in T_k.$$

It is not hard to verify that for this $\widehat{v}_j$ we have $\sum\limits_{j \in T_k} \widehat{v}_j b_j = \overline{v}_k \overline{b}_k$. Hence $\widehat{v} \in D(\overline{v})$ and the claim follows.

Propositions 2.1, 2.2, together with (2.12), yield immediately the following expression for the a priori bound:

Corollary 2.1. *For $d_{ij} = p_i t_j$ we have*

$$\overline{z} - z^* \leq \varepsilon^a(W) \equiv \max_{x \in W} \sum_{i \in S} \sum_{j \in T} \delta^a_{ij} x_{ij},$$

where

$$\delta^a_{ij} \equiv \left(\frac{\overline{c}_{ik(j)}}{\overline{d}_{ik(j)}} - \frac{c_{ij}}{d_{ij}} \right) d_{ij},$$

and $k(j)$ is the cluster to which j belongs, $k(j) \in T^A$.

Note that if $d_{ij} = 1$ for all i, j, then $\varepsilon^a(W)$ coincides with the a priori bound (2.14).

Remark 2.2. Under the assumption $d_{ij} = p_i t_j$ we have $\sum_{j \in T} t_j x_{ij} \le a_i/p_i$, $i \in S$ in the restrictions of the original GTP (2.1). This means that the "gain" along the arc (i, j) is the same for all sources i connected with a destination j. We can treat this case as an intermediate one between the classical TP, where the gains are the same for all arcs, and a general form of the GTP, where the gains vary from arc to arc. For example, in financial networks, sources of investments may be concentrated in one country or region, while destinations may be located in various countries/regions. Then t_j can be treated as a "cross-the-boarder" fee which is the same for all sources.

Interpreting the GTP as the machine loading problem (see, e.g., [1]), producing 1 unit of product j on machine i consumes d_{ij} hours of the machine's time. If p_i is a production "velocity" of the machine i and t_j is a "length" of the product j measured in suitable units, then the assumption $d_{ij} = p_i t_j$ also holds.

A priori localizations can be defined, for example, by manipulating the constraints of the original problem. Let

$$W_b = \{x_{ij} \mid 0 \le x_{ij} \le r_{ij}\ \forall i, j\},\ \ r_{ij} = \min\{b_j,\ a_i/d_{ij}\},$$

$$W_s = \{x_{ij} \mid \sum_{j \in T} d_{ij} x_{ij} \le a_i,\ \ i \in S,\ \ x_{ij} \ge 0\ \forall i, j\},$$

$$W_d = \{x_{ij} \mid \sum_{i \in S} x_{ij} = b_j,\ \ j \in T,\ \ x_{ij} \ge 0\ \forall i, j\},$$

where the lower indices b, s, and d show that the restrictions defining the respective localization are associated with the bounds, sources and destinations. Obviously W_s, W_d are localizations as well as $W_b \cap W_s$ and $W_b \cap W_d$. For these four localizations the optimization problem associated with $\varepsilon^a(W)$ can be solved analytically. Due to the decomposable structure of the localizations, computation of $\varepsilon^a(W_s), \varepsilon^a(W_d)$ results in a number of independent single-constrained continuous knapsack problems, which immediately gives

$$\varepsilon^a(W_s) = \sum_i a_i \max_j [\delta^a_{ij}]^+ / d_{ij},$$

$$\varepsilon^a(W_d) = \sum_j b_j \max_i \delta^a_{ij}.$$

Similarly, calculation of $\varepsilon^a(W_b \cap W_s), \varepsilon^a(W_b \cap W_d)$ is reduced to the solution of independent knapsack problems with upper bounds for the variables. Respective solutions can be obtained the same way as in (1.21).

Remark 2.3. Formally, the dual variables are not used in the calculation of the a priori bound $\varepsilon^a(W)$. Meanwhile, if W is defined by some, but not all constraints of the GTP, respective bound can still be improved by searching the duals similar to Section 1.4.2. We briefly discuss the dual searching approach for $\varepsilon^a(W_b \cap W_d)$. The other bounds can be treated similarly.

Denote by W_0 the localization defined by all constraints of the original problem (2.1). By definition of W_b, W_d, and W_s we have $W_0 = W_b \cap W_d \cap W_s$. By the weak Lagrangian duality we obtain for any $u \geq 0$:

$$\varepsilon^a(W_0) \equiv \max_{x \in W_b \cap W_d \cap W_s} \sum_{i \in S} \sum_{j \in T} \delta^a_{ij} x_{ij} \leq$$

$$\max_{x \in W_b \cap W_d} \left\{ \sum_{i \in S} \sum_{j \in T} \delta^a_{ij} x_{ij} - \sum_{i \in S} u_i (\sum_{j \in T} d_{ij} x_{ij} - a_i) \right\} \equiv \varepsilon^a(W_b \cap W_d, u).$$

Obviously, $\varepsilon^a(W_b \cap W_d) = \varepsilon^a(W_b \cap W_d, 0)$, and we may try to decrease the upper bound $\varepsilon^a(W_b \cap W_d)$ by changing the duals starting from $u = 0$.

Suppose the bound $\varepsilon^a(W_b \cap W_d)$ has been calculated and let $\widehat{x}$ be a unique optimal solution of the associated maximization problem. By the marginal value theorem [15] in this case $\varepsilon^a(W_b \cap W_d, u)$ is differentiable at $u = 0$, such that

$$\nabla_{u_i} \varepsilon^a(W_b \cap W_d, 0) = -(\sum_{j \in T} d_{ij} \widehat{x}_{ij} - a_i).$$

Taking from $u = 0$ a step of the projected gradient technique for the problem $\min_{u \geq 0} \varepsilon^a(W_b \cap W_d, u)$, we obtain

$$u(\rho) = \rho \Pi_+(-\nabla_u \varepsilon^a(W_b \cap W_d, 0)), \ \ \rho \geq 0.$$

The stepsize can be defined, for example, by the one-dimensional search: $\widetilde{\rho} = \arg\min\{\varepsilon^a(W_b \cap W_d, u(\rho)) \mid \rho \geq 0\}$. Note, that either $\varepsilon^a(W_b \cap W_d, u(\widetilde{\rho})) <$ $< \varepsilon^a(W_b \cap W_d)$ or $\varepsilon^a(W_b \cap W_d) = \varepsilon^a(W_0)$. Applying this process iteratively, similar to Section 1.4.1, we can sharpen $\varepsilon^a(W_b \cap W_d)$ down to $\varepsilon^a(W_0)$.

Suppose now the aggregated problem (2.9) was solved and its optimal primal and dual solutions $\overline{X}, \overline{z}, \overline{u}$ and $\overline{v} = \{\overline{v}_k, k = 1, ..., K\}$ are known.

The bound (2.12) is valid for any $u \geq 0, v = \{v_j\}$. To calculate the bound (2.12) based on the solution of the aggregated problem we can wish to construct these multipliers to meet equality (2.13). By Comments 2.1 this can be done in a nonunique way.

Suppose $\widetilde{W}$ is a localization and the constraints $\sum_{i\in S} x_{ij} = b_j,\ j \in T$ are used explicitly in its definition. In particular, W_d and $W_b \cap W_d$ are of that form. For the localization $\widetilde{W}$ nonunique duals v_j can be eliminated from the expression for the error bound. By Corollary 1.2 and (1.18) we obtain from (2.12):

$$\overline{z} - z^* \leq UB((u,v)^{\widetilde{W}}, \widetilde{W}) \equiv \overline{z} + \sum_{i\in S} a_i u_i + \max_{x\in\widetilde{W}} \sum_{i\in S} \sum_{j\in T} (-d_{ij}u_i - c_{ij})x_{ij},$$

where by definition $(u,v)^{\widetilde{W}} = u$.

Calculating this bound for $u = \overline{u}$ and using $\overline{z} = -\sum_{i\in S} a_i\overline{u}_i + \sum_{k\in T^A} \overline{b}_k\overline{v}_k$ yields

$$UB((\overline{u},v)^{\widetilde{W}}, \widetilde{W}) = \sum_{k\in T^A} \overline{b}_k\overline{v}_k + \max_{x\in\widetilde{W}} \sum_{i\in S} \sum_{j\in T} (-d_{ij}\overline{u}_i - c_{ij})x_{ij}. \qquad (2.16)$$

If the localization is not of the form $\widetilde{W}$, we may choose $\widehat{u} = \overline{u}, \widehat{v}_j = \overline{v}_k$, $j \in T_k$, $k = 1, ..., K$ as the multipliers resulted from the aggregated problem. By Comments 2.1 the equality (2.13) holds in this case, and by (2.12) we obtain

$$\overline{z} - z^* \leq \max_{x\in W} \sum_{i\in S} \sum_{j\in T} (-\widehat{u}_i d_{ij} + \widehat{v}_j - c_{ij})x_{ij} \equiv \varepsilon_Z^p(W).$$

For the four localizations $W_s, W_d, W_b \cap W_s$ and $W_b \cap W_d$ the maximization problem to calculate $\varepsilon_Z^p(W)$ (or (2.16)) can be solved analytically, similar to the bound $\varepsilon^a(W)$. We need only to use $-\widehat{u}d_{ij} + \widehat{v}_j - c_{ij}$ (or $-d_{ij}\overline{u}_i - c_{ij}$ in the case of W_d, $W_b \cap W_d$) instead of δ_{ij}^a.

After the a posteriori bound has been calculated for some fixed values of the duals, the dual searching approaches to tightening the bound can be applied. Considered in Sections 1.2, 1.4 they differ basically in the search direction used. Let $\widehat{u}, \widehat{v}$ be changed in the direction s, q with the stepsize ρ, such that $\widehat{u} + \rho s \geq 0$. The best (i.e., bound minimizing) stepsize in this direction is defined from the problem $\min_\rho\{UB(\widehat{u} + \rho s, \widehat{v} + \rho q, W) \mid \widehat{u} + \rho s \geq$ $\geq 0\}$. Note that here we do not restrict the sign of ρ. For the Mendelssohn approach we have $s = \widehat{u}, q = \widehat{v}$. The choice of Shetty-Taylor results in $s = \{a_i, i \in S\}, q = \{b_j, j \in T\}$. Denote by $\widehat{x}$ an optimal solution of the maximization problem to calculate $\varepsilon_Z^p(W)$, which is assumed to be unique. Then by (2.12) and the marginal value theorem we have $\nabla_{u_i} UB(\widehat{u}, \widehat{v}, W) = = a_i - \sum_{j \in T} d_{ij} \widehat{x}_{ij}$, $\nabla_{v_j} UB(\widehat{u}, \widehat{v}, W) = \sum_{i \in S} \widehat{x}_{ij} - b_j$ and the (projected) gradient can be used as the search direction as described in Section 1.4. We will denote the bounds, obtained by these dual-searching approaches as $\varepsilon_M^p(W), \varepsilon_{ST}^p(W)$ and $\varepsilon_\nabla^p(W)$, respectively.

2.3. Numerical Indications

As it was noted in [24, 26], there are two principal questions associated with the estimation of the quality of the error bounds. The first one is addressed to the quantitative quality of the bound, i.e. the difference of an error bound from the actual error. From this point of view, the closer the error bound to the actual error, the better. The second question is how to compare different aggregation strategies based on the error bound. It is not obvious that the aggregation yielding the tightest error bound has the smallest aggregation error. Moreover, different bound computation techniques may result in different correspondence between the error bound and the actual error.

We briefly discuss the results of a numerical study realized in [18] to test the bounds derived for the GTP and to explore the correspondence between the error bound and the actual error. Three classes of the problems were studied: the general GTP (GGTP), the specific GTP with $d_{ij} = p_i t_j$ (SGTP), and the classical transportation problem (TP). For each class three problem sizes were considered, specified by (number of sources)$\times$(number of destinations): small (30$\times$50), medium (50$\times$100), and large (100$\times$150). For each problem size 10 problem were randomly generated. The destinations are clustered to 80, 60, 40, 20 and 10 percents of the total number of destinations. Each cluster has $\lfloor |T| \,/\, |T_A| \rfloor$ destinations sorted by their index, except from the last cluster which contains all other destinations. Here $\lfloor \alpha \rfloor$ denotes the integer part of α. Thus there are a total of (3 problem sizes)$\times$(5 cluster levels)$\times$(10 problems)=150 observations for each class of the problems. Four localizations were used to calculate the aggregation bounds: $W_s, W_d, W_b \cap W_s$ and $W_b \cap W_d$. For each localization the a priori bound $\varepsilon^a(W)$ and four a posteriori bounds $\varepsilon_Z^p(W), \varepsilon_M^p(W), \varepsilon_{ST}^p(W)$, and $\varepsilon_\nabla^p(W)$ were calculated. These bounds were calculated for each problems class, except for the a priori bound, which was calculated only for the TP and SGTP.

The dimension of the original problem allows exact solution for all problem sizes. Similar to Table 1.1 the upper and the lower bounds for z^* were characterized by $(\overline{z} - z^*)/z^*$ and $(z^* - LB)/z^*$. Here $LB = \overline{z} - \varepsilon(W)$. The average values of these indicators for each bound calculation approach and localization are represented in Tables 2.1–2.3 along with the corresponding standard deviations. The correlations between the a priori error bound $\varepsilon^a(W)$ and actual error $\overline{z} - z^*$are given in Table 2.4 for the SGTP. For these correlations all p-values were less than 0.001 In all tables we use W_{bs} and W_{bd} to denote $W_b \cap W_s$ and $W_b \cap W_d$, respectively.

These preliminary computational results show that the error bounds calculated for the localization $W_b \cap W_d$ are the tightest, compared with three other localizations. This takes place for all types of problems and problems sizes, all bounds calculation approaches and types of the error bounds, a priori and a posteriori. We would like to stress, that the a priori bound $\varepsilon^a(W_b \cap W_d)$ for the SGTP is reasonably high and statistically significant correlated with the actual error. That is: a) we may expect that the actual error varies the same way the error bound does, and b) if the customers in the SGTP are aggregated using the strategy outlined in this study, the bound $\varepsilon^a(W_b \cap W_d)$ is an appropriate guide in selecting the clustering level.

The results provided in [18] show that the a posteriori error bounds are much more correlated with the actual error than the a priori bounds. Moreover, the dual searching approaches result in a higher correlation with the actual error compared with $\varepsilon^p(W)$. In particular, Mendelssohn type bound $\varepsilon^p_M(W)$ provides the highest correlation.

Concerning the quantitative characteristics, the a posteriori bounds provide much more tight approximation of the actual error than the a priori bounds. Again, the tightest approximation takes place for the localization $W_b \cap W_d$. Comparing various techniques to calculate the a posteriori bounds, the tightest bounds were obtained by the Mendelssohn approach and it is less sensitive to the type of the localization used. This differs from the experimental results reported for the continuous multidimensional knapsack problem [14, 17], where the tightest bounds were obtained by Shetty-Taylor and the gradient projection approaches.

Based on this experimental study we may conclude that the choice of the localization plays an important role in aggregation bounds computations.

	Problem size											
Bounds (in %)	Small				Medium				Large			
Upper bound	**154** 54				**163** 61				**128** 26			
Lower bounds	W_d	W_s	W_{db}	W_{sb}	W_d	W_s	W_{db}	W_{sb}	W_d	W_s	W_{db}	W_{sb}
A priori	**162**	**441**	**138**	**373**	**102**	**675**	**89**	**572**	**74**	**323**	**64**	**265**
	43	131	34	63	96	630	81	463	32	108	23	91
Zipkin	**29**	**293**	**17**	**242**	**27**	**479**	**23**	**399**	**20**	**252**	**16**	**229**
	13	156	8	131	11	136	11	153	5	43	2	21
Mendelssohn	**15**	**29**	**9**	**15**	**10**	**22**	**7**	**12**	**10**	**14**	**6**	**8**
	4	1	3	1	3	1	1	2	1	1	1	0.7
Shetty-Taylor	**23**	**261**	**16**	**182**	**25**	**408**	**22**	**371**	**19**	**240**	**15**	**226**
	8	120	10	108	12	162	11	120	3	38	3	18
Gradient	**14**	**156**	**8**	**96**	**18**	**207**	**14**	**141**	**14**	**195**	**9**	**124**
	6	53	3	24	4	57	5	38	5	12	3	21

Table 2.1. Bounds in % for the TP-**average** and standard deviation.

	Problem size											
Bounds (in %)	Small				Medium				Large			
Upper bound	**167** 68				**222** 93				**125** 37			
Lower bounds	W_d	W_s	W_{db}	W_{sb}	W_d	W_s	W_{db}	W_{sb}	W_d	W_s	W_{db}	W_{sb}
A priori	**193**	**440**	**160**	**371**	**160**	**401**	**117**	**342**	**88**	**611**	**71**	**375**
	66	217	63	140	52	208	51	194	24	452	35	116
Zipkin	**26**	**485**	**19**	**328**	**98**	**713**	**92**	**478**	**17**	**850**	**14**	**498**
	14	39	11	46	69	506	74	325	6	86	2	14
Mendelssohn	**13**	**62**	**7**	**51**	**15**	**51**	**9**	**41**	**10**	**63**	**5**	**58**
	4	8	3	11	1	18	1	23	1	5	1	4
Shetty-Taylor	**25**	**483**	**16**	**314**	**73**	**682**	**69**	**435**	**15**	**834**	**13**	**462**
	14	96	12	52	52	513	41	378	6	27	4	21
Gradient	**21**	**228**	**14**	**128**	**79**	**446**	**75**	**256**	**14**	**428**	**9**	**239**
	14	66	5	31	61	328	55	153	3	31	3	27

Table 2.2. Bounds in % for the SGTP-**average** and standard deviation.

	Problem size											
Bounds (in %)	Small				Medium				Large			
Upper bound	**164**				**163**				**191**			
	61				48				57			
Lower bounds	W_d	W_s	W_{db}	W_{sb}	W_d	W_s	W_{db}	W_{sb}	W_d	W_s	W_{db}	W_{sb}
Zipkin	**25**	**500**	**17**	**326**	**28**	**1028**	**23**	**554**	**65**	**685**	**60**	**496**
	13	121	7	23	14	96	11	41	23	121	24	35
Mendelssohn	**13**	**61**	**7**	**51**	**10**	**64**	**6**	**54**	**13**	**66**	**8**	**61**
	5	12	2	6	2	5	1	6	1	8	1	14
Shetty-Taylor	**23**	**486**	**16**	**307**	**27**	**1011**	**21**	**513**	**51**	**661**	**47**	**472**
	11	83	5	93	12	105	9	27	21	213	19	114
Gradient	**13**	**254**	**9**	**142**	**18**	**552**	**15**	**232**	**54**	**39**	**49**	**218**
	4	13	3	16	6	23	4	24	28	14	16	37

Table 2.3. Bounds in % for the GGTP-**average** and standard deviation.

Problem size											
Small				Medium				Large			
W_d	W_s	W_{db}	W_{sb}	W_d	W_s	W_{db}	W_{sb}	W_d	W_s	W_{db}	W_{sb}
0.5999	0.4873	0.6743	0.5957	0.7417	0.2305	0.7839	0.3693	0.4344	0.1861	0.5755	0.4905

Table 2.4. Correlation between ε^a and the actual error for the SGTP.

The proper localization not only improve the value of the error bound, but can also increase the correlation between the aggregation bound and the actual error.

§3. Variable Aggregation in Nonlinear Programming

In this section we show how to transform some results of linear programming aggregation to the nonlinear context.

Consider the following nonlinear programming problem:

$$z^* = \max f(x), \tag{3.1}$$

$$\text{s.t. } F(x) \leq 0, \ x \geq 0,$$

where $x = \{x_j\}$ is an n-vector of decision variables, $F(x) = \{F_i(x)\}$ is an m-vector-function and $f(x)$ is a scalar function. It is assumed that (3.1) has a finite optimal solution.

The corresponding variable aggregated problem is defined by a partition S of the set of variables indices into a set of clusters $S_k, k = 1, ..., K$, such that $S_k \cap S_p = \varnothing, k \neq p$ and $\cup_{k=1}^{K} S_k = \{1, ..., n\}$. Let $|S_k| = n_k$, $\sum_k n_k = n$. Consider an $n \times K$ weighting matrix $g = blockdiag\{g^k, k = 1, ..., K\}$, where $g^k = \{g_l^k, l \in S_k\}$ is a n_k-vector, fulfilling the following normalizing conditions:

$$g \in G = \left\{ g \mid \sum_{l \in S_k} g_l^k = 1, \ g_l^k \geq 0, \ k = 1, ..., K, \ l \in S_k \right\}.$$

Fixing $g \in G$, the aggregated problem, corresponding to (3.1), is stated as follows:

$$\overline{z} = \max f(gX), \tag{3.2}$$

$$\text{s.t. } F(gX) \leq 0, \ X = \{X_k\} \geq 0,$$

where X is a K-vector of aggregated variables. If (3.1) is a linear programming problem, then (3.2) coincides with the column aggregated problem considered in Section 1.1.

Remark 3.1. Huberman [10] used a slightly different definition of the aggregated problem, considering an arbitrary weighting matrix. This choice results in the constraints $gX \geq 0$ for (3.2). We use a nonnegative normalized weighting matrix to simplify this condition. Moreover, the choice of normalized weights guarantees us one-to-one correspondence with aggregation results for linear programming.

Clear, that $\overline{z} \leq z^*$, since any feasible solution of the aggregated problem always can be disaggregated into a feasible solution of the original problem. The simplest way is to use a fixed-weight disaggregation $x = gX$.

Having the lower bound $\overline{z} \leq z^*$, we want to generate an upper bound as well. To this end we impose some additional assumption for (3.1). Suppose, that $F_i(x)$ and $-f(x)$ are convex continuously differentiable functions. It is not hard to verify, that the same holds for $F_i(gX)$ and $-f(gX)$ with respect to the aggregated variables X for any fixed $g \in G$. Let the Slater regularity condition be satisfied for (3.2), that is $\exists \ \widetilde{X} \geq 0$, such that $F_i(g\widetilde{X}) < 0$,

$i = 1, .., m$. Then the Slater condition holds for (3.1) and the Kuhn-Tucker and saddle point theorems are applicable to both (3.1) and (3.2).

Following Section 1.4, we assume that a localization $W \subseteq \mathbf{R}^n$ is known, such that $x^* \in W$ and W is convex, closed and bounded. Moreover, if not specified otherwise, the condition $x \geq 0$ is always included in the definition of W. Let $\overline{X}$ be an optimal solution of (3.2), $\overline{x} = g\overline{X}$ be the fixed-weight disaggregated solution, and $\overline{u} = \{\overline{u}_i\}$ be an optimal vector of the Lagrange multipliers (duals) for (3.2). Denote by $u = \{u_i\} \geq 0$ a vector of the Lagrange multipliers for (3.2).

Proposition 3.1. *For any fixed $u \geq 0$ we have*

$$z^* \leq UB_1(u, W) \leq UB_2(u, W),$$

where

$$UB_1(u, W) = \max_{x \in W} \left\{ f(x) - \sum_i u_i F_i(x) \right\}, \tag{3.3}$$

$$UB_2(u, W) = f(\overline{x}) - \sum_i u_i F_i(\overline{x}) - \sum_j \overline{x}_j \left[\frac{\partial f}{\partial x_j}(\overline{x}) - \sum_i u_i \frac{\partial F_i}{\partial x_j}(\overline{x}) \right]$$

$$+ \max_{x \in W} \left\{ \sum_j x_j \left[\frac{\partial f}{\partial x_j}(\overline{x}) - \sum_i u_i \frac{\partial F_i}{\partial x_j}(\overline{x}) \right] \right\}. \tag{3.4}$$

Moreover,

$$z^* = \min_{u \geq 0} UB_1(u, W) \tag{3.5}$$

and for $u = \overline{u}$,

$$UB_2(\overline{u}, W) = f(\overline{x}) + \max_{x \in W} \left\{ \sum_j x_j \left[\frac{\partial f}{\partial x_j}(\overline{x}) - \sum_i \overline{u}_i \frac{\partial F_i}{\partial x_j}(\overline{x}) \right] \right\}. \tag{3.6}$$

Proof. By definition of the localization, we can attach the condition $x \in W$ to the constraints of (3.1) without changing its optimal solution, such that

$$z^* = \max f(x), \tag{3.7}$$

$$\text{s.t. } F(x) \leq 0, \; x \in W.$$

Let $L(x,u) = f(x) - \sum_i u_i F_i(x)$ be the Lagrangian function for (3.7). Since the set W is a convex compact, the saddle point theorem holds (see, e.g., [12]), such that

$$z^* = \max_{x\in W}\min_{u\geq 0} L(x,u) = \min_{u\geq 0}\max_{x\in W} L(x,u) \leq \max_{x\in W} L(x,u) \equiv UB_1(u,W) \quad (3.8)$$

and the upper bound (3.3) holds for any $u \geq 0$.

A differentiable function $\varphi : \mathbf{R}^n \to \mathbf{R}$ is concave if and only if

$$\varphi(\tau) \leq \varphi(\xi) - \sum_{j=1}^{n}(\xi_j - \tau_j)\frac{\partial\varphi}{\partial\xi_j}(\xi) \quad \text{for all } \tau,\xi \in \mathbf{R}^n.$$

Using this inequality for $\varphi(\cdot) = L(\cdot,u)$, $\tau = x$, $\xi = \overline{x} \equiv g\overline{X}$ we get

$$L(x,u) \leq L(\overline{x},u) - \sum_{j=1}^{n}(\overline{x}_j - x_j)\left[\frac{\partial f}{\partial x_j}(\overline{x}) - \sum_i u_i \frac{\partial F_i}{\partial x_j}(\overline{x})\right]$$

and taking the maximum of both parts we obtain

$$\max_{x\in W} L(x,u) \leq L(\overline{x},u) - \sum_{j=1}^{n}\overline{x}_j\left[\frac{\partial f}{\partial x_j}(\overline{x}) - \sum_i u_i \frac{\partial F_i}{\partial x_j}(\overline{x})\right]$$

$$+\max_{x\in W}\left\{\sum_{j=1}^{n} x_j\left[\frac{\partial f}{\partial x_j}(\overline{x}) - \sum_i u_i \frac{\partial F_i}{\partial x_j}(\overline{x})\right]\right\} \equiv UB_2(u,W),$$

which by the inequality (3.8) gives the upper bound (3.4). Note also, that the bound (3.4) remains valid, if we substitute in (3.4) any $\widetilde{x} \in \mathbf{R}^n$ instead of $\overline{x}$

To prove (3.5) note, that by (3.3) and the definition of the Lagrangian function $L(x,u)$ we have $UB_1(u,W) = \max_{x\in W} L(x,u)$. Then by (3.8) we obtain

$$z^* = \min_{u\geq 0}\max_{x\in W} L(x,u) = \min_{u\geq 0} UB_1(u,W)$$

as required.

Now let $u = \overline{u}$. The complementarity condition for the aggregated problem (3.2) gives

$$0 = \overline{u}_i F_i(g\overline{X}) = \overline{u}_i F_i(\overline{x}), \ i = 1,...,m$$

and hence for $u = \overline{u}$ the second term in the right-hand side of (3.4) is zero.

Denote $q(X) \equiv f(gX)$, $R_i(X) \equiv F_i(gX)$. The complementarity condition for the dual to (3.2) is

$$0 = \overline{X}_k \left[\frac{\partial q}{\partial X_k}(\overline{X}) - \sum_i \overline{u}_i \frac{\partial R_i}{\partial X_k}(\overline{X}) \right], \quad k = 1, ..., K. \tag{3.9}$$

Since $\overline{x}_l^k = g_l^k \overline{X}_k$, then by differentiation we get

$$\frac{\partial q}{\partial X_k}(\overline{X}) = \sum_{l \in S_k} g_l^k \frac{\partial f}{\partial x_l^k}(\overline{x})$$

and

$$\overline{X}_k \frac{\partial q}{\partial X_k}(\overline{X}) = \sum_{l \in S_k} \overline{x}_l^k \frac{\partial f}{\partial x_l^k}(\overline{x}).$$

Doing the same for the second term in (3.9), we obtain that (3.9) is equivalent to

$$0 = \sum_{l \in S_k} \overline{x}_l^k \left[\frac{\partial f}{\partial x_l^k}(\overline{x}) - \sum_i \overline{u}_i \frac{\partial F_i}{\partial x_l^k}(\overline{x}) \right], \quad k = 1, ..., K.$$

Summing up these equalities, we get

$$0 = \sum_k \sum_{l \in S_k} \overline{x}_l^k \left[\frac{\partial f}{\partial x_l^k}(\overline{x}) - \sum_i \overline{u}_i \frac{\partial F_i}{\partial x_l^k}(\overline{x}) \right] = \sum_j \overline{x}_j \left[\frac{\partial f}{\partial x_j}(\overline{x}) - \sum_i \overline{u}_i \frac{\partial F_i}{\partial x_j}(\overline{x}) \right].$$

The last equality holds since $\cup_{k=1}^K S_k = \{1, ..., n\}$ and clustering is simply a rearrangement of the indices of the original variables. Thus we way conclude, that for $u = \overline{u}$ the third term in the right-hand side of (3.4) is zero and hence (3.6) holds.

Note that if $f(x), F_i(x)$ are linear functions, then $UB_1(u, W)$ coincides with $UB_2(u, W)$.

Corollary 3.1. *Let some original constraints be included in the definition of W, such that*

$$W = \{x \mid F_i(x) \leq 0, \ i \in I \subseteq \{1, ..., m\} \equiv M, \ x \in \widetilde{W}\},$$

where $\widetilde{W}$ is another localization of x^*. Then for any fixed $u \geq 0$ we have

$$UB_1(u, W) \equiv \max_{x \in W} \left\{ f(x) - \sum_{i \in M} u_i F_i(x) \right\} \geq$$

$$\geq \max_{x \in W} \left\{ f(x) - \sum_{i \in M \setminus I} u_i F_i(x) \right\} \geq z^*.$$

Proof. The last inequality is derived similar to (3.8) if we define the Lagrangian in the form $L(x,u) = f(x) - \sum_{i \in M \setminus I} u_i F_i(x)$, including in $L(x,u)$ only those constraints which are not used in the definition of W. The first inequality holds since

$$\sum_{i \in M} u_i F_i(x) = \sum_{i \in I} u_i F_i(x) + \sum_{i \in M \setminus I} u_i F_i(x)$$

and $\sum_{i \in I} u_i F_i(x) \leq 0$ for any $u \geq 0$, $x \in W$.

Assume the bound $UB_1(u,W)$ or $UB_2(u,W)$ has been calculated for a fixed $u \geq 0$, say $u = \overline{u}$. Then it is natural to look for another dual u to decrease the upper bound. Below we concentrate on the bound $UB_2(u,W)$. The case $UB_1(u,W)$ may be treated in a similar way.

Denote by $\theta(u)$ the objective-value function of the maximization problem in (3.4):

$$\theta(u) \equiv \max_{x \in W} \left\{ \sum_{j=1}^{n} x_j \left[\frac{\partial f}{\partial x_j}(\overline{x}) - \sum_i u_i \frac{\partial F_i}{\partial x_j}(\overline{x}) \right] \right\}. \qquad (3.10)$$

Assume (3.10) has a unique optimal solution $\widehat{x}$ for $u = \overline{u}$. Then by the marginal value theorem (see, e.g., [15]) $\theta(u)$ is differentiable at $u = \overline{u}$, such that

$$\nabla_{u_i} \theta(\overline{u}) = -\sum_{j=1}^{n} \widehat{x}_j \frac{\partial F_i}{\partial x_j}(\overline{x}).$$

Respectively,

$$\nabla_{u_i} UB_2(\overline{u}, W) = -F_i(\overline{x}) + \sum_j \overline{x}_j \frac{\partial F_i}{\partial x_j}(\overline{x}) - \sum_{j=1}^{n} \widehat{x}_j \frac{\partial F_i}{\partial x_j}(\overline{x}).$$

Now, similar to Section 1.4.1, we can take one step of the projected gradient technique for the problem $\min_{u \geq 0} UB_2(u,W)$, starting from $u = \overline{u}$. Obtaining new u in the form $u = \Pi_+(\overline{u} - \widetilde{\rho} \nabla_u UB_2(\overline{u}, W))$ for a certain $\widetilde{\rho} \geq 0$, we can proceed this way iteratively. Note that for the bound UB_1we have $z^* = \min_{u \geq 0} UB_1(u,W)$, while for the bound UB_2 we can guarantee only $z^* \leq \min_{u \geq 0} UB_2(u,W)$.

§4. Sharpening Localization by Valid Inequalities in Integer Programming

In this section we apply the results of Section 1 to derive an upper bound for the optimal objective value of an integer linear program. The main idea is to use the valid inequalities (the constraints which are utilized to make LP-relaxation closer to the integer problem) to sharpen the localization.

Consider the following linear integer problem:

$$z_I^* = \max cx,$$

$$\text{s.t. } Ax \leq b, \quad 0 \leq x \leq r, \tag{4.1}$$

$$x \text{ integer},$$

where $c = \{c_j\}, r = \{r_j\}$ are n-vectors, $A = [a_{ij}]$ is an $m \times n$ matrix having columns a_j. Denote by x_I^* an optimal integer solution of (4.1).

In many integer programming techniques it is important to get an upper bound for z_I^* (see, e.g., [23]). Frequently, an upper bound is obtained from LP-relaxed problem

$$z_{LP}^* = \max cx, \tag{4.2}$$

$$\text{s.t. } Ax \leq b, \quad 0 \leq x \leq r,$$

such that $z_I^* \leq z_{LP}^*$. Denote x_{LP}^* as the optimal solution of (4.2).

Let W_I be a convex and bounded set, such that $x_I^* \in W_I$. Throughout this section we assume, that conditions $0 \leq x \leq r$ are always included in the definition of the localization W_I. Consider the problem

$$z_R^* = \max cx, \tag{4.3}$$

$$\text{s.t. } Ax \leq b, \ x \in W_I.$$

By the definition of W_I, (4.3) is the relaxation of (4.1), such that $z_I^* \leq z_R^*$.

To derive an upper bound for z_R^* we can apply the results of Section 1.4 to obtain

$$z_I^* \leq z_R^* \leq UB(u^{W_I}, W_I) \text{ for any } u \geq 0, \tag{4.4}$$

where $UB(u^{W_I}, W_I)$ is of the form (1.18) since the original restrictions $x \leq r$ have been included in the definition of W_I. The bound (4.4) holds for any localization W_I of an integer solution x_I^*. It seems natural to define W_I by a valid inequality, which preserves integer solutions of (4.1), but cut off a part of the feasible region of the relaxed problem (4.2). Below we specify the choice of the valid inequality, associated with the solution of the aggregated problem.

For a fixed partition of the variables and a normalized weighting matrix $g \in G$ we state the aggregated problem for (4.2) as was done in Section 1.1:

$$\overline{z}_{LP} = \max cgX, \tag{4.5}$$

$$\text{s.t. } AgX \leq b, \; gX \leq r, \; X \geq 0.$$

Let $\overline{u}$ be an optimal dual vector for (4.5), $\overline{X}$ optimal to (4.5), and $\overline{x} = g\overline{X}$ the respective disaggregated solution.

By definition, $\overline{x}$ is feasible to (4.2), but generally speaking, noninteger. Let the linear constraint $dx \leq p$ be a valid inequality for the problem (4.1) and cuts off $\overline{x}$. That is, $dx \leq p$ for any integer $0 \leq x \leq r$, $Ax \leq b$, but $d\overline{x} > p$.

Let W_I be defined by

$$W_I = \{x \mid dx \leq p, \; 0 \leq x \leq r\}. \tag{4.6}$$

By the definition of W_I, $x_I^* \in W_I$. The choice of the localization W_I is motivated as follows. If $\overline{x}$ is "close" enough to x_{LP}^*, then we may expect that cutting off $\overline{x}$, the point x_{LP}^* is cut as well. In this case $z_R^* < z_{LP}^*$ and estimating z_R^* by (4.4), a better upper bound than z_{LP}^* could be obtained. As an initial guess of u^{W_I} in (4.4) we can take $u^{W_I} = \overline{u}^{W_I}$, where by the definition of W_I, $\overline{u}^{W_I}$ is an optimal dual vector for the constraints $AgX \leq b$ in (4.5). Then the bound can be improved by changing u^{W_I} the way it was described in section 1.4.

To illustrate this approach of sharpening the localization, suppose that (4.1) is a multidimensional boolean knapsack problem. That is, all the data involved are nonnegative integers, $b > 0$, and r is a unit vector of an appropriate dimension. One of the possible ways to get a valid inequality is to use a minimal cover. A minimal cover Q_i of the ith knapsack constraint is a subset of $\{1, ..., n\}$ such that $\sum_{j \in Q_i} a_{ij} > b_i$ and $\sum_{j \in S_i^l} a_{ij} \leq b_i$ for all proper subsets S_i^l of Q_i. The extension $E(Q_i)$ of a minimal cover Q_i is defined by $E(Q_i) = Q_i \cup \{\widetilde{j} \in \{1, 2..., n\} \setminus Q_i \mid a_{i\widetilde{j}} \geq a_{ij} \text{ for any } j \in Q_i\}$. Then

$$\sum_{j \in E(Q_i)} x_j \leq |Q_i| - 1 \tag{4.7}$$

is a valid inequality for the knapsack polytope. It may also be possible to "lift" the inequality to make it stronger [23]. Note, that lifting can result in a valid inequality with the coefficients greater than 1.

Now we need to identify a valid inequality which is violated by a given point $\overline{x}$. It is a typical separation problem, arising when minimal cover inequalities are used in cutting plane algorithms. The measure of the violation is $\sum_{j\in E(Q_i)} x_j - |Q_i| + 1$ and the separation problem to maximize the violation may be stated as follows:

$$\zeta = \min \sum_j (1 - \overline{x}_j)\alpha_j, \tag{4.8}$$

$$\text{s.t.} \sum_j a_{ij}\alpha_j > b_i, \;\; \alpha_j \in \{0,1\},$$

where $\alpha_j = 1$ if $j \in E(Q_i)$, and $\alpha_j = 0$ otherwise. The problem (4.8) is often solved approximately, based on its LP-relaxation.

Now assume the minimal cover inequality (4.7), which cuts off $\overline{x}$ has been constructed. Then the localization W_I of the form (4.6) is as follows:

$$W_I = \left\{ \sum_{j\in E(Q_i)} x_j \le |Q_i| - 1, \;\; 0 \le x_j \le 1, \;\; j = 1,...,n \right\}.$$

Note, that this W_I is defined after the aggregated problem has been solved. Hence we may treat W_I as an example of a posteriori localization.

The principal step of the approach above is separation - generation of a valid inequality, that cuts off the disaggregated solution. We would like to stress, that a minimal cover is just an example of a valid inequality, which seems suitable for the boolean knapsack problem. But this approach is applicable for other problems as well, if a certain procedure is available to obtain a valid inequality that is violated by the disaggregated solution.

The other approach to identify the valid inequalities was proposed in [8] for the problem (4.1), where r is a unit vector e of an appropriate dimension.

Let $\overline{u} = (\overline{u}^1, \overline{u}^2)$ be an optimal dual solution for the aggregated problem (4.5). Here $\overline{u}^1$ corresponds to the linear constraints in (4.5), while $\overline{u}^2$ is associated with the upper bounds restrictions $gX \le e$. Let $\sigma' = \{S'_k,$ $k = 1,...,K'\}$ be any partition of the variables, not necessary the same as the partition used to define the aggregated problem.

Bearing in mind the valid inequalities (4.7), suppose that integers $p_k \leq |S'_k|, k = 1, ..., K'$ are known such that an optimal solution x_I^* fulfils the restrictions

$$\sum_{j \in S'_k} x_j \leq p_k, \ k = 1, ..., K'. \tag{4.9}$$

Define the localization

$$\widetilde{W}_I = \{x \mid \sum_{j \in S'_k} x_j \leq p_k, \ 0 \leq x_j \leq 1, \ j \in S'_k, \ k = 1, ..., K'\}$$

and suppose that the indices $j \in S'_k$ are rearranged such that $\delta_j \geq \delta_{j+1}$, where $\delta_j = c - \overline{u}^1 a^j - \overline{u}_j^2$. Then, calculating the bound (1.17) by (1.21) for $u = \overline{u}$, $W = W_I$, we obtain

$$z_I^* \leq z_R^* \leq UB(\overline{u}, \widetilde{W}_I) = \overline{z}_{LP} + \max_{x \in \widetilde{W}_I} \sum_{j=1}^{n} \delta_j x_j = \overline{z}_{LP} + \sum_{k=1}^{K'} \sum_{j=1}^{p_k} (\delta_j)^+.$$

Given a dual solution $\overline{u}$, a very general optimization problem can be formulated, the solution of which yields the best (i.e., minimizing $UB(\overline{u}, \widetilde{W}_I)$) set of inequalities (4.9). The problem is as follows:

$$\min_{\sigma', p} \sum_{k=1}^{K'} \sum_{j=1}^{p_k} (\delta_j)^+,$$

$$\text{s.t. } Ax \leq b,$$

$$0 \leq x_j \leq 1 \Longrightarrow \sum_{j \in S'_k} (x_I^*)_j \leq p_k,$$

$$x_j \text{ integer}, \ k = 1, ..., K'.$$

If the inequalities (4.9) are restricted to the type (4.7), this optimization problem is specialized to determine a set of disjoint valid inequalities (4.7). This is indeed a difficult optimization problem, in general, much more difficult than the original problem (4.1). It is really a clustering problem, since we wish to group together variables with approximately the same value of δ_j.

One possible heuristic procedure for performing such a clustering is to rank the variables according to δ_j, and then to start from the largest value and include one after the other until the set of variables is a cover for one of the constraints. The cover is then reduced to a minimal by a standard technique [23] and procedure is repeated to form the next set of variables, until no more covers can be found. The partition σ' is then defined by the set of covers $\{Q_1, Q_2, ...\}$ and the singleton sets of variables that are not included in a cover. A formal statement of the procedure can be found in [7].

We would like to stress the difference between the two considered approaches. In the first case a valid inequality is generated to cut off the disaggregated solution and the aim is to find the "strongest" inequality of this type. In the second approach a set of disjoint valid cuts is generated aimed to minimize an aggregation bound.

Consider the following illustrative example [8]:

$$\begin{aligned} z_I^* = \max\ & 19x_1 + 20x_2 + 21x_3 + 39x_4 + 40x_5 + 41x_6, \\ \text{s.t.}\ & 5x_1 + 6x_2 + 9x_3 + 14x_4 + 15x_5 + 16x_6 \le 19, \\ & 5x_1 + 5x_2 + 7x_3 + 11x_4 + 12x_5 + 13x_6 \le 13, \\ & x_j \in \{0, 1\}. \end{aligned}$$

The optimal solution is

$$x_I^* = (0,1,1,0,0,0) \text{ or } (0,0,0,0,0,1)$$

with $z_I^* = 41$. The LP-relaxation to the integer problem has the optimal solution

$$x_{LP}^* = (1,1,0,\frac{3}{11},0,0), \quad z_{LP}^* = 49\frac{7}{11} \simeq 49.64.$$

The aggregation is defined by clustering the variables in two groups: $K = 2$, $S_1 = \{1,2,3\}$, $S_2 = \{4,5,6\}$. The weighting matrix $g \in G$ is chosen as follows:

$$g = \frac{1}{3}\begin{bmatrix} 1\,1\,1\,0\,0\,0 \\ 0\,0\,0\,1\,1\,1 \end{bmatrix}^T.$$

The aggregated problem is then

$$\begin{aligned} \bar{z}_{LP} = \max\ & 20X_1 + 40X_2, \\ \text{s.t.}\ & 6\tfrac{2}{3}X_1 + 15X_2 \le 19, \\ & 5\tfrac{2}{3}X_1 + 12X_2 \le 13, \\ & 0 \le X_1 \le 3 \ \text{(3 times)}, \\ & 0 \le X_2 \le 3 \ \text{(3 times)}. \end{aligned}$$

The optimal aggregated solution is $\overline{X}_1 = 2\frac{5}{17} \simeq 2.294$, $\overline{X}_2 = 0$, $\overline{z}_{LP} = 45\frac{15}{17} \simeq$ $\simeq 45.88$. For the dual aggregated solution we have $\overline{u}_1 = 0$, $\overline{u}_2 = 3\frac{9}{17} \simeq 3.529$, while all other components are zeros. The disaggregated solution is $\overline{x}_{LP} =$ $= \frac{1}{17}(13, 13, 13, 0, 0, 0)$.

Using standard techniques [23] two lifted minimal cover inequalities, which cut off the disaggregated solution $\overline{x}_{LP}$, have been generated: $x_1 + x_2 + x_3 +$ $+ 2x_4 + 2x_5 + 2x_6 \leq 2$ and $x_1 + x_2 + x_3 + x_4 + 2x_5 + 2x_6 \leq 2$. Respectively, we have two localizations of the form (4.6):

$$W_I^1 = \{x \mid x_1 + x_2 + x_3 + 2x_4 + 2x_5 + 2x_6 \leq 2,\ 0 \leq x_j \leq 1\},$$

$$W_I^2 = \{x \mid x_1 + x_2 + x_3 + x_4 + 2x_5 + 2x_6 \leq 2,\ 0 \leq x_j \leq 1\}.$$

Note, that since the restrictions $x_j \leq 1$ have been used in the definition of W_I^1 and W_I^2, then $u^{W_I^1} = u^{W_I^2} = \{u_1, u_2\}$. Computing the upper bound by (4.4) we obtain for $u = \overline{u}$:

$$z_I^* \leq UB(\overline{u}^{W_I^1}, W_I^1) = \overline{z}_{LP} + \max\{\sum_j \delta_j x_j \mid x \in W_I^1\} = 49\frac{10}{17} \simeq 49.588,$$
$$z_I^* \leq UB(\overline{u}^{W_I^2}, W_I^2) = \overline{z}_{LP} + \max\{\sum_j \delta_j x_j \mid x \in W_I^2\} = 49\frac{10}{17}, \tag{4.10}$$

where the vector δ of the coefficients δ_j, associated with the bound computation problem (1.18) is as follows:

$$\delta = \frac{1}{17}(23, 40, -63, 3, -40, -83).$$

Note, that here the bound $UB(\overline{u}^{W_I}, W_I)$ coincides with the bound $UB(\overline{u}, W_I)$ since the components of $\overline{u}$, associated with the restrictions $x_j \leq 1$ are zeros. In (4.10) we have obtained in both cases a stronger bound $z_I^* \leq 49.588$ than the ordinary LP-relaxation gives: $z_I^* \leq z_{LP}^* = 49.64$. Note, that for both lifted inequalities that we have generated to cut off the disaggregated solution $\overline{x}_{LP}$, the x_{LP}^* was cut off as well.

Now try to improve the bounds by changing the duals u^{W_I} in (4.4). Note, that in our case u^{W_I} has only two components, associated with the first two constraints of the problem. The unique optimal solution of the minimization problem in (4.10), the same for both problems, is $\widehat{x} = (1,1,0,0,0,0)$. The gradient computed by (1.22) is

$$\nabla_{u_1} UB(\overline{u}^{W_I^1}, W_I^1) = \nabla_{u_1} UB(\overline{u}^{W_I^2}, W_I^2) = b_1 - \sum_j a_{1j}\widehat{x}_j = 8,$$

$$\nabla_{u_2} UB(\overline{u}^{W_I^1}, W_I^1) = \nabla_{u_1} UB(\overline{u}^{W_I^2}, W_I^2) = b_2 - \sum_j a_{2j}\widehat{x}_j = 3.$$

The projected gradient step leads to $u^{W_I}(\rho)$ in the form

$$u^{W_I}(\rho) = \Pi_+(\overline{u}^{W_I} - \rho\nabla_u UB(\overline{u}^{W_I}, W_I)) = \Pi_+ \begin{pmatrix} 0 - 8\rho \\ 3\frac{9}{17} - 3\rho \end{pmatrix},$$

since $\overline{u}^{W_I} = (\overline{u}_1, \overline{u}_2) = (0, 3\frac{9}{17})$, the same for both localizations. Further, we will drop the superscript and write simply $u(\rho)$. We see, that $u(\rho) = (0,0)^t$ for $\rho > \frac{20}{17} \simeq 1.176$ and $u(\rho) = (0, 3\frac{9}{17} - 3\rho)^t$ for $0 \le \rho \le 1.176$. Substituting this $u(\rho)$ in (4.4) we get for the first localization W_I^1 :

$$UB(u(\rho), W_I^1) = \sum_{i=1}^{2} u_i(\rho)b_i + \max\left\{\sum_j \delta_j(\rho)x_j \mid x \in W_I^1\right\},$$

where $\delta_j(\rho) = c_j - \sum_{i=1,2} a_{ij}u_i(\rho)$. Applying a one-dimensional search we obtain

$$\widetilde{\rho} = \arg\min\{UB(u(\rho), W_I^1) \mid 0 \le \rho \le 1.176\} = 1\frac{10}{153} \simeq 1.06536.$$

For this $\widetilde{\rho}$ we have $\delta(\widetilde{\rho}) = (17\frac{1}{3}, 18\frac{1}{3}, 18\frac{2}{3}, 35\frac{1}{3}, 36, 36\frac{2}{3})$ such that $UB(u(\widetilde{\rho}), W_I^1) = 41\frac{1}{3}$ and hence $z_I^* \le 41\frac{1}{3}$. Note, that since the objective coefficients of the original problem are integer, this is in fact a sharp bound for $z_I^* = 41$.

Moreover, if we attach the first generated valid inequality to the LP-relaxation of the original integer problem to form a problem similar to (4.3), we obtain $z_R^* = 41\frac{1}{3}$ for its optimal value. Hence, the second inequality in (4.4) holds as equality $z_R^* = UB(u(\widetilde{\rho}), W_I^1)$ and the bound $UB(u^{W_I^1}, W_I^1)$ cannot be further improved by changing the duals.

For the bound $UB(u(\rho), W_I^2)$ the one-dimensional search gives

$$\widetilde{\rho} = \arg\min\{UB(u(\rho), W_I^2) \mid 0 \le \rho \le 1.176\} = \frac{10}{153} \simeq 0.06536.$$

For this $\widetilde{\rho}$ we have $\delta(\widetilde{\rho}) = \frac{1}{3}(7, 10, -7, 7, 0, -7)$ and $UB(u(\widetilde{\rho}), W_I^2) = 49$. This bound also cannot be improved by changing the duals.

The aggregation approach was used to obtain upper bounds for the integer program. But it is also possible to treat this approach as a way to generate "strong" cuts, based on the solution of the aggregated problem. To estimate how good are the cuts generated this way, a branch-and-bound routine from the standard package FortMP was used. Solving our problem starting from its LP-relaxation, it was necessary to trace 8 nodes on the branch-and-bound tree to get an optimal integer solution and 31 nodes to provide its optimality. If we attach both generated cuts to the LP-relaxed problem, then it takes 1 and 3 nodes, respectively.

Applying to the same example the heuristic procedure to generate a set of disjoint valid inequalities, we obtain three constraints: $x_2 + x_4 + x_5 + x_6 \leq 1$, $x_1 \leq 1$, $x_3 \leq 1$. This gives the upper bound

$$z_I^* \leq UB(\overline{u}, \widetilde{W}_I) = \overline{z}_{LP} + \delta_1 + \delta_2 = 49\frac{10}{17} \simeq 49.588,$$

the same as in (4.10). Meanwhile, the generated disjoint cuts are not violated by the disaggregated solution $\overline{x}_{LP} = \frac{1}{17}(13, 13, 13, 0, 0, 0)$. If we attach the disjoint cuts to the LP-relaxation of the original problem, then the respective optimal solution of that problem be $(1, 0.5, 0, 0.5, 0, 0)$ with the objective value 48.5. This means that the bound obtained by disjoint cuts heuristics cannot be sharpened stronger than $z_I^* \leq 48.5$ by changing the duals.

§5. The Generalized Assignment Problem

The generalized assignment problem (GAP) is closely related to the GTP. Comparing with (2.1), in the GAP $b_j = 1$, $j \in T$ and $x_{ij} \in \{0, 1\}$, $\forall i, j$. In this section a continuous aggregated problem is formulated for the GAP similar to Section 2.1. Feasible integer solutions are generated by optimal disaggregation coupled with some complimentary heuristics.

An important application of the GAP is allocation of jobs to machines. Suppose that there are $m = |S|$ machines and $n = |T|$ jobs, and each job has to be processed on exactly one machine. The cost of processing job i on machine j is c_{ij} and job j takes d_{ij} times units to process on machine i. The total available processing time on machine i is a_i. As decision variables we have $x_{ij} = 1$, if job j is processed on machine i, $x_{ij} = 0$ otherwise. The standard GAP is then to determine the least expensive allocation of jobs given the assumptions above.

In [9] the aggregation was studied for an equivalent formulation of the GAP:

$$z^* = \max \sum_{i \in S} \sum_{j \in T} c_{ij} x_{ij}, \tag{5.1}$$

$$\sum_{j \in T} d_{ij} \le a_i, \ i \in S, \tag{5.2}$$

$$\sum_{i \in S} x_{ij} \le 1, \ j \in T, \tag{5.3}$$

$$x_{ij} \in \{0, 1\} \quad \forall i, j. \tag{5.4}$$

The optimal solution to (5.1)–(5.4) will fulfill the constraints (5.3) with equality if the standard GAP (the minimization problem with equality conditions in (1.3)) has a feasible solution. The reason for using this formulation instead the standard GAP is that the disaggregation will then always generate a feasible solution to the problem (5.1)–(5.4), but not necessary to the standard GAP.

In many practical applications the number of jobs is much larger than the number of machines. Then it seems natural to aggregate the jobs to obtain an aggregated problem with the same number of machines, but with a smaller number of jobs.

Similar to Section 2 consider a partition of the set T of jobs into a set T^A of clusters T_k, $k = 1, ..., K$. For each k non-negative normalized weights g_j^k are introduced, such that $\sum_{j \in T_k} g_j^k = 1$, $k = 1, ..., K$.

By Comments 1.1, the column (variable) aggregated problem to the LP-relaxation of (5.1)–(5.4) can be obtained by fixing the weights and substituting $x_{ij} = g_j^k X_{ik}$, $j \in T_k$ into (5.1)–(5.3). This yields:

$$\overline{z}_{LP} = \max \sum_{i \in S} \sum_{k \in T^A} \overline{c}_{ik} X_{ik},$$

$$\sum_{k \in T^A} \overline{d}_{ik} X_{ik} \le a_i, \ i \in S,$$

$$g_j^k \sum_{i \in S} X_{ik} \le 1, \; j \in T_k, k = 1, ..., K,$$

$$X_{ik} \ge 0,$$

where $\overline{c}_{ik} = \sum_{j \in T_k} g_j^k c_{ij}$, $\overline{d}_{ik} = \sum_{j \in T_k} g_j^k d_{ij}$.

Eliminating redundant constraints, we obtain

$$\overline{z}_{LP} = \max \sum_{i \in S} \sum_{k \in T^A} \overline{c}_{ik} X_{ik}, \tag{5.5}$$

$$\sum_{k \in T^A} \overline{d}_{ik} X_{ik} \le a_i, \; i \in S,$$

$$\sum_{i \in S} X_{ik} \le n_k, \; k = 1, ..., K,$$

$$X_{ik} \ge 0,$$

where $n_k = \min_{j \in T_k}(1/g_j^k)$. If all g_j^k are equal for each $k = 1, ..., K$, then $n_k = |T_k|$.

The problem (5.5) will be treated as an aggregated problem, corresponding to (5.1)–(5.4).

Remark 5.1. Similar to Section 2, the choice of the weights in the form g_j^k (independent on i) is aimed to achieve the redundancy in the column aggregated problem. This way a specific structure of the constraints of the GAP is preserved in the aggregated problem.

The fixed-weight disaggregated solution $\overline{x}_{ij} = g_j^k \overline{X}_{ik}$ is, generally speaking, non integer and hence is not feasible to the GAP (5.1)–(5.4). To obtain a feasible solution of the GAP based on an optimal aggregated solution $\overline{X}$, an approach similar to the optimal disaggregation (Section 1.1) can be used. Suppose that for each $k = 1, ..., K$ the following subproblem was solved:

$$z_k^*(\overline{X}_k) = \max \sum_{i \in S} \sum_{j \in T_k} c_{ij} x_{ij}, \tag{5.6}$$

$$\sum_{j \in T_k} d_{ij} x_{ij} \le \overline{d}_{ij} \overline{X}_{ik}, \; i \in S,$$

$$\sum_{i \in S} x_{ij} \le 1, \; j \in T_k,$$

$$x_{ij} \in \{0, 1\}, \; i \in S, \; j \in T_k.$$

Then the solution, composed of the optimal subproblem´s solutions, is feasible to the GAP (5.1)–(5.4) and $z^* \geq \sum_{k \in T^A} z_k^*(\overline{X}_k)$.

It may be possible to improve this feasible solution by some heuristics [9]. A common situation is that some jobs are not assigned to any of the machines while at the same time, some capacity is left in the machines. A simple heuristic is just to assign these jobs (as many as possible) in such a way that it will not lead to any reallocation. Moreover, if the number of jobs not assigned are just a few, it can be worthwhile to solve another generalized assignment problem, calculating the "best" additional assignment.

Finally, if the capacity left after the disaggregation is used up and some of the jobs have still not being assigned, an exchange heuristic can be carried out. This is done in such a way that the amount of unused capacity will be increased at the same time as the set of jobs assigned will be unchanged. This will hopefully lead to an increased number of jobs assigned. Note, that the aim is to assign all the jobs, since the standard GAP has equality semi assignment constraints.

To obtain the upper bound on z^* we can use the results of Section 2.2, remembering that in (5.1)–(5.4) we maximize the objective and $z^* \leq z_{LP}^*$, where z_{LP}^* is the optimal value of the LP-relaxation to (5.1)–(5.4). Similar to the bound $\varepsilon^p(W)$ in Section 2.2 we get

$$z^* \leq \overline{z}_R + \max_{x \in W} \sum_{i \in S} \sum_{j \in T} \delta_{ij}^p x_{ij} \equiv UB(\overline{u}, \overline{v}, W),$$

where W is a localization for the LP-relaxation to (5.1)–(5.4), $\delta_{ij}^p = c_{ij} - d_{ij}\overline{u}_i - v_j$, and by Comments 2.1 $v \in \{v_j \geq 0,\ j \in T \mid \sum_{j \in T} v_j = \sum_{k \in T^A} n_k \overline{v}_k\}$. Here $\overline{u}_i, \overline{v}_k$ are the optimal dual multipliers to the aggregated problem (5.5).

In particular, defining the localization by

$$W_0 = \{x_{ij} \geq 0 \mid \sum_{i \in S} x_{ij} \leq 1,\ j \in T\},$$

we obtain the upper bound provided in [9]:

$$z^* \leq UB(\overline{u}, v, W_0) = \overline{z}_{LP} + \sum_{j \in T} \max_{i \in S} (\delta_{ij}^p)^+.$$

Similar to (2.16) the multipliers v_j can be excluded from the expression for the upper bound:

$$z^* \le UB((\overline{u}, v)^{W_0}, W_0) = \overline{z}_{LP} - \sum_{k \in T^A} n_k \overline{v}_k + \sum_{j \in T} \max_{i \in S}(c_{ij} - d_{ij}\overline{u}_i)^+.$$

To illustrate the aggregation/disaggregation procedure, consider an example [9] with 2 machines and 20 jobs, such that

$$\{a_i\} = \begin{pmatrix} 60 \\ 55 \end{pmatrix},$$

$$[d_{ij}] = \begin{bmatrix} 5 & 3 & 13 & 9 & 9 & 5 & 6 & 9 & 6 & 12 & 8 & 3 & 4 & 12 & 3 & 10 & 15 & 4 & 11 & 7 \\ 8 & 3 & 4 & 12 & 6 & 9 & 10 & 4 & 6 & 5 & 11 & 8 & 3 & 1 & 7 & 11 & 8 & 7 & 9 & 3 \end{bmatrix},$$

$$[c_{ij}] = 100 + [\widetilde{c}_{ij}], \text{ where}$$

$$[\widetilde{c}_{ij}] = \begin{bmatrix} 1 & 1 & 1 & 6 & 4 & 1 & 1 & 1 & 9 & 1 & 1 & 7 & 2 & 1 & 10 & 4 & 1 & 5 & 1 & 1 \\ 1 & 2 & 6 & 1 & 1 & 1 & 3 & 3 & 1 & 2 & 10 & 1 & 1 & 9 & 1 & 1 & 2 & 1 & 1 & 3 \end{bmatrix}.$$

The reason for the objective function coefficient having this special distribution, is that the example originates from a standard GAP and has been transformed into a problem with structure (5.1)–(5.4).

The optimal solution to this problem can be written in the form (1, 2, 2, 1, 2, 1, 2, 2, 1, 2, 2, 1, 1, 2, 1, 1, 2, 1, 1, 2), which means that the jobs 1, 4, 6, 9, 12, 13, 15, 16, 18 and 19 are processed on machine 1, and the remaining jobs on machine 2. The optimal objective function value is $z^* = 2087$ the optimal objective function value to the LP-relaxation is $z^*_{LP} = 2089.46$.

The first question which arises when an aggregation technique is to be applied, is in what way the jobs should be partitioned. Bearing in mind results of Section 2, each job and machine can be characterized by the quotient $c_{ij}a_i/d_{ij}$. A large quotient represents a good machine-job assignment since it combines Then the partition is obtained by simply merging the jobs which are close to each other in the sense of this quotient. This yields (see

[9] for the details) the following partition:

Group	Jobs
1	4, 11, 16, 17, 19
2	1, 6, 7, 9, 18
3	12, 15
4	2, 13
5	3, 5, 8, 10
6	20
7	14

Taking the normalized weights equal inside each cluster, the aggregated problem (5.5), corresponding to 2 machines and 7 aggregated jobs, was obtained. the aggregated data are as follows:

$$\{a_i\} = \begin{pmatrix} 60 \\ 55 \end{pmatrix},$$

$$[\overline{d}_{ik}] = \begin{bmatrix} 10.60 & 5.20 & 3.00 & 3.50 & 10.75 & 7.00 & 12.00 \\ 10.20 & 8.00 & 7.50 & 3.00 & 4.75 & 3.00 & 1.00 \end{bmatrix},$$

$$[\overline{c}_{ik}] = \begin{bmatrix} 102.60 & 103.40 & 108.50 & 101.50 & 101.75 & 101.00 & 101.00 \\ 103.00 & 101.40 & 101.00 & 101.50 & 103.00 & 103.00 & 109.00 \end{bmatrix}.$$

The optimal solution to this aggregated problem is

$$[\overline{X}_{ik}] = \begin{bmatrix} 1.863 & 5 & 2 & 2 & 0 & 0 & 0 \\ 3.137 & 0 & 0 & 0 & 4 & 1 & 1 \end{bmatrix},$$

$$\overline{z}_{LP} = 2075.25,$$

$$\{\overline{u}_i\} = (0,\ 0.0392),$$

$$\{\overline{v}_k\} = (102.6,\ 103.4,\ 108.5,\ 101.5,\ 102.814,\ 108.961).$$

If the optimal disaggregation is applied, seven generalized assignment problems (5.6) will be solved. The optimal disaggregation provides $z^* \geq \geq \sum_{k \in T^A} z_k^*(\overline{X}_k) = 2084$ and the respective disaggregated solution can be represented in the form (1, 1, 2, 1, 2, 1, 1, 2, 1, 2, 2, 1, 1, 2, 1, 1, 2, 1, 2, 2).

Calculating the upper bound we obtain $z^* \leq UB((\overline{u}, v)^{W_0}, W_0) = 2090.23$. Note, that the choice $\widehat{v}_j = \overline{v}_k$, $j \in T_k$ yields $UB(\overline{u}, \widehat{v}, W_0) = 2075.25 + +24.182 = 2099.43$. Hence in this case we have a strong inequality in Corollary 1.2: $UB((\overline{u}, \widehat{v})^{W_0}, W_0) < UB(\overline{u}, \widehat{v}, W_0)$. The upper bound $UB((\overline{u}, v)^{W_0}, W_0)$ can be tightening by searching the duals u. For this example the vector u has only two components. Applying the gradient projection scheme similar to Section 2.2 we obtain that the upper bound $UB((u, v)^{W_0}, W_0)$ can be strengthening to z_{LP}^*. This yields $2084 \leq z^* \leq z_{LP}^* = 2089.46$. Using the heuristics mentioned earlier, the optimal disaggregated solution can be improved to the feasible solution with the objective function value 2087 (in this case the optimal solution).

Comments and References to Chapter 1

Various aspects of aggregation/disaggregation and bound calculations are summarized in the paper by Rogers D.F. et al. [26]. This excellent reference source covers a broad range of aggregation topics and is highly recommended as an introduction to the subject. Aggregation bounds for the case of primal or (and) dual degeneracy in linear programs are studied, in particular, in the book by Leisten R. [13]. Applications of the aggregation bounds in stochastic programming one can find in the works by Birge J.R. [3], Edirisinghe N.C.P. and Ziemba W.T. [4], Rosa C.H. and Takriti S. [27], Wright S.E. [32] and the references therein. An original approach to aggregation bounds in location problems is studied in the paper by Francis R.L., Lowe T.J., and Tamir A. [6]. The works by Weintraub A. and Cholaky A. [30], and Weintraub A., Saez G., and Yadlin M. [31] are focused on computational aspects of aggregation in large scale forest planning models. Some recent trends in aggregation/disaggregation are presented in the special issue of *Computers & Operations Research* (1999, vol. 26, no. 10/11). Section 1 is based on the papers by Zipkin P.H [33], Leisten R. [14], Litvinchev I.S. [16], and Litvinchev I.S. and Rangel S. [17]. Aggregation in transportation problems, presented in Section 2, is studied in Evans J.R. [5], Norman S.K., Rogers D.F., and Levy M.S. [24], Zipkin P.H. [35], Litvinchev I.S., Rangel S., and Andrade I.G. [18], and Alvarez A. et al. [2]. Sections 3 and 4 are based on the cited papers by Litvinchev I.S. [16], and Litvinchev I.S. and Rangel S. [17], and the work by Hallefjord A. and Storøy S. [8]. Aggregation in the generalized assignment problem of the last section is studied in Hallefjord A., Jörnsten K, and Varbrand P. [9].

References to Chapter 1

[1] Ahuja R.K., Magnati T.L. and Orlin J.B., Network Flows: Theory, Algorithms and Applications, New York: Prentice Hall, 1993.

[2] Alvarez A., Chacon O., Litvinchev I.S., and Rangel S., Aggregation in the Generalized Transportation problem, *J. Comput. Syst. Scis. Intl.*, 2001, vol. 40, no. 6, pp. 923–929.

[3] Birge J.R., Aggregation Bounds in Stochastic Linear Programming, *Math. Progr.*, 1985, vol. 31, pp. 25–41.

[4] Edirisinghe N.C.P. and Ziemba W.T., Tight Bounds for Stochastic Convex Programs, *Oper. Res.*, 1992, vol. 40, no. 4, pp. 660–677.

[5] Evans J.R., Aggregation in the Generalized Transportation Problem, *Computers Oper. Res.*, 1979, vol. 6, pp. 199–204.

[6] Francis R.L., Lowe T.J., and Tamir A., Aggregation Error Bounds for a Class of Location Models, *Oper. Res.*, 2000, vol. 48, no. 2, pp. 294–307.

[7] Hallefjord A. and Storøy S., Aggregation and Disaggregation in Integer Programming Problems, CMI Working Paper 862090-8, Chr. Michelsens Inst., Norway,1986.

[8] Hallefjord A. and Storøy S., Aggregation and Disaggregation in Integer Programming Problems, *Oper. Res.*, 1990, vol. 38, pp. 619–623.

[9] Hallefjord A., Jörnsten K, and Varbrand P., Solving Large Scale Generalized Assignment Problems – An Aggregation/Disaggregation Approach, *Eur. J. Oper. Res.*, 1993, vol. 64, pp. 103–114.

[10] Huberman G., Error Bounds for the Aggregated Convex Programming Problems, *Math. Progr.*, 1983, vol. 26, pp. 100–108.

[11] Knolmayer G., Computing a Tight a Posteriori Bound for Column Aggregated Linear Programs, *Methods Oper. Res.*, 1986, vol. 53, pp. 103–114.

[12] Lasdon L., Optimization Theory for Large Systems, New York: MacMillan, 1970.

[13] Leisten R., Iterative Aggregation und Mehrstufige Entscheidungsmodelle, Heidelberg: Physica–Verlag, 1995.

[14] Leisten R., A Posteriori Bounds in Linear Programming Aggregation, *Computers & Oper. Res.*, 1997, vol. 24, no. 1, pp. 1–16.

[15] Levitin E.S., *O differentsiryemosti po parametry optimal'nogo znacheniya parametricheskih zadach matematicheskogo programmirovaniya* (On Differentiability with Respect to a Parameter of Optimal Value in Parametric Problems of Mathematical Programming), *Kibernetika*, 1976, no.1, pp. 44–60.

[16] Litvinchev I.S., Bounds on the Suboptimalization of Aggregation in Convex Programming, *Comput. Maths. Math. Phys.*, 1993, vol. 33, no. 8, pp. 1007–1015.

[17] Litvinchev I.S. and Rangel S., Localization of the Optimal Solution and a Posteriori Bounds for Aggregation, *Computers & Oper. Res.*, 1999, vol. 26, no. 10–11, pp. 967–988.

[18] Litvinchev I.S., Rangel S., and Andrade I.G., Error Bounds for the Aggregation in the Generalized Transportation Problem, Technical Report TR 2000-001-OC, San Paulo State Univ. (UNESP), Dept. Comput. Maths. Statist., S.J. Rio Preto, Brasil, 2000.

[19] Litvinchev I.S., Ellipsoids Containing Optimal Solutions of the Linear Programming Problem, *Comput. Maths. Math. Phys.*, 2000, vol. 40, no. 2, pp. 178–188.

[20] Litivnchev I.S., On Improved Choi-Goldfarb Solution-Containing Ellipsoids in Linear Programming, *Oper. Res. Let.*, 2000, vol. 27, no.1, pp. 29–37.

[21] Mendelssohn R., Improved Bounds for Aggregated Linear Programs, *Oper. Res.*, 1980, vol. 28, pp.1450–1453.

[22] Murty K., Linear and Combinatorial Programming, New York: Wiley, 1976.

[23] Nemhauser G. and Wolsey L., Integer and Combinatorial Optimization, New York: Wiley, 1988.

[24] Norman S.K., Rogers D.F., and Levy M.S., Error Bound Comparisons for Aggregation/Disaggregation Techniques Applied to the Transportation Problem, *Computers & Oper. Res.*, 1999, vol. 26, no. 10–11, pp. 1003–1014.

[25] Osborne M.R., Finite Algorithms in Optimization and Data Analysis, New York: Wiley, 1987.

[26] Rogers D.F., Plante R.D., Wong R.T., and Evans J.R., Aggregation and Disaggregation Techniques and Methodology in Optimization, *Oper. Res.*, 1991, vol. 39, pp. 553–582.

[27] Rosa C.H. and Takriti S., Improving Aggregation Bounds for Two-Stage Stochastic Programs, *Oper. Res. Let.*, 1999, vol. 24, pp. 127–137.

[28] Shetty C.M. and Taylor R.W., Solving Large-Scale Linear Programs by Aggregation, *Computers & Oper. Res.*, 1987, vol. 14, pp. 385–393.

[29] Storøy S., Optimal Weights and Degeneracy in Variable Aggregated Linear Programs, *Oper. Res. Lett.*, 1996, vol.19, pp. 29–31.

[30] Weintraub A. and Cholaky A., A Hierarchical Approach to Forest Planning, *Forest Sci.*, 1991, vol. 37, no. 2, 439–460.

[31] Weintraub A., Saez G., and Yadlin M., Aggregation Procedures in Forest Management Planning Using Cluster Analysis, Technical Report, Univ. of Chile, Dept. Ind. Eng., 1997.

[32] Wright S.E., Primal-Dual Aggregation and Disaggregation for Stochastic Linear Programs, *Math. Oper. Res.*, 1994, vol. 19, no. 4, pp. 893–908.

[33] Zipkin P., Bounds on the Effect of Aggregating Variables in Linear Programs, *Oper. Res.*, 1980, vol. 28, no. 2, pp. 403–418.

[34] Zipkin P.H., Bounds for Row-Aggregation in Linear Programming, *Oper. Res.*, 1980, vol. 28, no. 4, pp. 903–916.

[35] Zipkin P.H., Bounds for Aggregating Nodes in Network Problems, *Math. Progr.*, 1980, vol. 19, pp. 155–177.

[36] Zipkin P.H., Aggregation and Disaggregation in Convex Network Problems, *Networks,* 1982, vol. 12, pp. 101–107.

Chapter 2

Iterative Aggregation-Decomposition in Optimization Problems

In the previous chapter aggregation was used to construct and analyze approximate solutions of optimization problems. The parameters of aggregation, such as weights and clustering, were fixed. In this chapter we focus on iterative methods aimed to construct a sequence of aggregated problems and update aggregation parameters to get an optimal solution of the original problem.

Section 1 is concerned with the case of linear aggregation in finite-dimensional convex problems. An iterative procedure is suggested and analyzed; at its one level, a macroproblem with a small number of aggregated variables is solved, while at the other, the weights of aggregation are adjusted. The latter problem decomposes into a number of independent problems under a certain choice of aggregation matrix.

In Section 2 we study finite-dimensional convex problems with constraints of special structure, such as block, block-separable, or block with coupling variables. It is shown how the special structure of the original problem can be used in constructing iterative decomposition algorithms based on aggregation.

In the first two sections the decomposition effect was achieved for the problem of updating the aggregation weights. The structural properties of the aggregated problem were not taken into account. In Section 3 we study a special aggregation for block-separable problems, resulting in decomposition of the macroproblem also.

In the next section we consider an iterative clustering technique, where not only the weights are updated, but the variables are reassigned to clusters.

In Section 5 two procedures of iterative aggregation of constraints are presented. One uses a certain correspondence between aggregation of variables in primal setting of the original problem and aggregation of constraints in its dual formulation. Another approach differs from the aggregation provided by Lagrange multipliers and works for problems for which duality does not hold.

Section 6 extends the results of Section 2 to aggregation of controls in dynamical problems with cross connections.

Section 7 considers linear-quadratic optimal control problems. Aggregation reduces the dimension with respect to the state variables, and the weights are recalculated by the method of successive approximations.

In Section 8, the iterative aggregation method is applied to problems where control systems are described by partial differential equations arising in mathematical physics.

The last section describes numerical tests performed with the use of the aggregation method for a number of optimal control problems.

§1. Linear Aggregation in Finite-Dimensional Problems

In this section, a procedure for iterative aggregation in convex mathematical programming problems is described. Each iteration is constructed according to a two-level scheme. At one level, a problem in macrovariables, which depends on disaggregation parameters (weights), is solved. At the other, the values of these parameters are adjusted in order to obtain a disaggregated solution optimal to the original problem. A relative freedom in the choice of the macroproblem is used to reduce the recalculation of the disaggregation parameters to the solution of a number of independent subproblems of lower dimensions. As a result, we obtain an iterative process for constructing disaggregated solutions feasible to the original problem, and this process is monotone with respect to the objective functional. Importantly, the decomposition into independent subproblems mentioned above is determined by the properties of aggregation and disaggregation, and it is not related to the presence of special structural properties of the original problem, such as those of being block or block-separable. This allows us to consider a wide class of extremal problems with cross connections from a unified point of view. Our approach is more general than the known methods of iterative aggregation in block-separable problems, where the macrovariables are formed from unknowns that belong to the same [42] or different [53, 54] blocks, and the recalculation of the disaggregation parameters is decomposed into subproblems due to the block-separable structure of the constraints and the objective functional. The iterative aggregation approach developed in this section aims at solving nonseparable optimization problems of various types.

Consider the mathematical programming problem

$$f(x) \to \max, \quad g_k(x) \leq 0, \quad k = 1, 2, \ldots, K, \quad x \geq 0, \tag{1.1}$$

where $x \in \mathbf{R}^n$. It is assumed that (1.1) has a solution and the scalar functions $-f(x)$ and $g_k(x)$ are convex.

Suppose that

$$x = D(\alpha, X), \tag{1.2}$$

is a transformation of variables depending on parameters $\alpha \in \mathbf{R}^r$; in (1.2), $X \in \mathbf{R}^m$ and $\alpha \in A$, where A is a closed convex set in the parameter space. We refer to the family of transformations (1.2) as a disaggregation. Fixing $\alpha \in A$ and substituting (1.2) into the original problem, we obtain

$$\begin{aligned} &f[D(\alpha, X)] \to \max, \quad g_k[D(\alpha, X)] \leq 0, \\ &k = 1, 2, \ldots, K, \quad D(\alpha, X) \geq 0. \end{aligned} \tag{1.3}$$

If $n > M$, then we call (1.3) an aggregated problem, or a macroproblem.

Let us introduce a continuous vector function $H\colon \mathbf{R}^n \to \mathbf{R}^M$ such that, for any fixed $\alpha \in A$,

$$X = H[D(\alpha, X)] = H(x). \tag{1.4}$$

For a transformation $H(\cdot)$ satisfying (1.4), we use the term aggregation.

In what follows, we are interested in finding an $\alpha^* \in A$ such that, if X^* is an optimal solution to the macroproblem (1.3) for $\alpha = \alpha^*$, then the disaggregated solution $x^* = D(\alpha^*, x^*)$ is optimal to the original problem (1.1). The iterative procedures for constructing such an α^* that involve two levels, one to solve the macroproblem for a fixed $\alpha \in A$ and the other to adjust the aggregation parameters, are known as the iterative aggregation methods [5].

Consider the auxiliary problem

$$\theta(\alpha) \to \max, \quad \alpha \in A, \tag{1.5}$$

where $\theta(\alpha)$ is an optimal value of the objective function of (1.3) depending on $\alpha \in A$; it is assumed that problem (1.3) has a finite optimal solution for each $\alpha \in A$. Note that $\theta(\alpha)$ is not concave in general. The following theorem establishes a relation between problems (1.1), (1.3), and (1.5).

Theorem 1.1. *Suppose that $\alpha^* \in A$ is a local maximum in (1.5) and $P(\alpha^*)$ is the set of optimal solutions to the macroproblem (1.3) for $\alpha = \alpha^*$. For $X^* \in P(\alpha^*)$, consider the disaggregated solution $x^* = D(\alpha^*, X^*)$ defined by (1.2). Let $U(x^*, \delta)$ be the δ-neighborhood of x^*. Suppose that there exists $\alpha(x)$ an A-valued mapping to A such that it is continuous at x^*, $\alpha(x^*) = \alpha^*$, and $x = D[\alpha(x), H(x)]$. Then the disaggregated solution $x^* = D(\alpha^*, X^*)$ is optimal for the original problem (1.1).*

Proof. Theorem 1.1 is proved similarly to the corresponding assertions in [29, 30]. Let $V(\alpha^*, \varepsilon)$ be the ε-neighborhood of α^*. Since α^* is a point of the local maximum in (1.5), there exists an $\varepsilon^* > 0$ such that $\theta(\alpha^*) \geq \theta(\alpha)$ for all $\alpha \in V(\alpha^*, \varepsilon^*) \cap A$.

By virtue of the continuity of $\alpha(x)$, we can choose $\delta^* > 0$ such that $\alpha(x)$ is defined and $\|\alpha(x^*) - \alpha(x)\| \leq \varepsilon^*$ for any $x \in U(x^*, \delta^*)$. Let P be the set of the feasible solutions to problem (1.1). By the definition of the macroproblem (1.3), the disaggregated solution $x^* = D(\alpha^*, X^*)$ belongs to P. Since P is convex, we have $U(x^*, \delta^*) \cap P \neq 0$.

Let $\tilde{x} \in U(x^*, \delta^*) \cap P$. According to the above considerations, $\alpha(\tilde{x}) \in V(\alpha^*, \varepsilon^*) \cap A$.

By assumption, we have $f(x^*) = f[D(\alpha^*, X^*)] = \theta(\alpha^*) \geq \theta(\alpha(\tilde{x}))$. Put $\tilde{X} = H(\tilde{x})$. The definition of $\alpha(x)$ implies that $\tilde{x} = D[\alpha(\tilde{x}), H(\tilde{x})] = D[\alpha(\tilde{x}), \tilde{X}]$. Since $\tilde{x}$ is feasible to the original problem (1.1), we have $g_k[D(\alpha(\tilde{x}), \tilde{X})] \leq 0$ for $k = 1, 2, \ldots, m$, and $D[\alpha(\tilde{x}), \tilde{X}] \geq 0$. Therefore, the element $\tilde{X}$ is feasible to the aggregated problem at $\alpha = \alpha(\tilde{x})$, and $\theta(\alpha(\tilde{x})) \geq f[D(\alpha(\tilde{x}), \tilde{X})] = f(\tilde{x})$. Thus, there exists a $\delta^* > 0$ such that $f(x^*) \geq f(\tilde{x})$ for any $\tilde{x} \in U(x^*, \delta^*) \cap P$. Taking into account the convexity of the function $-f(x)$, we obtain the optimality of x^* for the original problem (1.1).

Theorem 1.1 implies that solving the original problem can be reduced to finding a local maximum of the implicit function $\theta(\alpha)$ on the set A. To evaluate $\theta(\mathring{\alpha})$ at a point $\mathring{\alpha} \in A$, it is necessary to solve the macroproblem (1.3) with respect to M variables X. There is a certain freedom in the choice of the aggregation $H(\cdot)$, the disaggregation $D(\cdot, \cdot)$, and the set A satisfying the conditions of Theorem 1.1. Below, for the case of linear aggregation, we choose them to simplify problem (1.5) and endow it with decomposition properties.

Consider the case of the linear disaggregation

$$x = \alpha X, \tag{1.6}$$

where $\alpha = [\alpha_{lm}]$ is a rectangular $n \times M$ matrix ($l = 1, 2, \ldots, n$; $m = 1, 2, \ldots, M$). Let us define the convex set A by

$$A = \{\alpha \colon C\alpha = E\}, \tag{1.7}$$

where C is a fixed rectangular $M \times n$ matrix of rank M and E is the identity matrix of order M. According to (1.6) and (1.7), we have

$$X = Cx \tag{1.8}$$

for all $\alpha \in A$. Thus, condition (1.4) is obviously fulfilled and the matrix C determines the linear aggregation. Note that the set A is nonempty since

$$\alpha = C^{\mathrm{T}}(CC^{\mathrm{T}})^{-1} \in A.$$

Consider the original problem (1.1) without the assumption that the variables x are nonnegative:

$$f(x) \to \max, \quad g_k(x) \leq 0, \quad k = 1, 2, \ldots, K. \tag{1.9}$$

The corresponding aggregated problem is

$$F(X) \to \max, \quad G_k(X) \leq 0, \quad k = 1, 2, \ldots, K, \tag{1.10}$$

where

$$F(X) = f\left(\sum_m \alpha_{1m}X_m, \sum_m \alpha_{2m}X_m, \ldots, \sum_m \alpha_{nm}X_m\right) \equiv f(\alpha X),$$

$$G_k(X) = g_k\left(\sum_m \alpha_{1m}X_m, \ldots, \sum_m \alpha_{lm}X_m, \ldots, \sum_m \alpha_{nm}X_m\right) \equiv g_k(\alpha X)$$

for each fixed $\alpha \in A$. It is easy to verify that $F(X)$ and $G(X)$ are convex with respect to the variables X.

Now, suppose that the functions $f(x)$ and $g_k(x)$ ($k = 1, 2, \ldots, K$) are continuously differentiable and their partial derivatives satisfy the Lipschitz condition. We assume that, for any $\alpha \in A$, the aggregated problem (1.10) has a finite optimal solution and (as well as (1.9)) satisfies the Slater regularity condition.

The problem dual to (1.10) is

$$\begin{aligned} & F(X) - \sum_k \lambda_k G_k(X) \to \min, \\ & \frac{\partial F(X)}{\partial X_m} - \sum_k \lambda_k \frac{\partial G_k(x)}{\partial X_m} = 0, \quad m = 1, 2, \ldots, M, \\ & \lambda_k \geq 0, \quad k = 1, 2, \ldots, K. \end{aligned} \tag{1.11}$$

Let us denote the sets of optimal solutions to problems (1.10) and (1.11) for $\alpha \in A$ by $P(\alpha)$ and $Q(\alpha)$, respectively. The following theorem gives the necessary and sufficient conditions for the optimality of a disaggregated solution in terms of the macroproblem.

Theorem 1.2. *Suppose that $\mathring{\alpha} \in A$ and $\mathring{X} \in P(\mathring{\alpha})$. Then the disaggregated solution $\mathring{x} = \mathring{\alpha}\mathring{X}$ is optimal for problem (1.9) if and only if there exists $\mathring{\lambda} = (\mathring{\lambda}_1, \ldots, \mathring{\lambda}_K) \in Q(\mathring{\alpha})$ such that*

$$\Delta_l(\mathring{\alpha}) = \left(\frac{\partial f}{\partial x_l} - \sum_k \mathring{\lambda}_k \frac{\partial g_k}{\partial x_l} \right) \Bigg|_{x = \mathring{x}} = 0, \quad l = 1, 2, \ldots, n. \tag{1.12}$$

Proof. Necessity. Suppose that (1.12) holds. Consider the problem dual to (1.9) in the form

$$\begin{aligned} & f(x) - \sum_k \nu_k g_k(x) \to \min, \\ & \frac{\partial f(x)}{\partial x_l} - \sum_k \nu_k \frac{\partial g_k(x)}{\partial x_l} = 0, \quad l = 1, 2, \ldots, m, \\ & \nu = (\nu_1, \ldots, \nu_K) \geq 0. \end{aligned} \tag{1.13}$$

By the definition of the macroproblem (1.10), the disaggregated solution $\mathring{x}$ is feasible to problem (1.9). The complementary slackness conditions for the aggregated problem (1.10) imply $0 = \sum_k \mathring{\lambda}_k G_k(\mathring{X}) = \sum_k \mathring{\lambda}_k g_k(\mathring{x})$. Thus, we have a pair $(\mathring{x}, \mathring{\lambda})$ such that $\mathring{x}$ is feasible for (1.9), $(\mathring{x}, \mathring{\lambda})$ is feasible for the dual problem (1.13) (by the assumptions of the theorem), and the complementary slackness relations hold. Therefore, by the Kuhn–Tucker theorem, $\mathring{x}$ is optimal for (1.9).

Sufficiency. Suppose that $\overset{\circ}{x} = \overset{\circ}{\alpha}\overset{\circ}{X}$ is optimal for problem (1.9). By the Kuhn–Tucker theorem for the pair of problems (1.9), (1.13) there exists a $\overset{\circ}{\nu} = (\overset{\circ}{\nu}_1, \dots, \overset{\circ}{\nu}_k) \geq 0$ such that

$$\left(\frac{\partial f}{\partial x_l} - \sum_k \frac{\partial g_k}{\partial x_l} \overset{\circ}{\nu}_k \right) \Bigg|_{x=\overset{\circ}{x}} = 0, \quad l = 1, 2, \dots, n. \tag{1.14}$$

Since

$$\left(\frac{\partial F}{\partial X_m} - \sum_k \overset{\circ}{\nu}_k \frac{\partial G_k}{\partial X_m} \right) \Bigg|_{x=\overset{\circ}{x}} = \sum_{l=1}^{n} \overset{\circ}{\alpha}_{lm} \left(\frac{\partial f}{\partial x_l} - \sum_k \overset{\circ}{\nu}_k \frac{\partial g_k}{\partial x_l} \right) \Bigg|_{x=\overset{\circ}{x}},$$

the pair $\overset{\circ}{X}, \overset{\circ}{\lambda}$ with $\overset{\circ}{\lambda} = \overset{\circ}{\nu}$ satisfies the constraints of problems (1.10) and (1.11). Moreover, by the complementary slackness conditions for problem (1.9), we have $0 = \sum_k \overset{\circ}{\nu}_k g_k(\overset{\circ}{x}) = \sum_k \overset{\circ}{\nu}_k G_k(\overset{\circ}{X})$. Therefore, the pair $(\overset{\circ}{x}, \overset{\circ}{\lambda})$ is optimal to problem (1.11), $\overset{\circ}{\lambda} \in Q(\overset{\circ}{\alpha})$, and condition (1.12) holds by virtue of (1.14). This completes the proof of the theorem.

In fact, Theorem 1.2 gives the necessary and sufficient conditions for an $\alpha \in A$ to be an optimal solution to problem (1.5). Thus, although the function $\theta(\alpha)$ is not convex in general, problem (1.5) has one property similar to convex programming problems. Namely, the necessary optimality condition is simultaneously sufficient. Thus we may expect that, when $\theta(\alpha)$ is differentiable, the first-order optimality conditions for problem (1.5) are also sufficient.

Consider the differential properties of the function $\theta(\alpha)$. Suppose that the sets $P(\alpha)$ and $Q(\alpha)$ are uniformly bounded with respect to $\alpha \in A$. It is easy to verify that, under the above assumptions about the differentiability of $f(x)$ and $g_k(x)$, the family of the macroproblems (1.10) depending on the parameter $\alpha \in A$ satisfies all conditions of the marginal-value theorem (see, e.g., [10, 27]). In particular, the uniform convergence of the derivatives of G_k with respect to the parameter α is derived from the Lipschitz continuity of the derivatives of g_k.

Let $\overset{\circ}{\alpha} \in A$. By the marginal-value theorem, the function $\theta(\alpha)$ has a right derivative in any direction $s = [s_{lm}]$ at the point $\overset{\circ}{\alpha}$, and it is given by the formula

$$\lim_{\rho \to +0} \frac{\theta(\overset{\circ}{\alpha} + \rho s) - \theta(\overset{\circ}{\alpha})}{\rho} = \max_{\overset{\circ}{X} \in P(\overset{\circ}{\alpha})} \min_{\overset{\circ}{\lambda} \in Q(\overset{\circ}{\alpha})} \frac{\partial L(\overset{\circ}{X}, \overset{\circ}{\lambda}, 0)}{\partial \rho},$$

where $L(\overset{\circ}{X}, \overset{\circ}{\lambda}, \rho) = f(\alpha \overset{\circ}{X}) - \sum_k \overset{\circ}{\lambda}_k g_k(\alpha \overset{\circ}{X})$ is the Lagrange function of the aggregated problem (1.10) at $\alpha = \overset{\circ}{\alpha} + \rho s$ and $\partial L(\overset{\circ}{X}, \overset{\circ}{\lambda}, 0)/\partial \rho$ is the derivative of the Lagrange function at $\rho = 0$.

Differentiation yields

$$\frac{\partial L(\overset{\circ}{X}, \overset{\circ}{\lambda}, 0)}{\partial \rho} = \sum_l \sum_m s_{lm} \overset{\circ}{X}_m \left(\frac{\partial f}{\partial x_l} - \sum_k \overset{\circ}{\lambda}_k \frac{\partial g_k}{\partial x_l} \right) \Bigg|_{x=\overset{\circ}{x}},$$

where $\overset{\circ}{x} = \overset{\circ}{\alpha} \overset{\circ}{X}$ is a disaggregated solution.

If the aggregated problem (1.10) and its dual problem (1.11) have unique solutions for $\overset{\circ}{\alpha} \in A$, then the function $\theta(\alpha)$ is differentiable [27], and its partial derivatives are

$$\frac{\partial \theta}{\partial \alpha_{ml}} = \overset{\circ}{X}_m \left(\frac{\partial f}{\partial x_l} - \sum_k \overset{\circ}{\lambda}_k \frac{\partial g_k}{\partial x_l} \right) \Bigg|_{x=\overset{\circ}{x}}. \tag{1.15}$$

Suppose that the sets $P(\alpha)$ and $Q(\alpha)$ are singletons for any $\alpha \in A$. The following theorem shows that any stationary point of problem (1.5) (i.e., a point satisfying the first-order necessary optimality conditions) is optimal to problem (1.5).

Theorem 1.3. *Suppose that a point $\alpha^* \in A$ satisfies the necessary optimality conditions $(\nabla \theta(\alpha^*), \alpha - \alpha^*) \le 0$, for all $\alpha \in A$, where the components of the gradient $\nabla \theta(\alpha^*)$ are defined according to (1.15). If X^* is an optimal solution to the macroproblem (1.10) for $\alpha = \alpha^*$ and not all its components are zero, then the disaggregated solution $x^* = \alpha^* X^*$ is optimal to the original problem (1.9).*

Proof. As is shown in [57], the optimality condition mentioned in the statement of the theorem is equivalent to the fulfillment of the equality $\alpha^* = \Pi_A(\alpha^* + \rho \nabla \theta(\alpha^*))$ for all $\rho > 0$, where Π_A is the projector onto the set $A = \{\alpha\colon C\alpha = E\}$. By assumption, the matrix C is of full rank, and the projection $\Pi_A(\alpha)$ is defined as $\Pi_A(\alpha) = \alpha - C^{\mathrm{T}}(CC^{\mathrm{T}})^{-1}(C\alpha - E)$. Using this expression and the condition $\alpha^* \in A$, we can write the necessary optimality condition in the form

$$\nabla \theta(\alpha^*) = C^{\mathrm{T}}(CC^{\mathrm{T}})^{-1} C \nabla \theta(\alpha^*). \tag{1.16}$$

Let $X^* = (X_1^* \dots X_M^*)$ be a column vector from $P(\alpha^*)$. Consider the row vector $\Delta^* = (\Delta_1^* \dots \Delta_l^* \dots \Delta_n^*)$ with the components

$$\Delta_l^* = \left(\frac{\partial f}{\partial x_l} - \sum_k \lambda_k^* \frac{\partial g_k}{\partial x_l} \right) \Bigg|_{x=x^*},$$

where $\lambda^* \in Q(\alpha^*)$ and $x^* = \alpha^* X^*$. By virtue of (1.15), the components δ_{lm}^* of the matrix $\nabla\theta(\alpha^*) = [\delta_{lm}^*]$ are defined by $\delta_{lm}^* = X_m^* \Delta_l^*$ ($m = 1, 2, \dots, M$; $l = 1, 2, \dots, n$) or in the matrix form, $\nabla\theta(\alpha^*) = \Delta^{*\mathrm{T}} X^{*\mathrm{T}}$. Since the pair (X^*, λ^*) is optimal for the problem dual to the aggregated problem (1.11), we have

$$0 = \left(\frac{\partial F(X)}{\partial X_m} - \sum_k \lambda_k^* \frac{\partial G_k(X)}{\partial X_m} \right) \Bigg|_{X=X^*} =$$
$$= \sum_{l=1}^{n} \alpha_{lm}^* \left(\frac{\partial f}{\partial x_l} - \sum_l \lambda_k^* \frac{\partial g_k}{\partial x_l} \right) \Bigg|_{x=x^*}, \qquad m = 1, 2, \dots, M,$$

or, in the matrix form, $\Delta^* \alpha^* = 0$. Multiplying both sides of (1.16) from the left by $\alpha^{*\mathrm{T}}$ while taking into account the expression for $\nabla\theta(\alpha^*)$ and the relation $C\alpha^* = E$, we obtain the following sequence of equalities:

$$0 = \alpha^{*\mathrm{T}} \nabla\theta(\alpha^*) - \alpha^{*\mathrm{T}} C^{\mathrm{T}} (CC^{\mathrm{T}})^{-1} C \nabla\theta(\alpha^*) =$$
$$= \alpha^{*\mathrm{T}} \Delta^{*\mathrm{T}} \; X^{*\mathrm{T}} \; - (CC^{\mathrm{T}})^{-1} C \nabla\theta(\alpha^*) = -(CC^{\mathrm{T}})^{-1} C \nabla\theta(\alpha^*), \qquad (1.17)$$

where we have used the relation $(\Delta^* \alpha^*)^{\mathrm{T}} = \alpha^{*\mathrm{T}} \Delta^{*\mathrm{T}}$. Formulas (1.16) and (1.17) give $\nabla\theta(\alpha^*) = 0$. If $X_m^* \neq 0$ for at least one subscript $m = 1, 2, \dots, M$, then the expression $\delta_{lm}^* = X_m^* \Delta_l^*$ for the elements of the matrix $\nabla\theta(\alpha^*)$ gives $\Delta_l^* = 0$ for $l = 1, 2, \dots, n$.

Therefore, by Theorem 1.2, the disaggregated solution $x^* = \alpha^* X^*$ is optimal to problem (1.9).

The stationary points of problem (1.5) can be found by the first-order methods, such as projected gradient, feasible directions, conditional gradient techniques, etc. [57]. We use the projected gradient method. Applying to (1.5), it consists of constructing a sequence of points $\{\alpha_\nu\} \in A$ such that $\alpha_{\nu+1} = \Pi_A(\alpha_\nu + \rho_\nu \nabla\theta(\alpha_\nu))$, where $\nu = 0, 1, \dots$ is the number of the iteration and the step length ρ_ν maximizes the function $\xi(\rho) = \theta(\Pi_A(\alpha_\nu + \rho\nabla\theta(\alpha_\nu))$ over $\rho \geq 0$. As the initial approximation α_0, we can take any element from the set A (say, $\alpha_0 = C^{\mathrm{T}}(CC^{\mathrm{T}})^{-1}$).

Below, we describe one iteration of the method. Let $\overset{\circ}{\alpha} \in A$ be the current solution.

Step A. The aggregated problem (1.10) and its dual are solved for $\alpha = \overset{\circ}{\alpha}$. Suppose that $\overset{\circ}{X}$ and $\overset{\circ}{\lambda}$ are the optimal solutions to (1.10) and its dual, respectively, and $\overset{\circ}{x} = \overset{\circ}{\alpha}\overset{\circ}{X}$ is the disaggregated solution.

Step B. The components $\overset{\circ}{\delta}_{ml}$ $(m = 1, 2, \dots, M;\ l = 1, 2, \dots, n)$ of the gradient $\nabla\theta(\overset{\circ}{\alpha}) = [\overset{\circ}{\delta}_{ml}]$ are calculated by the formula (1.15). If $\overset{\circ}{\Delta}_l = 0$ for $l = 1, 2, \dots, n$, then the disaggregated solution $\overset{\circ}{x} = \overset{\circ}{\alpha}\overset{\circ}{X}$ is optimal for problem (1.9), and the process terminates.

Step C. If $\overset{\circ}{\Delta}_l \neq 0$ for at least one subscript l, then the next approximation is calculated by the formula

$$\alpha = \Pi_A(\overset{\circ}{\alpha} + \overset{\circ}{\rho}\nabla\theta(\overset{\circ}{\lambda})) = \\ = \overset{\circ}{\alpha} + \overset{\circ}{\rho}\nabla\theta(\overset{\circ}{\alpha}) - C^{\mathrm{T}}(CC^{\mathrm{T}})^{-1}\big[C(\overset{\circ}{\alpha} + \overset{\circ}{\rho}\nabla\theta(\overset{\circ}{\alpha})) - E\big], \qquad (1.18)$$

where $\overset{\circ}{\rho}$ is determined by maximizing the function $\xi(\rho)$ defined above. Other procedures for evaluating $\overset{\circ}{\rho}$ are described in [20, 57].

Thus, the original problem (1.9) having n variables, is reduced to iterative solution of lower–dimensional problems. Namely, the macroproblem having M aggregated variables to be solved at step A and the maximization problem in one variable ρ to be solved at Step C. To update α by (1.18) we need to invert the nondegenerate matrix $(CC)^{\mathrm{T}}$ of order M, which is not changed from iteration to iteration. The convergence of the projected gradient method is proved in [57] under the assumption that the gradient of the objective function satisfies the Lipschitz condition. The fulfillment of the Lipschitz condition for the gradient of $\theta(\alpha)$ follows from (1.15), the α-uniform boundedness of the sets $P(\alpha)$ and $Q(\alpha)$, and the assumption that the derivatives of $f(x)$ and $g_k(x)$ $(k = 1, 2, \dots, K)$ are Lipschitz.

The disaggregated solution $\overset{\circ}{x}_\nu = \overset{\circ}{\alpha}_\nu\overset{\circ}{X}_\nu$ obtained at the ν-th iteration of the method described above is feasible to problem (1.19). Moreover, the condition $\theta(\overset{\circ}{\alpha}_\nu) \leq \theta(\overset{\circ}{\alpha}_{\nu+1})$ implies that the sequence $\{f(\overset{\circ}{x}_\nu),\ \nu = 0, 1, \dots\}$ is monotonically nondecreasing, and if $\overset{\circ}{x}_\nu$ is not optimal for (1.9), then $f(\overset{\circ}{x}_{\nu+1}) > f(\overset{\circ}{x}_\nu)$.

Note that we have a relatively large freedom to choose the aggregation matrix. For example, we may choose the matrix C to endow the method of iterative aggregation with decomposition properties. Indeed, note that the constraints of the problem $\{\theta(\alpha)|\alpha \in A\}$ do not depend on the structure of the original problem and are fully determined by the structure of the set A. Suppose that A is represented in the form of a direct product of independent subsets as $A = A_1 \times A_2 \times \ldots \times A_I$. Then the problem of projecting the gradient $\nabla\theta(\alpha)$ onto the set A also decomposes into I independent projection subproblems of lower dimensions. A similar effect occurs when the other first-order methods are used. For example, in the method of conditional gradient [25, 83], the ascent direction is determined by maximizing the linear increment of the function $\theta(\alpha)$ over the set A. This problem also decomposes into I independent subproblems due to the additive separability of the linear increment and to the block structure of A.

The desirable structure of the set A specified above can be obtained by choosing a suitable block-diagonal aggregation matrix C. Suppose that the components of the vectors x and X are divided into I disjoint subsets as

$$\begin{aligned} &x = (x_1 \ldots x_i \ldots x_I), \quad x_i \in \mathbf{R}^{J_i}, \\ &x_i = \{x_1^i \ldots x_j^i \ldots x_{J_i}^i\}, \quad \sum_i J_i = n, \\ &X = (X^1 \ldots X^i \ldots X^I), \quad X^i \in \mathbf{R}^{M_i}, \\ &M_i \le J_i, \quad \sum_i M_i = M, \quad M < n. \end{aligned} \tag{1.19}$$

Define the linear disaggregation by

$$x_i = \alpha_i X^i, \quad i = 1, 2, \ldots, I, \tag{1.20}$$

where α_i are rectangular $J_i \times M_i$ matrices, and specify the set A by

$$A = A_1 \times \ldots \times A_i \cdots \times A_I, \quad \alpha_i = \{\alpha_i \colon C_i\alpha_i = E_i\}, \quad i = 1, 2, \ldots I, \tag{1.21}$$

where C_i are rectangular $M_i \times J_i$ matrices and E_i are the identity matrices of orders M_i. Then $X^i = C_i x_i$ $(i = 1, 2, \ldots, I)$ for any $\alpha \in A$. If all the matrices C_i are of full rank, then projecting onto the set A by formula (1.18) involves inverting the matrices $C_i C_i^{\mathrm{T}}$ of order M_i for $i = 1, 2, \ldots, I$, while in the general case, a matrix C of order M is inverted.

Thus, the decomposition effect is achieved by choosing a suitable aggregation matrix rather than due to the block-separable structure of the original problem. In Section 2, we show that, if the original problem has block structure, then this additional property can also be taken into account in choosing the aggregation matrix.

Another approach to the choice of the aggregation matrix C in (1.7) aims at maximally simplifying the aggregated problem (1.10).

Consider the problem

$$\max\{f(x) \mid Cx \leq b\}, \tag{1.22}$$

where $x \in \mathbf{R}^n$, $b \in \mathbf{R}^m$, $\operatorname{rank} C = m$, and $-f(x)$ is strongly convex. Let $A = \{\alpha\colon C\alpha = E\}$, where α is a rectangular $n \times m$ matrix. Considering vectors of aggregated variables $X \in \mathbf{R}^m$ and the linear disaggregation $x = \alpha X$, we obtain $X = Cx$ for all $\alpha \in A$. The aggregated problem for (1.22) has the form

$$\theta(\mathring{\alpha}) = \max\{f(\mathring{\alpha}X) \mid X \leq b\} \tag{1.23}$$

for a fixed $\mathring{\alpha} \in A$. Using the condition $\operatorname{rank} C = m$, we can easily verify that $-f(\alpha X)$ is strongly convex with respect to X. Thus, the aggregated problem (1.23) has a unique optimal solution for any $\alpha \in A$. Suppose that the function $f(x)$ is continuously differentiable and its partial derivatives satisfy the Lipschitz condition. Writing out the constraints of the problem dual to (1.23), we obtain that, for the solution $\mathring{X}$ to the aggregated problem at $\alpha = \mathring{\alpha}$, the dual problem has a unique solution, namely, $\mathring{\lambda} = (\partial F/\partial X)_{x=\mathring{x}}$, where $F(X) = f(\mathring{\alpha}X)$. Therefore, the function $\theta(\alpha)$ is differentiable, its partial derivatives are given by (1.15), Theorem 1.3 is valid, and the problem $\max\{\theta(\alpha) \mid \alpha \in A\}$ can be solved by the projected gradient method as described above. The constraints of the aggregated problem (1.23) have a very simple structure, and this problem can be solved with the use of, e.g., projected gradient technique.

§2. Problems with Constraints of Special Structure

In Section 1, the macroproblem (1.3) was constructed in such a way that the constraint $x \geq 0$ in the original problem resulted in the condition $\alpha X \geq 0$ in the macroproblem providing the aggregation $x = \alpha X$ is used with $\alpha \in A = \{\alpha\colon C\alpha = E\}$. This section concentrates mainly on the case where the conditions $\alpha X \geq 0$ can be transformed so that the aggregated problem retains the constraint $X \geq 0$, while the condition $\alpha \geq 0$ is taken into account in the definition $A = \{\alpha\colon C\alpha = E,\ \alpha \geq 0\}$ used in the problem $\max\{\theta(\alpha) \mid \alpha \in A\}$. It is not always possible to realize this approach using the results of Section 1, at least because the system $C\alpha = E$, $\alpha \geq 0$ has a solution not for all aggregation matrices C. Below, we consider a special case of linear aggregation meeting the above requirement and suggest how to use the structural properties of the constrains of the original problem in updating the aggregation parameters.

Consider the problem

$$f(x) \to \max, \quad g_k(x) \leq 0, \quad k = 1, 2, \dots, K, \quad x \geq 0, \tag{2.1}$$

where $x \in \mathbf{R}^n$ and the functions $-f(x)$ and $g_k(x)$ $(k = 1, 2, \dots, K)$ are convex. Suppose that problem (2.1) has a finite optimal solution and satisfies the Slater condition.

Let us partition the components of the vector x into I disjoint subsets:

$$x = (x_1 \dots x_i \dots x_I), \quad x_i \in \mathbf{R}^{J_i},$$

$$x_i = \{x_1^i \dots x_j^i \dots x_{J_i}^i\}, \quad \sum_{i=1}^{I} J_i = n.$$

Consider the simple case of the block aggregation (1.20), (1.21) (see Section 1), where exactly one aggregated variable is associated with each subvector x_i. Let us introduce the vector $X = (X^1, \dots, X^i, \dots, X^I) \in \mathbf{R}^I$ of the aggregated variables and let $x = \alpha X$ be the linear disaggregation, such that the components of x are

$$x_j^i = \alpha_j^i X^i, \quad i = 1, 2, \dots, I, \quad j = 1, 2, \dots J_i, \tag{2.2}$$

where α_j^i are the elements of the block–diagonal matrix α. Here, the i-th block of the matrix α is the column–vector $(\alpha_1^i, \alpha_2^i, \dots, \alpha_{J_i}^i)$.

Define the set A by

$$\begin{aligned} A &= A_1 \times A_2 \times \dots A_I, \\ A_i &= \left\{ a_j^i \mid \sum_j \alpha_j^i = 1,\ \alpha_j^i \geq 0,\ j = 1, 2, \dots J_i \right\}. \end{aligned} \tag{2.3}$$

It follows from (2.2) and (2.3) that $X^i = \sum_j x_j^i$ for $i = 1, 2, \dots, I$. Therefore, the aggregation matrix can be written as $C = \operatorname{block\,diag}\{C_i, i = 1, 2, \dots$ $\dots, I\}$, where C_i is the row of J_i ones.

The aggregated problem for (2.1) has the form

$$F(X) \to \max, \quad G_k(X) \leq 0, \quad k = 1, 2, \ldots, K, \quad X \geq 0, \tag{2.4}$$

where $F(X) = f(\alpha X)$ and $G_k(X) = g_k(\alpha X)$ for every $\alpha \in A$. We assume that problem (2.4) has a finite solution for any $\alpha \in A$. Obviously, if $\mathring{X}$ is an optimal solution to (2.4) obtained for $\mathring{\alpha} \in A$, then the disaggregated solution $\mathring{x} = \mathring{\alpha}\mathring{X}$ is feasible to the original problem (2.1) since the conditions $\mathring{x} \geq 0$ in (2.4) and $\alpha \geq 0$ in (2.3) imply $\mathring{x} = \mathring{\alpha}\mathring{X} \geq 0$.

Let $\theta(\alpha)$, $\alpha \in A$ be the objective function value for (2.4) and consider the problem

$$\theta(\alpha) \to \max, \quad \alpha \in A. \tag{2.5}$$

The following theorem establishes a relation between problems (2.1), (2.4), and (2.5) providing that the set A is defined by (2.3).

Theorem 2.1. *Let $\mathring{x} \in A$ be a local maximum point in (2.5). Denote by $P(\mathring{\alpha})$ the set of optimal solutions to macroproblem (2.4) for $\alpha = \mathring{\alpha}$. Let $\mathring{X} \in P(\mathring{\alpha})$ and $\mathring{X}$ is componentwise positive. Then the disaggregated solution $\mathring{x} = \mathring{\alpha}\mathring{X}$ is optimal for the original problem (2.1).*

The proof of this theorem is obtained similar to the proof of Theorem 1.1 given in Section 1 by defining the mapping $\alpha(x)$ in a neighborhood of the point $\mathring{x}$ as

$$\alpha_j^i(x) = \begin{cases} x_j^i \Big(\sum\limits_j x_j^i\Big)^{-1} & \text{if } \sum\limits_j x_j^i > 0, \\ \mathring{\alpha}_j^i & \text{if } \sum\limits_j x_j^i = 0, \quad i = 1, 2, \ldots, I, \ j = 1, 2, \ldots J_i. \end{cases}$$

Clearly, (2.2) yields $\alpha(\mathring{x}) = \mathring{\alpha}$. Moreover, the condition $\mathring{X}^i = \sum\limits_j \mathring{x}_j^i$ ($i = 1, 2, \ldots, I$) implies that the function $\alpha(x)$ is continuous at $x = \mathring{x}$. Hence there exists a neighborhood $U(\mathring{x}, \delta^*)$ such that, for all $x \in U(\mathring{x}, \delta^*)$, we have $\sum\limits_j x_j^i = X^i > 0$ ($i = 1, 2, \ldots, I$) and $x_j^i = \alpha_j^i(x) X^i$ ($i = 1, 2, \ldots, I$; $j = 1, 2, \ldots, J_i$). Now, the required assertion is proved by the same argument as that used in the proof of Theorem 1.1 from Section 1.

Next, suppose that the functions $f(x)$ and $g_k(x)$ ($k = 1, 2, \ldots, K$) are continuously differentiable and the aggregated problem (2.4) satisfies the Slater regularity conditions for any $\alpha \in A$.

The problem dual to (2.4) has the form

$$\begin{aligned} & F(X) - \sum_k \lambda_k G_k(X) \to \min, \\ & \frac{\partial F(X)}{\partial X^i} - \sum_k \lambda_k \frac{\partial G_k(X)}{\partial X^i} \le 0, \quad X^i = 0, \\ & \frac{\partial F(X)}{\partial X^i} - \sum_k \lambda_k \frac{\partial G_k(X)}{\partial X^i} = 0, \quad X^i > 0, \\ & i = 1, 2, \dots I, \quad \lambda_k \ge 0, \quad k = 1, 2, \dots, K. \end{aligned} \tag{2.6}$$

Denote by $P(\alpha)$ and $Q(\alpha)$ the sets of optimal solutions to the macroproblem and its dual, respectively. The following theorem gives the optimality criterion for the disaggregated solution.

Theorem 2.2. *Suppose that $\mathring{\alpha} \in A$, $\mathring{X} \in P(\mathring{\alpha})$, and $\mathring{X}$ is componentwise positive. The disaggregated solution $\mathring{x} = \mathring{\alpha}\mathring{X}$ is optimal for the original problem (2.1) if and only if there exists $\mathring{\lambda} = (\mathring{\lambda}_1, \dots, \mathring{\lambda}_K) \in Q(\mathring{\alpha})$ such that*

$$\begin{aligned} & \mathring{\Delta}(ij) \equiv \left(\frac{\partial f(x)}{\partial x_j^i} - \sum_k \mathring{\lambda}_k \frac{\partial g_k(x)}{\partial x_j^i} \right) \Bigg|_{\mathring{x}} \le 0, \quad \mathring{x}_j^i = 0; \\ & \mathring{\Delta}(ij) = 0, \quad \mathring{x}_j^i > 0; \quad i = 1, 2, \dots, I, \ j = 1, 2, \dots, J_i. \end{aligned} \tag{2.7}$$

Proof. Necessity. Suppose that conditions (2.7) hold. Consider the following problem dual to (2.1):

$$\begin{aligned} & f(x) - \sum_k \nu_k g_k(x) \to \min, \\ & \frac{\partial f(x)}{\partial x_j^i} - \sum_k \nu_k \frac{\partial g_k(x)}{\partial x_j^i} = 0, \quad x_j^i > 0; \quad i = 1, 2, \dots, I, \\ & \frac{\partial f(x)}{\partial x_j^i} - \sum_k \nu_k \frac{\partial g_k(x)}{\partial x_j^i} \le 0, \quad x_j^i = 0; \\ & j = 1, 2, \dots, J_i, \quad \nu_k \ge 0, \quad k = 1, 2, \dots K. \end{aligned} \tag{2.8}$$

The complementary slackness conditions for the aggregated problem (2.4) and its dual (2.6) give

$$\begin{aligned} 0 &= \sum_k \mathring{\lambda}_k G_k(\mathring{X}) = \sum_k \mathring{\lambda}_k g_k(\mathring{x}) = 0, \\ 0 &= \mathring{X}^i \left(\frac{\partial f(X)}{\partial X^i} - \sum_k \mathring{\lambda}_k \frac{\partial G_k(X)}{\partial X^i} \right) \Bigg|_{\mathring{X}} = \\ &= \mathring{X}^i \sum_j \mathring{\alpha}^i_j \mathring{\Delta}(ij) = \sum_j \mathring{x}^i_j \mathring{\Delta}(ij). \end{aligned} \tag{2.9}$$

Thus, the element $\mathring{x}$ is feasible to the original problem (2.1), the pair $(\mathring{x}, \mathring{\lambda})$ is feasible to (2.8) by virtue of (2.7), and the complementary slackness conditions (2.9) hold. Therefore, by the Kuhn–Tucker theorem, $\mathring{x}$ is optimal for (2.1).

Sufficiency. Now, suppose that $\mathring{x} = \mathring{\alpha}\mathring{X}$ is optimal for (2.1). By the Kuhn–Tucker theorem for the pair of problems (2.1), (2.8), there exists $\mathring{\nu} = (\mathring{\nu}_1, \dots, \mathring{\nu}_K)$ satisfying the constraints of the problem (2.8). Then we have

$$\begin{aligned} 0 = \sum_j \mathring{x}^i_j \left(\frac{\partial f(x)}{\partial x^i_j} - \sum_k \mathring{\nu}_k \frac{\partial g_k(x)}{\partial x^i_j} \right) \Bigg|_{\mathring{x}} &= \\ = \mathring{X}^i \sum_j \mathring{\alpha}^i_j \left(\frac{\partial f(x)}{\partial x^i_j} - \sum_k \mathring{\nu}_k \frac{\partial g_k(x)}{\partial x^i_j} \right) \Bigg|_{\mathring{x}} &= \\ = \mathring{X}^i \left(\frac{\partial F}{\partial X^i} - \sum_k \mathring{\nu}_k \frac{\partial G_k}{\partial X^i} \right) \Bigg|_{\mathring{X}} &. \end{aligned} \tag{2.10}$$

Put $\mathring{\lambda} = \mathring{\nu}$. By assumption, we have $\mathring{X}^i > 0$ for $i = 1, 2, \dots, I$, and (2.10) implies that the pair $(\mathring{x}, \mathring{\lambda})$ satisfies the constraints of the dual problem (2.6). It is easy to show that the complementary slackness conditions are also fulfilled, and the application of the Kuhn–Tucker theorem to the pair of problems (2.4), (2.6) gives $\mathring{\lambda} \in Q(\mathring{\alpha})$. Since $\mathring{\lambda} = \mathring{\nu}$, conditions (2.7) are met too, which completes the proof of the theorem.

Suppose that the derivatives of the functions $f(x)$ and $g_k(x)$ $(k = 1, 2, \dots, K)$ satisfy the Lipschitz condition and the sets $P(\alpha)$ and $Q(\alpha)$ are bounded uniformly with respect to $\alpha \in A$.

First, consider the case when the sets $P(\alpha)$ and $Q(\alpha)$ are singletons for all $\alpha \in A$. By the marginal-value theorem [27], the function $\theta(\alpha)$ in the problem (2.5) is differentiable, and its partial derivatives are

$$\left.\frac{\partial \theta}{\partial \alpha_j^i}\right|_{\mathring{\alpha}} = \mathring{X}^i \left(\frac{\partial f(x)}{\partial x_j^i} - \sum_k \mathring{\lambda}_k \frac{\partial g_k(x)}{\partial x_j^i} \right) \Bigg|_{\mathring{x}} = \mathring{X}^i \mathring{\Delta}(ij), \tag{2.11}$$

where $\mathring{\alpha} \in A$, $\mathring{X} \in P(\mathring{\alpha})$, $\mathring{\lambda} \in Q(\mathring{\alpha})$, and $\mathring{x} = \mathring{\alpha}\mathring{X}$.

The following theorem is similar to Theorem 1.3 proved in Section 1.

Theorem 2.3. *Suppose that the first-order necessary optimality conditions, $(\nabla\theta(\mathring{\alpha}), \alpha - \mathring{\alpha}) \le 0$ for any $\alpha \in A$, hold at a point $\mathring{\alpha} \in A$. If $\mathring{X} = P(\mathring{\alpha})$ is componentwise positive, then the disaggregated solution $\mathring{x} = \mathring{\alpha}\mathring{X}$ is optimal for the original problem (2.1).*

Proof. Using the expression (2.11) for the derivatives, we can write the necessary optimality conditions for the point $\mathring{\alpha}$ as

$$S(\mathring{\alpha}) = \max\left\{ \sum_i \sum_j \alpha_j^i \mathring{X}^i \mathring{\Delta}(ij) \mid \alpha \in A \right\} = 0. \tag{2.12}$$

According to Theorem 2.2, $\mathring{x}$ is optimal for (2.2) if and only if $\mathring{\Delta}(ij) \le 0$ for $\mathring{x}_j^i = 0$ and $\mathring{\Delta}(ij) = 0$ for $\mathring{x}_j^i > 0$.

Suppose that condition (2.12) holds but there exists a pair of indices (i_0, j_0) such that $\mathring{\Delta}(i_0 j_0) > 0$. The complementary slackness conditions for the dual aggregated problem (2.6) give $0 = \mathring{X}^i \sum_j \mathring{\alpha}_j^i \mathring{\Delta}(ij)$ for $i = 1, 2, \ldots, I$. Since $\mathring{X}^i > 0$ and $\sum_j \mathring{\alpha}_j^i = 1$, then for any fixed i all $\mathring{\Delta}(ij)$ cannot be strictly positive or strictly negative. Therefore, for each i, there exists a j_i such that $\mathring{\Delta}(i, j_i) \ge 0$. Let us define $\tilde{\alpha} \in A$ by setting

$$\tilde{\alpha}_{j_0}^{i_0} = 1, \quad \tilde{\alpha}_j^{i_0} = 0, \quad j \ne j_0; \quad \tilde{\alpha}_{j_i}^i = 1, \quad \tilde{\alpha}_j^i = 0, \quad j \ne j_i, \quad i \ne i_0.$$

Since $\mathring{\Delta}(i_0, j_0) > 0$, we have $\sum_i \sum_j \mathring{X}^i \tilde{\alpha}_j^i \mathring{\Delta}(ij) > 0$, which contradicts (2.12). Therefore, if (2.12) holds, then $\mathring{\Delta}(ij) \le 0$. In addition, from the relations $0 = \mathring{X}^i \sum_j \mathring{\alpha}_j^i \mathring{\Delta}(ij) = \sum_j \mathring{x}_j^i \mathring{\Delta}(ij)$ and $\mathring{x}_j^i \ge 0$, it follows that $\mathring{\Delta}(ij) = 0$ for

$\overset{\circ}{x}{}^i_j > 0$. Hence, (2.12) implies the optimality of $\overset{\circ}{x}$ for (2.1), which proves the theorem.

Now, suppose that the macroproblem (2.4) has a unique solution for any $\alpha \in A$, while its dual problem (2.6) may has a nonunique solution. In this case, the function $\theta(\alpha)$ is differentiable in any direction. Therefore, for an arbitrary direction α and a sufficiently small $\rho > 0$, the following expansion at the point $\overset{\circ}{\alpha}$ holds:

$$\theta(\overset{\circ}{\alpha} + \rho\alpha) = \theta(\overset{\circ}{\alpha}) +$$
$$+ \rho \min_{\lambda \in Q(\overset{\circ}{\alpha})} \sum_i \sum_j \overset{\circ}{X}{}^i \alpha^i_j \left(\frac{\partial f(x)}{\partial x^i_j} - \sum_k \lambda_k \frac{\partial g_k(x)}{\partial x^i_j} \right) \Bigg|_{\overset{\circ}{x}} + o_{\alpha,\overset{\circ}{\alpha}}(\rho),$$
$$\lim_{\rho \to 0} \frac{o_{\alpha,\overset{\circ}{\alpha}}(\rho)}{\rho} = 0.$$

A necessary condition for $\overset{\circ}{\alpha}$ to be a point of optimum in problem (2.5) when problem (2.6) has nonunique points of optimum $\overset{\circ}{\lambda}$ is the fulfillment of the equality

$$0 = \max_{\alpha \in A} \min_{\lambda \in Q(\overset{\circ}{\alpha})} \sum_i \sum_j \overset{\circ}{X}{}^i \alpha^i_j \left(\frac{\partial f(x)}{\partial x^i_j} - \sum_k \lambda_k \frac{\partial g_k(x)}{\partial x^i_j} \right) \Bigg|_{\overset{\circ}{x}}, \tag{2.13}$$

where a saddle point exists since the sets A and $Q(\overset{\circ}{\alpha})$ are bounded. Below, we show that the necessary condition (2.13) is sufficient.

Theorem 2.4. *Suppose that condition (2.13) holds at a point $\overset{\circ}{\alpha} \in A$. If $\overset{\circ}{X}$ is componentwise positive, then the disaggregated solution $\overset{\circ}{x} = \overset{\circ}{\alpha}\overset{\circ}{X}$ is optimal for the original problem (2.1).*

Proof. Let $\overset{\circ}{\lambda} \in Q(\overset{\circ}{\alpha})$ be the component of a saddle point in (2.13). By the definition of the saddle point, we have

$$S(\overset{\circ}{\alpha}, \overset{\circ}{\lambda}) = \max_{\alpha \in A} \sum_i \sum_j \overset{\circ}{X}{}^i \alpha^i_j \left(\frac{\partial f(x)}{\partial x^i_j} - \sum_k \overset{\circ}{\lambda}_k \frac{\partial g_k(x)}{\partial x^i_j} \right) \Bigg|_{\overset{\circ}{x}} = 0. \tag{2.14}$$

Problem (2.14) is similar to (2.12). The same argument as that used in the proof of Theorem 2.3 shows that $(\overset{\circ}{x}, \overset{\circ}{\lambda})$ satisfies conditions (2.7). Therefore, the disaggregated solution $\overset{\circ}{x} = \overset{\circ}{\alpha}\overset{\circ}{X}$ is optimal for the original problem (2.1) by Theorem 2.2.

The stationary points for problem (2.5) can be found, as in Section 1, by the first-order techniques, such as the methods of conditional gradient, feasible directions, etc. (see, e.g., [57]). If the aggregated problem and its dual have unique solutions, then the auxiliary problem of determining the ascent directions decomposes into I independent subproblems, since $A = A_1 \times A_2 \times \cdots \times A_I$.

Below, we describe one iteration of the conditional gradient method. Let $\overset{\circ}{\alpha} \in A$ be the current solution, and let the sets $P(\overset{\circ}{\alpha})$ and $Q(\overset{\circ}{\alpha})$ be singletons.

Step 1. The aggregated problem (2.4) and its dual (2.6) are solved for $\alpha = \overset{\circ}{\alpha}$. Suppose that $\overset{\circ}{X} \in P(\overset{\circ}{\alpha})$, $\overset{\circ}{X} > 0$, $\overset{\circ}{\lambda} \in Q(\overset{\circ}{\alpha})$, and $\overset{\circ}{x} = \overset{\circ}{\alpha}\overset{\circ}{X}$.

Step 2. The I independent subproblems

$$S_i(\overset{\circ}{\alpha}) = \max\left\{ \sum_j \overset{\circ}{X}^i \alpha_j^i \overset{\circ}{\Delta}(i,j) \mid \alpha_i \in A_i \right\}, \quad i = 1, 2, \ldots, I \tag{2.15}$$

are solved. Let $\bar{\alpha} = (\bar{\alpha}_1, \ldots, \bar{\alpha}_I)$ be a solution to problems (2.15). If $\bar{\alpha} = \overset{\circ}{\alpha}$, then $\overset{\circ}{x} = \overset{\circ}{\alpha}\overset{\circ}{X}$ is optimal to (2.1), and the process terminates. Otherwise, go to step 3 with $s = \bar{\alpha} - \overset{\circ}{\alpha}$.

Step 3. The step length $\overset{\circ}{\rho}$ is determined by solving the problem

$$\theta(\overset{\circ}{\alpha} + \overset{\circ}{\rho} s) = \max\{\theta(\overset{\circ}{\alpha} + \rho s) \mid 0 \le \rho \le 1\}.$$

The next approximation for α is chosen to be $\alpha = \overset{\circ}{\alpha} + \overset{\circ}{\rho} s$.

The convergence of the conditional gradient method is proved in [57], where the gradient of the objective function is required to satisfy the Lipschitz condition. If the aggregated problem and its dual have unique solutions for any $\alpha \in A$, then the gradient of $\theta(\alpha)$ does satisfy this condition by virtue of (2.11), the α-uniform boundedness of $P(\alpha)$ and $Q(\alpha)$, and the Lipschitz continuity of the derivatives of $f(x)$ and $g_k(x)$ $(k = 1, 2, \ldots, K)$. Note that the set A is bounded.

If a solution to the dual aggregated problem (2.6) is nonunique, then the choice of the direction s at step 2 depends on the solution of the maximin problem (2.13). Let $\bar{\alpha}$ be the component of a saddle point, then set $s = \bar{\alpha} - \overset{\circ}{\alpha}$. The convergence of the conditional gradient method in this case is proved similar to [7].

The maximin problem (2.13) can be solved, for example, by Brown's method of fictitious play (see [7]). Since the function

$$\varphi(\lambda) = \max_{\alpha \in A} \sum_i \sum_j \alpha_j^i \overset{\circ}{X}{}^i \left(\frac{\partial f(x)}{\partial x_j^i} - \sum_k \lambda_k \frac{\partial g_k(x)}{\partial x_j^i} \right) \Bigg|_{\overset{\circ}{x}}$$

is convex and piecewise linear, the problem

$$\min_{\lambda \in Q(\overset{\circ}{\alpha})} \varphi(\lambda)$$

can also be solved by special algorithms [7], and, for a fixed $\overset{\circ}{\lambda} \in Q(\overset{\circ}{\alpha})$, the maximization problem decomposes into I independent subproblems.

The nonuniqueness of the dual estimates in the aggregated problem complicates the solution, since the directional derivative of the function $\theta(\alpha)$ cannot be expressed explicitly. It is natural to expect that nonuniqueness occurs when the number I of the aggregated variables is small compared with the number K of constraints. For example, in the linear programming problems, the dual estimates are nonunique if more than I inequality constraints are active at an optimal solution.

The determination of the ascent direction at Step 2 uses only the properties of the set A. As a result, the corresponding problem decomposes into independent subproblems. The structural properties of the original problem, such as the block structure and/or separability, are not taken into account. Below, we consider an approach that makes it possible to exploit the special features related to the structure of the original problem (2.1).

Let $p_r(x)$ $(r = 1, 2, \dots, R)$ be sufficiently smooth convex functions. Define $\Omega = \{x \colon p_r(x) \le 0,\ x \ge 0,\ r = 1, 2, \dots, R\}$.

Assumption 2.1.

(1) Ω is a closed bounded set;

(2) $\Omega \supseteq P$, where P is the set of feasible solutions to the original problem (2.1);

(3) for any $\alpha \in A$, the optimal solution to the aggregated problem is uniformly bounded from below, i.e., $\overset{\circ}{X}{}^i \ge \varepsilon > 0$ for $i = 1, 2, \dots, I$.

Let us modify the ascent direction problem to be solved at Step 2 of the algorithm. For every fixed $\overset{\circ}{\alpha} \in A$, consider the auxiliary problem

$$\delta(\overset{\circ}{\alpha}) = \max\{x, \overset{\circ}{\Delta}) \mid x \in \Omega\}, \tag{2.16}$$

where $\overset{\circ}{\Delta} = \{\overset{\circ}{\Delta}(ij)\}$. We assume that the sets $P(\overset{\circ}{\alpha})$ and $Q(\overset{\circ}{\alpha})$ are singletons and $\overset{\circ}{\Delta}(ij)$ is defined by (2.7).

Theorem 2.5. *Suppose that Assumption 2.1 holds and $\hat{x}$ is an optimal solution to (2.16) such that $\hat{X}^i = \sum_j \hat{x}^i_j > 0$ for $i = 1, 2, \ldots, I$. Put*

$$\hat{\alpha}^i_j = \hat{x}^i_j / \hat{X}_i, \quad \hat{s}^i_j = \hat{X}^i(\hat{\alpha}^i_j - \overset{\circ}{\alpha}{}^i_j)/\overset{\circ}{X}{}^i, \quad i = 1, 2, \ldots, I, \quad j = 1, 2, \ldots J_i.$$

Then the following statements hold:

(a) If $\delta(\overset{\circ}{\alpha}) > 0$, then there exists $\rho > 0$ such that $\theta(\overset{\circ}{\alpha}) < \theta(\overset{\circ}{\alpha} + \rho\hat{s})$; i.e., $\hat{s}$ is an ascent direction for problem (2.5);

(b) If $\delta(\overset{\circ}{\alpha}) = 0$, then the disaggregated solution $\overset{\circ}{x} = \overset{\circ}{\alpha}\overset{\circ}{X}$ is optimal for the original problem (2.1);

(c) For each $\overset{\circ}{\alpha} \in A$, there exists $T > 0$ such that $S(\overset{\circ}{\alpha}) \geq T\delta(\overset{\circ}{\alpha})$, where $S(\overset{\circ}{\alpha})$ is defined by (2.12).

Proof. By Assumption 2.1, $\Omega \supseteq P$. Since the element $\overset{\circ}{x}$ is feasible to the original problem (2.1), we have $\overset{\circ}{x} \in \Omega$. The complementary slackness conditions for the aggregated problem give

$$0 = \overset{\circ}{X}{}^i \sum_j \overset{\circ}{\alpha}{}^i_j \overset{\circ}{\Delta}(ij) = \sum \overset{\circ}{x}{}^i_j \overset{\circ}{\Delta}(ij) \quad (i = 1, 2, \ldots, I);$$

therefore, $(\overset{\circ}{x}, \overset{\circ}{\Delta}) = 0$. This implies $\delta(\overset{\circ}{\alpha}) \geq 0$.

Put $\theta(\rho) = \theta(\overset{\circ}{\alpha} + \rho\hat{s})$. Since $P(\overset{\circ}{\alpha})$ and $Q(\overset{\circ}{\alpha})$ are singletons, $\theta(\rho)$ is differentiable at $\rho = 0$, and the marginal-value theorem implies

$$(\theta(0))' = \partial\left[F(\overset{\circ}{X}) - \sum_k \overset{\circ}{\lambda}_k G_k(\overset{\circ}{X})\right] \Big/ \partial\rho \bigg|_{\rho=0}.$$

Recalling the definition of F and G_k and differentiating, we ob- tain

$$(\theta(0))' = \sum_i \sum_j \hat{s}^i_j \overset{\circ}{X}{}^i \overset{\circ}{\Delta}(ij) =$$

$$= \sum_i \sum_j \frac{\hat{X}^i}{\overset{\circ}{X}{}^i} (\hat{\alpha}^i_j - \overset{\circ}{\alpha}{}^i_j) \overset{\circ}{X}{}^i \overset{\circ}{\Delta}(ij) = \sum_i \sum_j \hat{x}^i_j \overset{\circ}{\Delta}(ij) = \delta(\overset{\circ}{\alpha}). \qquad (2.17)$$

(The next to the last equality in (2.17) follows from $(\overset{\circ}{x}, \overset{\circ}{\Delta}) = 0$.) Thus, if $\delta(\overset{\circ}{\alpha}) > 0$, then $(\theta(0))' > 0$, which proves part (a) of Theorem 2.5.

Suppose that $\delta(\overset{\circ}{\alpha}) = 0$, but there exists a pair of indices (i_0, j_0) such that $\overset{\circ}{\Delta}(i_0, j_0) > 0$. Consider the same $\tilde{\alpha} \in A$ as in the proof of Theorem 2.3. By the definition of $\tilde{\alpha}$, we have $\sum_j \tilde{\alpha}_j^{i_0} \overset{\circ}{\Delta}(i_0, j) > 0$ and $\sum_j \tilde{\alpha}_j^i \overset{\circ}{\Delta}(ij) \geq 0$, $i \neq i_0$. According to Assumption 2.1, the optimal solution $\tilde{X}$ to the aggregated problem for $\alpha = \tilde{\alpha}$ is componentwise strictly positive, i.e., $\tilde{X}^i > 0$ for $i = 1, 2, \ldots, I$.

Since $\tilde{x} = \tilde{\alpha}\tilde{X} \in P$ and $\Omega \supseteq P$, we have $\tilde{x} \in \Omega$. Taking into account the definition of $\tilde{\alpha}$ and the condition $\tilde{X} > 0$, we obtain

$$0 < \sum_i \tilde{X}^i \sum_j \tilde{\alpha}_j^i \overset{\circ}{\Delta}(ij) = (\tilde{x}, \overset{\circ}{\Delta}), \tag{2.18}$$

which contradicts the assumption $\delta(\overset{\circ}{x}) = 0$, because $\tilde{x} \in \Omega$. Therefore, if $\delta(\overset{\circ}{\alpha}) = 0$, then $\overset{\circ}{\Delta}(ij) \leq 0$ for $i = 1, 2, \ldots, I$ and $j = 1, 2, \ldots, J$. The relations $0 = \overset{\circ}{X}{}^i \sum_j \overset{\circ}{\alpha}{}_j^i \overset{\circ}{\Delta}(ij) = \sum_j \overset{\circ}{x}{}_j^i \overset{\circ}{\Delta}(ij)$ and $\overset{\circ}{\Delta}(ij) \leq 0$ imply $\overset{\circ}{\Delta}(ij) = 0$ at $\overset{\circ}{x}{}_j^i > 0$. Thus, according to Theorem 2.2, $\overset{\circ}{x}$ is optimal for the original problem (2.1), which proves part (b) of Theorem 2.5.

The last part of Theorem 2.5 follows from the following chain of inequalities

$$S(\overset{\circ}{\alpha}) = \max_{\alpha \in A} \sum_i \overset{\circ}{X}{}^i \sum_j \alpha_j^i \overset{\circ}{\Delta}(ij) = \sum_i \overset{\circ}{X}{}^i \max_{\alpha_i \in A_i} \sum_j \alpha_j^i \overset{\circ}{\Delta}(ij) =$$

$$= \sum_i \frac{\overset{\circ}{X}{}^i}{\hat{X}^i} \hat{X}^i \max_{\alpha_i \in A_i} \sum_j \alpha_j^i \overset{\circ}{\Delta}(ij) \geq \left(\min_i \frac{\overset{\circ}{X}{}^i}{\hat{X}^i} \right) \sum_i \hat{X}^i \max_{\alpha_i \in A_i} \sum_j \alpha_j^i \overset{\circ}{\Delta}(ij) \geq$$

$$\geq \frac{\varepsilon}{B} \sum_i \hat{X}^i \sum_j \hat{\alpha}_j^i \overset{\circ}{\Delta}(ij) = \frac{\varepsilon}{B} \delta(\overset{\circ}{\alpha}) = T\delta(\overset{\circ}{\alpha}), \quad T > 0, \tag{2.19}$$

where $\hat{X}^i \leq B$ for $i = 1, 2, \ldots, I$ and $B > 0$. Such a number B exists since Ω is bounded and $\hat{X}^i = \sum_j \hat{x}_j^i$ for $i = 1, 2, \ldots, I$. The second equality in (2.19) follows from the decomposition $A = A_1 \times A_2 \times \cdots \times A_I$. The first inequality in (2.19) is valid since $\max\left\{ \sum_j \alpha_j^i \overset{\circ}{\Delta}(ij) \mid \alpha_i \in A_i \right\} \geq 0$, which is implied by

$$0 = \overset{\circ}{X}{}^i \sum_j \overset{\circ}{\alpha}{}_j^i \overset{\circ}{\Delta}(ij) \quad \text{and} \quad \overset{\circ}{X}{}^i > 0 \quad (i = 1, 2, \ldots, I), \quad \overset{\circ}{\alpha} \in A.$$

The last inequality in (2.19) holds by virtue of the relation

$$\left\{\sum_j \alpha_j^i \overset{\circ}{\Delta}(ij) \mid \alpha_i \in A_i\right\} \geq \sum_j \hat{\alpha}_j^i \overset{\circ}{\Delta}(ij),$$

which is valid for $\hat{\alpha} \in A$. This completes the proof of Theorem 2.5.

If a solution of the dual aggregated problem is nonunique, then, using the same argument as in Theorem 2.4, we can easily show that all statements of Theorem 2.5 remain valid if $\overset{\circ}{\lambda} \in Q(\overset{\circ}{\alpha})$ in the problem (2.16) is the component of the saddle point of the maximin problem

$$\max_{x \in \Omega} \min_{\lambda \in Q(\overset{\circ}{\alpha})} \sum_i \sum_j x_j^i \left(\frac{\partial f}{\partial x_j^i} - \sum_k \lambda_k \frac{\partial g_k}{\partial x_j^i} \right) \Bigg|_{\overset{\circ}{x}}. \tag{2.20}$$

Problem (2.20) has a saddle point since the convex set Ω is closed and bounded and $Q(\alpha)$ is uniformly bounded with respect to α.

In what follows, we assume for simplicity that the aggregated problem and its dual have unique solutions for any $\alpha \in A$.

Consider the iterative process $\alpha_{\nu+1} = \alpha_\nu + \rho_\nu s_\nu$, where ν is the number of iteration, $\alpha_\nu \in A$, s_ν is defined as in Theorem 2.5, and the step length ρ is determined by the condition

$$\theta(\alpha_\nu + \rho_\nu s_\nu) = \max\{\theta(\alpha_\nu + \rho s_\nu) \mid 0 \leq \rho \leq 1\}.$$

It is shown in [44] that, if the statements (a)–(c) of Theorem 2.5 hold and the derivatives of $\theta(\alpha)$ satisfy the Lipschitz condition, then this iterative process converges and $\{\alpha_\nu\}$ is a maximizing sequence for problem (2.5). In our case the Lipschitz continuity of the derivatives of $\theta(\alpha)$ is implied by (2.11), the differential properties of $f(x)$ and $g_k(x)$, and the uniform boundedness of $P(\alpha)$ and $Q(\alpha)$ with respect to α. The convergence of the iterative process can also be proved directly based on Theorem 2.5, as in [52].

Thus, in the conditional gradient method for problem (2.5), we can replace Step 2 with the following modified Step 2′:

Step 2′. Calculate $\overset{\circ}{\Delta}(ij)$ for $1 = 1, 2, \ldots, I$; $j = 1, 2, \ldots, J_i$. Solve problem (2.16) and find its optimal solution $\hat{x}$. It is assumed that the assumptions of Theorem 2.5 hold and $\hat{X}_i = \sum_j \hat{x}_j^i > 0$ for $i = 1, 2, \ldots, I$. If $\delta(\overset{\circ}{\alpha}) = 0$ in (2.16), then $\overset{\circ}{x} = \overset{\circ}{\alpha}\overset{\circ}{X}$ is optimal and the process terminates. Otherwise, put

$\hat{\alpha}^i_j = \hat{x}^i_j/\hat{X}^i$ and $\hat{s}^i_j = \hat{X}^i(\hat{\alpha}^i_j - \overset{\circ}{\alpha}{}^i_j)/\overset{\circ}{X}{}^i$ $(i = 1, 2, \ldots, I;\ j = 1, 2, \ldots, J_i)$ and go to Step 3 with $\overset{\circ}{s} = \hat{s}$.

A freedom in the choice of the set Ω in Theorem 2.5 can be used to take into account the structural properties of the original problem, if there are any. Below, we consider a number of typical statements of problems with cross connections [24] where the set Ω can be chosen to decompose the ascent direction problem (2.16) at Step 2′ into a number of independent subproblems, resulting thus in a decomposition algorithm for the original problem.

Suppose that problem (2.1) has block constraints corresponding to L subsystems and binding constraints representing the interconnections between the subsystems. The problem is then stated as follows:

$$\begin{aligned} &f(x_1 \ldots x_l \ldots x_L) \to \max, \\ &g^k_l(x_l) \leq 0, \quad x_l \geq 0, \quad k = 1, 2, \ldots, K, \quad l = 1, 2, \ldots, L, \\ &d^s(x_1 \ldots x_l \ldots x_L) \leq 0, \quad s = 1, 2, \ldots, S, \end{aligned} \tag{2.21}$$

where $x = (x_1 \cdots x_l \cdots x_L)$ and $n = \sum_l \dim x_l$. It is assumed that the scalar functions $-f(\cdot)$, $g^k_l(\cdot)$, and $d^s(\cdot)$ are convex and continuously differentiable, and their partial derivatives satisfy the Lipschitz condition.

Define

$$\Omega = \{x \mid g^k_l(x_l) \leq 0, \quad x_l \geq 0, \quad l = 1, 2, \ldots, L, \quad k = 1, 2, \ldots, K_l\}. \tag{2.22}$$

Obviously, the set Ω is convex, and $\Omega \supseteq P$. If Ω is bounded, then, according to Theorem 2.5, the ascent direction problem at Step 2′ decomposes into the following L independent subproblems of lower dimensions:

$$\begin{aligned} &\max\big\{(x_l, \overset{\circ}{\Delta}_l) \mid g^k_l(x_l) \leq 0, \quad x_l \geq 0, \quad k = 1, 2, \ldots, K_l\big\}, \\ &l = 1, 2, \ldots, L, \end{aligned} \tag{2.23}$$

where the vector $\overset{\circ}{\Delta}$ is defined similar to (2.7) and $\overset{\circ}{\Delta}_l$ is the part of $\overset{\circ}{\Delta}$ corresponding to x_l.

Denote by $\overset{\circ}{v}$, $\overset{\circ}{u}$ the unique optimal duals for the aggregated problem, corresponding to block and binding constraints, respectively.

Now we apply the inequality $(\nabla\varphi(\overset{\circ}{x}), x) \le \varphi(x) - \varphi(\overset{\circ}{x}) + (\nabla\varphi(\overset{\circ}{x}), \overset{\circ}{x})$, which is valid for a convex differentiable function $\varphi(x)$, to transform the objective function of the ascent direction problem involved in Step 2′. For any $x \in \Omega$ we have

$$
\begin{aligned}
&(x, \overset{\circ}{\Delta}) = (x, \nabla f(\overset{\circ}{x})) - \sum_{k,l} \overset{\circ}{v}{}_l^k (x_l, \nabla g_l^k(\overset{\circ}{x}_l)) - \sum_s \overset{\circ}{u}_s (x, \nabla d^s(\overset{\circ}{x})) \ge \\
&\ge (x, \nabla f(\overset{\circ}{x})) - \sum_s \overset{\circ}{u}_s (x, \nabla d^s(\overset{\circ}{x})) - \sum_{k,l} \overset{\circ}{v}{}_l^k [g_l^k(x_l) - g_l^k(\overset{\circ}{x}_l) + (\overset{\circ}{x}_l, \nabla g_l^k(\overset{\circ}{x}_l))] \ge \\
&\qquad \ge (x, \nabla f(\overset{\circ}{x})) - \sum_{k,l} \overset{\circ}{v}{}_l^k (\overset{\circ}{x}_l, \nabla g_l^k(\overset{\circ}{x}_l)) - \sum_s \overset{\circ}{u}_s (x, \nabla d^s(\overset{\circ}{x})) \ge \\
&\quad (x, \nabla f(\overset{\circ}{x})) - \sum_s \overset{\circ}{u}_s [d^s(x) - d^s(\overset{\circ}{x}) + (\overset{\circ}{x}, \nabla d^s(\overset{\circ}{x}))] - \sum_{k,l} \overset{\circ}{v}{}_l^k (\overset{\circ}{x}_l, \nabla g_l^k(\overset{\circ}{x}_l)) = \\
&\quad = (x, \nabla f(\overset{\circ}{x})) - \sum_s \overset{\circ}{u}_s d^s(x) - \sum_s \overset{\circ}{u}_s (\overset{\circ}{x}, \nabla d^s(\overset{\circ}{x})) - \sum_{k,l} \overset{\circ}{v}{}_l^k (\overset{\circ}{x}_l, \nabla g_l^k(\overset{\circ}{x}_l)) \ge \\
&\ge f(x) - \sum_s \overset{\circ}{u}_s d^s(x) - f(\overset{\circ}{x}) + (\overset{\circ}{x}, \nabla f(\overset{\circ}{x}) - \sum_s \overset{\circ}{u}_s \nabla d^s(\overset{\circ}{x}) - \sum_{k,l} \overset{\circ}{v}{}_l^k \nabla g_l^k(\overset{\circ}{x}_l)) \ge \\
&\qquad\qquad \ge f(x) - \sum_s \overset{\circ}{u}_s d^s(x) - f(\overset{\circ}{x}). \qquad (2.24)
\end{aligned}
$$

The second inequality in (2.24) holds since $\sum_{k,l} \overset{\circ}{v}{}_l^k g_l^k(\overset{\circ}{x}_l) = 0$ by the complementarity conditions for the aggregated problem and $\sum_{k,l} \overset{\circ}{v}{}_l^k g_l^k(x_l) \le 0$ for $x \in \Omega$, $\overset{\circ}{v}{}_l^k \ge 0$. The first equality in (2.24) follows from $\sum_s \overset{\circ}{u}_s d^s(\overset{\circ}{x}) = 0$ by the complementarity conditions for the aggregated problem. Finally, the last equality in (2.24) is fulfilled since $(\overset{\circ}{x}, \nabla f(\overset{\circ}{x}) - \sum_s \overset{\circ}{u}_s \nabla d^s(\overset{\circ}{x}) - \sum_{k,l} \overset{\circ}{v}{}_l^k \nabla g_l^k(\overset{\circ}{x}_l)) = 0$ by the complementarity conditions to the dual aggregated problem. Note, that a special structure of the set Ω has not been assumed in deriving (2.24).

By (2.24) and the definition of $\delta(\overset{\circ}{\alpha})$ in (2.16) we have

$$
\delta(\overset{\circ}{\alpha}) \ge \max\{f(x) - \sum_s \overset{\circ}{u}_s d^s(x) \mid x \in \Omega\} - f(\overset{\circ}{x}) \equiv \pi(\overset{\circ}{\alpha}). \qquad (2.25)
$$

The following proposition shows that problem (2.25) can be used in the ascend direction problem at Step 2′.

Theorem 2.6. *Let $\hat{x}$ be an optimal solution to problem (2.25), such that $\hat{X}_i = \sum_i \hat{x}_j^i > 0$ for $i = 1, 2, \ldots, I$. Define $\hat{s}_j^i$ similar to Theorem 2.5. If*

$\pi(\overset{\circ}{\alpha}) = 0$, *then the disaggregated solution* $\overset{\circ}{x} = \overset{\circ}{\alpha}\overset{\circ}{X}$ *is optimal to (2.21). If* $\pi(\overset{\circ}{\alpha}) > 0$, *then* $\hat{s}$ *is the ascent direction at the point* $\overset{\circ}{\alpha}$ *for problem (2.5).*

Proof. Bearing in mind the definition of Ω, we have by the saddle point theorem for the Lagrangian function of problem (2.21):

$$\max_{x\in\Omega}\{f(x) - \sum_s \overset{\circ}{u}_s d^s(x)\} \geq \max_{x\in\Omega}\min_{u^s\geq 0}\{f(x) - \sum_s \overset{\circ}{u}_s d^s(x)\} = f(x^*) \geq f(\overset{\circ}{x}), \tag{2.26}$$

where x^* is the optimal solution to (2.21). From (2.26) we have $\pi(\overset{\circ}{\alpha}) \geq$ $\geq f(x^*) - f(\overset{\circ}{x}) \geq 0$. Hence, if $\pi(\overset{\circ}{\alpha}) = 0$, then $\overset{\circ}{x}$ is optimal to (2.21).

Define $\theta(\rho) = \theta[\overset{\circ}{\alpha} + \rho\hat{s}]$. Doing the same as in the proof of the part a) of Theorem 2.5 we obtain

$$(\theta(0))' = \sum_{i,j} \hat{x}^i_j \overset{\circ}{\Delta}(i,j).$$

This, together with (2.24), yields

$$(\theta(0))' = \sum_{i,j} \hat{x}^i_j \overset{\circ}{\Delta}(i,j) \geq f(\hat{x}) - \sum_s \overset{\circ}{u}_s d^s(\hat{x}) - f(\overset{\circ}{x}) = \pi(\overset{\circ}{\alpha}).$$

Thus, if $\pi(\overset{\circ}{\alpha}) > 0$, then $(\theta(0))' > 0$ and hence $\hat{s}$ is the ascend direction, which completes the proof.

Now suppose that the binding constraints and the objective function of (2.21) have the following additive-separable form:

$$f(x) = \sum_l f_l(x_l), \ d^s(x) = \sum_l d^s_l(x_l). \tag{2.27}$$

In this case, due to the block structure of the set Ω, problem (2.25) used to determine the ascent direction at Step 2$'$ decomposes into L independent subproblems of the form

$$\max\left\{ f_l(x_l) - \sum_s \overset{\circ}{u}_s d^s_l(x_l) \mid g^k_l(x_l) \leq 0, \ x_l \geq 0, \ k = 1,2,...,K_l \right\}, \tag{2.28}$$
$$l = 1,2,...,L.$$

Note that the decomposition of the problem of determining the ascent direction is independent on the way the aggregated problem is constructed. That is, subproblems (2.23) or (2.28) are fully determined by the block or block-separable structure of problem (2.21). Moreover, the structure of the aggregation, i.e., the number I of the aggregated variables and the partitioning of the components of a vector x into I disjoint subsets can vary in a wide range. The only requirements are the uniqueness of the optimal solutions to the macroproblem and its dual, and the positiveness of components of $\overset{\circ}{X}$, $\hat{X}$.

In [33, 34] the decomposition method for the block-separable problems of type (2.21), (2.27) is constructed using an aggregation of variables from different blocks, which is a particular case of the aggregation (2.2), (2.3). In the notations of problem (2.21), (2.27), the aggregation of the variables from the different blocks corresponds to $I = I_0 = \dim x_l$ and $X = \sum_l x_l$ $(l = 1, 2, ..., L)$. In [53] the aggregation parameters α are updated by the formula

$$\alpha_j^i(\rho) = \left[\overset{\circ}{x}{}_j^i + \rho(\hat{x}_j^i - \overset{\circ}{x}{}_j^i)\right] / \sum_j \left[\overset{\circ}{x}{}_j^i + \rho(\hat{x}_j^i - \overset{\circ}{x}{}_j^i)\right], \tag{2.29}$$

where ρ is determined by the one-dimensional maximization of $\theta[\alpha(\rho)]$ over the interval [0,1]. Relation (2.29) differs from our formula

$$\alpha_j^i(\rho) = \left\{\overset{\circ}{x}{}_j^i + \rho\left[\hat{x}_j^i / \sum_r \hat{x}_r^i - \overset{\circ}{x}{}_j^i / \sum_r \overset{\circ}{x}{}_r^i\right] \sum_r \hat{x}_r^i\right\} / \sum_r \overset{\circ}{x}{}_r^i$$

used at Step 2′ to update α. However, it is easy to verify that the expression for the derivative $(\alpha_j^i(0))'$ evaluated according to (2.29),

$$(\alpha_j^i(0))' = \left(\hat{x}_j^i - \overset{\circ}{\alpha}{}_j^i \sum_r \hat{x}_r^i\right) / \sum_r \overset{\circ}{x}{}_r^i$$

and determining the direction of variation of α in [53], is precisely the same as the expression for $\hat{s}$ used at Step 2′ of the algorithm considered above. The relation to the aggregation method used in [42] for block linear programming problems, where the aggregated variable is introduced as the sum of variables from a separate block, can be examined similarly.

Using relations (2.24) and Theorem 2.6 it is possible to formulate different independent subproblems based on the structure of problem (2.21). Suppose, for example, that in problem (2.21) the objective function is nonseparable, while the binding constraints are additive-separable: $d^s(x) = \sum_l d_l^s(x_l)$. Using the second equality in (2.24) we have

$$(x,\overset{\circ}{\Delta}) \geq (x,\nabla f(\overset{\circ}{x})) - \sum_s \overset{\circ}{u}_s d^s(x) + (\overset{\circ}{x},\nabla f(\overset{\circ}{x})) - -(\overset{\circ}{x},\nabla f(\overset{\circ}{x})) -$$
$$- \sum_{k,l} \overset{\circ}{v}{}_l^k(\overset{\circ}{x}_l,\nabla g_l^k(\overset{\circ}{x}_l)) - \sum_s \overset{\circ}{u}_s(\overset{\circ}{x},\nabla d^s(\overset{\circ}{x})) =$$
$$= (x,\nabla f(\overset{\circ}{x})) - \sum_s \overset{\circ}{u}_s d^s(x) - -(\overset{\circ}{x},\nabla f(\overset{\circ}{x})) \geq f(x) - \sum_s \overset{\circ}{u}_s d^s(x) - f(\overset{\circ}{x}). \quad (2.30)$$

Here the equality was obtained by the complementarity theorem for the dual aggregated problem. From (2.30) it follows that

$$\delta(\overset{\circ}{\alpha}) \geq \max\left\{(x,\nabla f(\overset{\circ}{x})) - \sum_s \overset{\circ}{u}_s d^s(x) \mid x \in \Omega\right\} - (\overset{\circ}{x},\nabla f(\overset{\circ}{x})) \equiv \pi_f(\overset{\circ}{\alpha}) \geq \pi(\overset{\circ}{\alpha}). \quad (2.31)$$

Since $\pi(\overset{\circ}{\alpha}) \geq 0$, then for $\pi_f(\overset{\circ}{\alpha}) = 0$ we get by (2.31) and Theorem 2.6 the optimality of the disaggregated solution $\overset{\circ}{x}$. If $\pi_f(\overset{\circ}{\alpha}) > 0$ and $\hat{x}$ is an optimal solution of the maximization problem in (2.31), then using the same argument as in Theorem 2.6 we can construct the ascent direction $\hat{s}$. Due to the block structure of the set Ω, the maximization problem in (2.31) decomposes into L independent subproblems of the form

$$\max\left\{(x_l,\nabla_l f(\overset{\circ}{x})) - \sum_s \overset{\circ}{u}_s d_l^s(x_l) \mid g_l^k(x_l) \leq 0,\ x_l \geq 0,\ k = 1,2...,K_l\right\},$$
$$l = 1,2,....,L, \quad (2.32)$$

where $\nabla_l f(\overset{\circ}{x})$ is the part of the vector $\nabla f(\overset{\circ}{x})$ corresponding to x_l.

Let now in problem (2.21) the binding constraints be nonseparable, while the objective function be additive-separable, $f(x) = \sum_l f_l(x_l)$. Similar to (2.30) we have

$$(x,\overset{\circ}{\Delta}) \geq (x,\nabla f(\overset{\circ}{x})) - \sum_{k,l} \overset{\circ}{v}{}_l^k(\overset{\circ}{x}_l,\nabla g_l^k(\overset{\circ}{x}_l)) - \sum_s \overset{\circ}{u}_s(x,\nabla d^s(\overset{\circ}{x})) \geq$$
$$\geq f(x) - f(\overset{\circ}{x}) + (\overset{\circ}{x},\nabla f(\overset{\circ}{x})) - \sum_s \overset{\circ}{u}_s(x,\nabla d^s(\overset{\circ}{x})) - \sum_{k,l} \overset{\circ}{v}{}_l^k(\overset{\circ}{x}_l,\nabla g_l^k(\overset{\circ}{x}_l)) =$$
$$= [f(x) - \sum_s \overset{\circ}{u}_s(x,\nabla d^s(\overset{\circ}{x})] - [f(\overset{\circ}{x}) - -(\overset{\circ}{x},\sum_s \overset{\circ}{u}_s \nabla d^s(\overset{\circ}{x}))]+$$

$$+ [\overset{\circ}{x}, \nabla f(\overset{\circ}{x}) - \sum_{k,l} \overset{\circ}{v}{}_l^k \nabla g_l^k(\overset{\circ}{x}_l) - \sum_s \overset{\circ}{u}_s \nabla d^s(\overset{\circ}{x}))] =$$

$$= [f(x) - \sum_s \overset{\circ}{u}_s(x, \nabla d^s(\overset{\circ}{x})] - [f(\overset{\circ}{x}) - -(\overset{\circ}{x}, \sum_s \overset{\circ}{u}_s \nabla d^s(\overset{\circ}{x}))] \geq$$

$$\geq f(x) - \sum_s \overset{\circ}{u}_s d^s(x) - f(\overset{\circ}{x}).$$

This yields

$$\delta(\overset{\circ}{\alpha}) \geq \max \left\{ f(x) - \sum_s \overset{\circ}{u}_s(x, \nabla d^s(\overset{\circ}{x}) \mid x \in \Omega \right\} -$$
$$- f(\overset{\circ}{x}) + (\overset{\circ}{x}, \sum_s \overset{\circ}{u}_s \nabla d^s(\overset{\circ}{x})) \equiv \pi_d(\overset{\circ}{\alpha}) \geq \pi(\overset{\circ}{\alpha}). \qquad (2.33)$$

The optimality criterion is $\pi_d(\overset{\circ}{\alpha}) = 0$. If $\pi_d(\overset{\circ}{\alpha}) > 0$, then the decent direction is constructed the same as before. The maximization problem in (2.33) decomposes into L subproblems of the form

$$\max \left\{ f_l(x_l) - \sum_s \overset{\circ}{u}_s(x_l, \nabla_l d^s(\overset{\circ}{x}) \mid g_l^k(x_l) \leq 0,\ x_l \geq 0,\ k = 1, 2..., K_l \right\},$$
$$l = 1, 2..., L, \qquad (2.34)$$

where $\nabla_l d^s(\overset{\circ}{x})$ is the part of the vector $\nabla d^s(\overset{\circ}{x})$ corresponding to x_l.

Note that if the objective function and (or) binding constraints are separable, then the independent subproblems can be formulated either in the form (2.23) or in the form (2.28), (2.32), (2.34). The latter is preferable since only the duals for the binding constraints of the macroproblem are utilized. Taking into account this observation it is worth to exclude the duals for the block constraints block constraints form the subproblems formulations in the nonseparable case also. Using the second inequality in (2.24) we have

$$(x, \overset{\circ}{\Delta}) \geq (x, \nabla f(\overset{\circ}{x})) - \sum_{k,l} \overset{\circ}{v}{}_l^k (\overset{\circ}{x}_l, \nabla g_l^k(\overset{\circ}{x}_l)) -$$
$$- \sum_s \overset{\circ}{u}_s(x, \nabla d^s(\overset{\circ}{x})) \geq (x, \nabla f(\overset{\circ}{x})) - \sum_s \overset{\circ}{u}_s(x, \nabla d^s(\overset{\circ}{x}) +$$
$$+ \left[\left(\overset{\circ}{x}, \nabla f(\overset{\circ}{x}) - \sum_s \overset{\circ}{u}_s \nabla d^s(\overset{\circ}{x}) - \sum_{k,l} \overset{\circ}{v}{}_l^k \nabla g_l^k(\overset{\circ}{x}_l) \right) \right] -$$

$$-(\overset{\circ}{x}, \nabla f(\overset{\circ}{x})) + \left(\overset{\circ}{x}, \sum_s \overset{\circ}{u}_s \nabla d^s(\overset{\circ}{x})\right) =$$

$$= \left(x, \nabla f(\overset{\circ}{x}) - \sum_s \overset{\circ}{u}_s \nabla d^s(\overset{\circ}{x})\right) - \left(\overset{\circ}{x}, \nabla f(\overset{\circ}{x}) - \sum_s \overset{\circ}{u}_s \nabla d^s(\overset{\circ}{x})\right) \geq$$

$$\geq f(x) - \sum_s \overset{\circ}{u}_s d^s(x) - f(\overset{\circ}{x}).$$

This yields

$$\delta(\overset{\circ}{\alpha}) \geq \max\left\{\left(x, \nabla f(\overset{\circ}{x}) - \sum_s \overset{\circ}{u}_s \nabla d^s(\overset{\circ}{x})\right) \mid x \in \Omega\right\} -$$

$$- \left(\overset{\circ}{x}, \nabla f(\overset{\circ}{x}) - \sum_s \overset{\circ}{u}_s \nabla d^s(\overset{\circ}{x})\right) \equiv \pi_{fd}(\overset{\circ}{\alpha}) \geq \pi(\overset{\circ}{\alpha}). \qquad (2.35)$$

The optimality criterion for the disaggregated solution is $\pi_{fd}(\overset{\circ}{\alpha}) = 0$, while maximization in (2.35) results in L independent subproblems using only duals to the binding constraints.

Note an additional property, taking place for various assumptions about separability of the functions involved. From (2.26) it follows that

$$\pi(\overset{\circ}{\alpha}) + f(\overset{\circ}{x}) \geq f(x^*) \geq f(\overset{\circ}{x}).$$

Moreover, $\pi_f(\overset{\circ}{\alpha}) \geq \pi(\overset{\circ}{\alpha})$, $\pi_d(\overset{\circ}{\alpha}) \geq \pi(\overset{\circ}{\alpha})$, and $\pi_{fd}(\overset{\circ}{\alpha}) \geq \pi(\overset{\circ}{\alpha})$. Thus in all forms of subproblems formulations we have in each iteration a lower and an upper bound for the optimal objective in (2.21). The lower bound increases monotonously iteration from iteration, coinciding with the upper bound for the optimal disaggregated solution.

Consider yet another class of block problems with coupling variables, which is frequently encountered in applications [24]:

$$\begin{aligned}
&f(x_1 \ldots x_l \ldots x_L) \to \max, \\
&g_l^k(x_l) + r_l^k(x_1 \ldots x_l \ldots x_L) \leq 0, \quad x_l \geq 0, \quad k = 1, 2, \ldots K_l, \\
&d^s(x_1 \ldots x_l \ldots x_L) \leq 0, \quad s = 1, 2, \ldots, S, \\
&l = 1, 2, \ldots L,
\end{aligned} \qquad (2.36)$$

where $-f(\cdot)$, $d^s(\cdot)$, and $g_l^k(\cdot) + r_l^k(\cdot)$ are continuously differentiable convex scalar functions with Lipschitz partial derivatives. Generally, $r_l^k(\cdot)$ may be nonconvex.

Suppose that, for all $x \geq 0$, $d^s(x) \leq 0$ $(s = 1, 2, \ldots, S)$, the terms $r_l^k(x)$ determining the cross connections between the subsystems have the lower bounds

$$r_l^k(x) \geq \xi_l^k, \quad k = 1, 2, \ldots, K_l, \quad l = 1, 2, \ldots L, \tag{2.37}$$

where ξ_l^k are given real numbers.

We put

$$\Omega = \Big\{x \mid g_l^k(x_l) \leq -\xi_l^k,\ x_l \geq 0,\ k = 1, 2, \ldots K_l,\ l = 1, 2, \ldots, L\Big\}.$$

Obviously, the set Ω is convex and $\Omega \supseteq P$. The problem of determining the ascent direction for $\theta(\alpha)$ involved in Step 2$'$ decomposes into L independent subproblems

$$\max\Big\{x_l, \overset{\circ}{\Delta}_l) \mid g_l^k(x_l) \leq -\xi_l^k,\ x_l \geq 0,\ k = 1, 2, \ldots, K_l\Big\},\ l = 1, 2, \ldots, L, \tag{2.38}$$

where $\overset{\circ}{\Delta}(ij)$ is the vector with the components

$$\overset{\circ}{\Delta}(ij) = \Bigg[\partial f(x)/\partial x_j^i - \sum_k \sum_l \overset{\circ}{\nu}{}_l^k \Big(\partial g_l^k(x_l)/\partial x_j^i + \partial r_l^k(x)/\partial x_j^i\Big) - \\ - \sum_s \overset{\circ}{u}_s (\partial d^s(x)/\partial x_j^i)\Bigg]\Bigg|_{\overset{\circ}{x}}$$

and $\overset{\circ}{\Delta}_l$ is the part of $\overset{\circ}{\Delta}$ corresponding to x_l.

In the case of convex functions $r_l^k(x)$, we have the following estimate similar to (2.24):

$$(x, \overset{\circ}{\Delta}) \geq f(x) - \sum_s \overset{\circ}{u}_s d^s(x) - \sum_k \sum_l \overset{\circ}{\nu}{}_l^k \left[g_l^k(x_l) + r_l^k(x)\right] - f(\overset{\circ}{x}).$$

If $f(x)$ and $d^s(x)$ have form (2.27) and $r_l^k(x)$ are additively separable with respect to the blocks, i.e., $r_j^k(x) = \sum_{t=1}^{L} r_{lt}^k(x_t)$, then (2.38) can be replaced by local subproblems of the type

$$\max\Bigg\{ f_l(x_l) - \sum_s \overset{\circ}{u}_s d_l^s(x_l) - \sum_k \overset{\circ}{\nu}{}_l^k g_l^k(x_l) - \\ - \sum_t \sum_k \overset{\circ}{\nu}{}_t^k r_{tl}^k(x_l) \mid g_l(x_l) \leq -\xi_l^k,\ x_l \geq 0\Bigg\} \quad l = 1, 2, \ldots, L.$$

In addition to (2.37), bounds of the form

$$r_l^k(x) \geq \varphi_l^k(x_l), \quad k = 1, 2, \ldots, K_l, \quad l = 1, 2, \ldots, L,$$

where $\phi_l^k(\cdot)$ are given convex functions, can be used; they also give a decomposition into independent subproblems at Step 2′.

It often happens that $r_l^k(x) \geq 0$ for $x \leq 0$. In this case, the local problems (2.38) correspond to isolated subsystems without cross connections. The interconnections are only involved in the objective function of (2.38) and in the constraints of the aggregated problem.

§3. Decomposition of the Macroproblem in the Block-Separable Case

In the previously considered approaches the decomposition effect was achieved due to the special properties of the ascent direction problem. The structural properties of the aggregated problem were not taken into account. Now we consider a way of introducing aggregated variables resulting in a decomposition of the macroproblem also.

The original problem is stated in the form

$$\max\left\{\sum_l f_l(x_l) \mid \sum_l d_l^k(x_l) \leq b^k,\ k = 1, 2, ..., K,\ x_l \in P_l,\ l = 1, 2, ..., L\right\}, \tag{3.1}$$

where $d_l^k(x_l)$, $-f_l(x_l)$ are convex functions, P_l is a convex closed and bounded set for all l.

Problem (3.1) is commonly treated as corresponding to a decentralized system composed of one high-level decision making unit and L low-level decision making units/divisions [16, 18]. The decision vector under influence of the division l is x_l, the binding constraints correspond to resources that are common to all subdivisions, while restrictions $x_l \in P_l$ are individual constraints realized by the division l. The target of the central unit is to allocate common resources, using some control mechanism, among different divisions so that the total profit of the organization is maximized. Bearing in mind resource-allocation nature of (3.1) we assume for simple consideration that the vector b is strictly positive componentwise. In contrast to the previous constructions, where the decision variables were aggregated, we focus here on aggregation of resources allocated by the central unit among the divisions. It is natural to refer to this approach as resource aggregation.

Similar to resource-directive approach [21], artificial variables r_l are introduced and (3.1) is reformulated as follows:

$$\max_{x,r} \left\{ \sum_l f_l(x_l) \mid d_l^k(x_l) \leq r_l^k,\ x_l \in P_l,\ l = 1, 2, ..., L, \right.$$
$$\left. \sum_l r_l^k = b^k,\ k = 1, 2, ..., K \right\}. \tag{3.2}$$

This problem is aimed to find the best allocation $\{r_1, ..., r_L\}$ of the overall resource b among the divisions, maximizing the total profit.

The aggregated variables are defined by $R^k = \sum_l r_l^k$, $k = 1, 2, ..., K$, while the set A of the disaggregation parameters is given by

$$A = \left\{ \alpha_l^k \mid \sum_l \alpha_l^k = 1,\ k = 1, 2, ..., K \right\}.$$

Note that we do not impose here nonnegativity restrictions for the weights α_l^k since the variables r_l^k are unrestricted in sign. Respectively, the disaggregated variables are defined in the form $r_l^k = \alpha_l^k R^k$.

Fixing $\mathring{\alpha} \in A$ and substituting the disaggregated variables into (3.2) we obtain the aggregated problem:

$$\theta(\mathring{\alpha}) = \max_{x,R} \left\{ \sum_l f_l(x_l) \mid d_l^k(x_l) \leq \mathring{\alpha}_l^k R^k,\ x_l \in P_l,\ l = 1, 2, ..., L, \right.$$
$$\left. \sum_l \mathring{\alpha}_l^k R^k = b^k,\ k = 1, 2, ..., K \right\}. \tag{3.3}$$

The Lagrangian function for this problem is as follows:

$$\mathcal{L} = \sum_l f_l(x_l) + \sum_{k,l} \lambda_l^k \left(\mathring{\alpha}_l^k R^k - d_l^k(x_l) \right) + \sum_k u^k \left(b^k - \sum_l \mathring{\alpha}_l^k R^k \right).$$

Since the aggregated variables R^k are unrestricted in sign, the optimality condition $\partial\mathcal{L}/\partial R^k = 0$ yield $\mathring{u}^k(\sum_l \mathring{\alpha}_l^k) = \sum_l \mathring{\lambda}_l^k \mathring{\alpha}_l^k$. Here $\mathring{u}, \mathring{\lambda}$ are the optimal duals for the binding and the block constraints of the aggregated problem (3.3). Since $\overset{\circ}{\alpha} \in A$, we finally have

$$\overset{\circ}{u}^k = \sum_l \overset{\circ}{\lambda}{}_l^k \overset{\circ}{\alpha}{}_l^k, \ k = 1, 2, ..., K. \tag{3.4}$$

That is, knowing the duals for the block constraints in (3.3), we can easily calculate the duals for the binding constraints.

Consider now how to solve the aggregated problem (3.3). The last restriction of (3.3) and the condition $\overset{\circ}{\alpha} \in A$ yield $\overset{\circ}{R}{}^k = b^k$, $k = 1, 2, ..., K$. Substituting this solution into the aggregated problem, we conclude that it decomposes into L independent subproblems of the form

$$\theta_l(\overset{\circ}{\alpha}_l) = \max_{x_l} \left\{ f_l(x_l) \mid d_l^k(x_l) \leq \overset{\circ}{\alpha}{}_l^k b^k, \ x_l \in P_l, \ k = 1, 2, ..., K \right\}. \tag{3.5}$$

Obviously, if $\overset{\circ}{x}_l$ is an optimal solution of (3.5), then $\overset{\circ}{x} = \{\overset{\circ}{x}_1, ..., \overset{\circ}{x}_L\}$ and $\overset{\circ}{R} = b$ is an optimal solution of (3.3) with $\theta(\overset{\circ}{\alpha}) = \sum_l \theta_l(\overset{\circ}{\alpha}_l)$. Moreover, the Lagrange multiplier $\overset{\circ}{\lambda}{}_l^k$ for the condition $d_l^k(x_l) \leq \overset{\circ}{\alpha}{}_l^k b^k$ in (3.5) coincides with the Lagrange multiplier for the condition $d_l^k(x_l) \leq \overset{\circ}{\alpha}{}_l^k R^k$ in (3.3). Thus we may conclude that to find the optimal (primal and dual) solutions to (3.3) we can solve L independent subproblems (3.5), find their primal and dual solutions $\overset{\circ}{x}_l, \overset{\circ}{\lambda}_l$, and then calculate $\overset{\circ}{u}$ by (3.4). That is, in the case of resource aggregation, the aggregated problem decomposes into independent subproblems associated with separate divisions.

Define the set Ω by the block constraints of (3.2):

$$\Omega = \{x_l, r_l \mid d_l^k(x_l) \leq r_l^k, \ x_l \in P_l, \ l = 1, 2, ..., L\}.$$

Bearing in mind resource-allocation nature of (3.2) we assume for simple consideration that $d_l^k(x_l) \geq 0$ for all $x_l \in P_l$. Then we can always make the set Ω bounded, attaching the restrictions $0 \leq r_l^k \leq b^k$ if necessary.

To find the search direction according to our basic iterative scheme we need to solve the problem

$$\max_{x,r} \left\{ \sum_l f_l(x_l) - \sum_k \overset{\circ}{u}^k (\sum_l r_l^k - b^k) \mid x, r \in \Omega \right\},$$

which decomposes into L independent subproblems

$$\max_{x_l, r_l} \left\{ f_l(x_l) - \sum_k \overset{\circ}{u}^k r_l^k \mid d_l^k(x_l) \le r_l^k, \ x_l \in P_l \right\}, \tag{3.6}$$

where $\overset{\circ}{u}^k$ is defined by (3.4). Let $\hat{x}_l, \hat{r}_l$ be the optimal solution of these subproblems such that $\hat{R}^k = \sum_l \hat{r}_l^k \neq 0, \ k = 1, ..., K$. If

$$\pi(\overset{\circ}{\alpha}) = \sum_l f_l(\hat{x}_l) - \sum_k \overset{\circ}{u}^k \Big(\sum_l \hat{r}_l^k - b^k \Big) - \theta(\overset{\circ}{\alpha}) = 0,$$

then $\overset{\circ}{x}_l, l = 1, ..., L$ is optimal for the original problem. Otherwise,

$$\hat{s}_l^k = \hat{R}^k (\hat{\alpha}_l^k - \overset{\circ}{\alpha}_l^k)/b^k, \ \hat{\alpha}_l^k = \hat{r}_l^k / \hat{R}^k$$

defines the search direction. Note that by the definition of $\overset{\circ}{u}^k$

$$\pi(\overset{\circ}{\alpha}) = \sum_l \pi_l(\overset{\circ}{\alpha}), \ \pi_l(\overset{\circ}{\alpha}) = \Big[f_l(\hat{x}_l) - \sum_k \overset{\circ}{u}^k \hat{r}_l^k \Big] - \Big[\theta_l(\overset{\circ}{\alpha}_l) - \sum_k \overset{\circ}{\lambda}_l^k \overset{\circ}{\alpha}_l^k b^k \Big].$$

Now we are ready to state the iterative aggregation procedure for the case of resource aggregation in (3.2). We assume that the divisions do not know the resource vector b, while the center does not know f_l, d_l, and P_l associated with the subdivisions. Moreover, it is assumed that there are no information interconnections between the divisions. Then each iteration of the planning scheme consists of the following steps:

Step 0. (Initialization, central unit's problem).

Define $\delta \geq 0$ - accuracy of optimization. Initialize normalized $\overset{\circ}{\alpha}_l^k$ and transfer $\overset{\circ}{\alpha}_l^k b^k$ to the divisions.

Step 1. (Subdivisions problem).

Solve the problems

$$\theta_l(\overset{\circ}{\alpha}_l) = \max_{x_l} \left\{ f_l(x_l) \mid d_l^k(x_l) \le \overset{\circ}{\alpha}_l^k b^k, \ x_l \in P_l, \ k = 1, 2, ..., K \right\}$$

for all subdivisions, obtaining the optimal primal and dual solutions $\overset{\circ}{x}_l, \overset{\circ}{\lambda}_l$. Transfer $\theta_l(\overset{\circ}{\alpha}_l)$ and $\overset{\circ}{\lambda}_l$ to the center.

Step 2. (Central unit's problem).

Compute average prices $\overset{\circ}{u}^k = \sum_l \overset{\circ}{\lambda}_l^k \overset{\circ}{\alpha}_l^k, \ k = 1, 2, ..., K$ and transfer to the divisions.

Step 3. (Subdivisions problems).
Solve the problems

$$\max_{x_l, r_l} \left\{ f_l(x_l) - \sum_k \overset{\circ}{u}{}^k r_l^k \mid d_l^k(x_l) \leq r_l^k, \; x_l \in P_l \right\}$$

for all divisions, obtaining the optimal solution $\hat{r}_l, \hat{x}_l$. Calculate

$$\pi_l(\overset{\circ}{\alpha}) = \left[f_l(\hat{x}_l) - \sum_k \overset{\circ}{u}{}^k \hat{r}_l^k] - [\theta_l(\overset{\circ}{\alpha}_l) - \sum_k \overset{\circ}{\lambda}{}_l^k \overset{\circ}{\alpha}{}_l^k b^k \right]$$

and transfer $\pi_l(\overset{\circ}{\alpha}), \hat{r}_l$ to the center.

Step 4. (Central unit's problem).
Calculate $\pi(\overset{\circ}{\alpha}) = \sum_l \pi_l(\overset{\circ}{\alpha})$. If $\pi(\overset{\circ}{\alpha}) \leq \delta$ then stop. Else compute $\hat{R}^k =$ $= \sum_l \hat{r}_l^k$ and $\hat{\alpha}_l^k = \hat{r}_l^k / \hat{R}^k$. Set $\hat{s}_l^k = \hat{R}^k (\hat{\alpha}_l^k - \overset{\circ}{\alpha}{}_l^k)/b^k$. Go to Step 1 with $\alpha = \overset{\circ}{\alpha} + \rho \hat{s}$, where $0 \leq \rho \leq 1$ is a properly chosen stepsize, such that $\sum_l \theta_l(\alpha_l) \geq \sum_l \theta_l(\overset{\circ}{\alpha}_l)$.

The central unit's problem is very simple - there is no an optimization problem associated with the higher level. It is necessary only to sum up the local quantities $\pi_l(\overset{\circ}{\alpha}), \hat{r}_l^k$ generated by the subdivisions and calculate the mean price using the weights α_l. It is not hard to verify, using the definition of u in (3.4) and the normalizing conditions for α, that u is the mean price defined as is customary. Let $v_l^k = \alpha_l^k b^k$ be an amount of the resource k given for the division l. Then by (3.4) we have

$$u^k = \sum_l \lambda_l^k \alpha_l^k = \sum_l \lambda_l^k (v_l^k / b^k) = \left(\sum_l \lambda_l^k v_l^k \right) / b^k.$$

That is, to calculate the respective component of u we need to divide the total cost of the allocated resource, calculated for the local prices, by the total amount of the resource available.

In the subproblem (3.5) the division takes its decision under the fixed resource allocation $v_l^k = \alpha_l^k b^k$, $k = 1, ..., K$. From this point of view (3.5) is similar to the subproblem arising in resource directive decomposition [21]. In the subproblem (3.6) resources available for the division are unlimited, but the division should pay for resources used. The price for the resource u^k is the weighted mean of the local prices λ_l^k, $l = 1, ..., L$ obtained for the fixed resource allocation. That is, the center uses two control mechanisms consequently. At first, the central unit applies resource directive strategy and then, based on its results, uses pricing control policy.

The value $\theta_l(\alpha_l) - \sum_k \lambda_l^k \alpha_l^k b^k$ is computed in the subdivision after (3.5) has been solved. It is the profit of the subdivision under the fixed resource allocation $v_l^k = \alpha_l^k b^k$ minus the cost of these resources calculated for the local prices λ_l^k. The value $f_l(\hat{x}_l) - \sum_k u^k \hat{r}_l^k$ is calculated in the division after (3.6) has been solved. It is the profit of the subdivision under unlimited resources minus the cost of resources used calculated for the mean price u^k. So the value $\pi_l(\alpha) = [f_l(\hat{x}_l) - \sum_k u^k \hat{r}_l^k] - [\theta_l(\alpha_l) - \sum_k \lambda_l^k \alpha_l^k b^k]$ is the division's measure to compare two control polices of the center - the pricing mechanism and the resource directive strategy. It is not hard to verify using the Lagrange duality that $\pi(\alpha) = \sum_l \pi_l(\alpha) = 0$ if and only if $\pi_l(\alpha) = 0$ for all subdivisions. Thus for the optimal solution both control polices of the center are equivalent from the subdivision's point of view.

Moreover, suppose that the current resource allocation is optimal. Then all local prices λ_l are equal to each other since otherwise it is possible to relocate resources and increase the overall objective. Alternatively, if λ_l are equal to each other, then relocation of resources does not increase the overall objective and the current allocation is optimal. Hence by the definition of the mean price $u^k = \sum_l \lambda_l^k \alpha_l^k$ it follows that in optimality the mean price coincides with local price and vice versa. Thus the pricing policy of the center is as follows: the center averages the local prices until the local prices coincide with the mean price.

If λ_l in (3.5) is nonunique, say $\lambda_l \in D_l$, then the mean price u is also nonunique and π depends on u. In this case it is necessary to find a saddle point $\{\hat{u}, \hat{r}\}$ solving the problem $\min\{\pi(u) \mid u^k = \sum_l \lambda_l^k \alpha_l^k,\ \lambda_l \in D_l,\ l = = 1, ..., L\}$. Then the $\hat{r}$-component of the saddle point is used at Step 4 to update the disaggregation parameters (weights).

Now consider the resource allocation policy of the center. Let $v_l^k = \alpha_l^k b^k$ be the amount of the resource k given to the division l at the current iteration. Then by the weight updating formula of Step 4 the division receives at the next iteration

$$(\alpha_l^k + \rho \hat{s}_l^k) b^k = [\alpha_l^k + \rho(\hat{R}^k/b^k)(\hat{\alpha}_l^k - \alpha_l^k)] b^k = v_l^k + \rho \hat{R}^k (\hat{\alpha}_l^k - \alpha_l^k) =$$
$$(1-\rho) v_l^k + \rho[\hat{r}_l^k + v_l^k(1 - \hat{R}^k/b^k)]$$

units of the resource. That is the division receives an amount of the resource from the interval $[v_l^k, \hat{r}_l^k + v_l^k(1 - \hat{R}^k/b^k)]$. Denote by γ^k the ratio $\hat{R}^k/b^k$. Then $1 - \gamma^k = (b^k - \hat{R}^k)/b^k$ is the specific excess of the total resource over the total demand per unit of the total resource. The division l received v_l^k units of the resource at the current iteration. Hence its respective portion of the excess is $v_l^k(1 - \gamma^k)$ and the resource demand $\hat{r}_l^k$ is adjusted respectively. That is, if the total resource demand does not coincide with the total resource available at the center, the respective difference is distributed between the divisions proportionally to the amount of the resource given to the division previously. Then the division receives an amount of the resource laying between the amount given previously and the corrected resource demand. Such a control mechanism is a typical policy in planning practice. Note also that the local demands are corrected such that $\sum_l [\hat{r}_l^k + v_l^k(1 - \gamma^k)] = \sum_l v_l^k = b^k$.

In [34] another weight updating formula was used in the iterative aggregation algorithm. The direction $\hat{s}$ was defined by $\hat{s}_l^k = \hat{\alpha}_l^k - \alpha_l^k$. Then at the next iteration division l receives

$$(\alpha_l^k + \rho \hat{s}_l^k) b^k = (1 - \rho) v_l^k + \rho \hat{r}_l^k (b^k / \hat{R}^k)$$

units of the resource. This mechanism to relocate resources can be interpreted similarly.

We see that if quantities γ^k and ρ are transferred down to the divisions, it is possible to compute new resource allocation at subdivisions level. To simplify comparison of the proposed approach with classical decomposition schemes we also present the planning scheme in the standard terms, refining the description from the aggregation terminology:

Step 0. (Initialization, central unit's problem).

Define $\delta \geq 0$ - accuracy of optimization. Initialize a feasible resource allocation v_l^k such that $\sum_l v_l^k = b^k$, $k = 1, ..., K$ and transfer these v_l^k to the divisions.

Step 1. (Subdivisions problem).

Solve the problems

$$\theta_l(v_l) = \max_{x_l} \left\{ f_l(x_l) \mid d_l^k(x_l) \leq v_l^k, \ x_l \in P_l, \ k = 1, 2, ..., K \right\}$$

for all subdivisions, obtaining the optimal primal and dual solutions x_l, λ_l. Transfer $\theta_l(v_l)$ and $p_l^k = \lambda_l^k v_l^k$ to the center.

Step 2. (Central unit's problem).

Compute mean prices $u^k = (\sum_l p_l^k)/b^k$, $k = 1, 2, ..., K$ and transfer to the divisions.

Step 3. (Subdivisions problems).

Solve the problems

$$\max_{x_l, r_l} \left\{ f_l(x_l) - \sum_k u^k r_l^k \mid d_l^k(x_l) \le r_l^k, \; x_l \in P_l \right\}$$

for all divisions, obtaining the optimal solution $\hat{r}_l, \hat{x}_l$. Calculate

$$\pi_l = \left[f_l(\hat{x}_l) - \sum_k u^k \hat{r}_l^k] - [\theta_l(v_l) - \sum_k p_l^k \right]$$

and transfer $\pi_l, \hat{r}_l$ to the center.

Step 4. (Central unit's problem).

Calculate $\pi = \sum_l \pi_l$. If $\pi \le \delta$ then stop. Else compute $\hat{R}^k = \sum_l \hat{r}_l^k$, $\gamma^k = \hat{R}^k / b^k$ and transfer γ^k down to the divisions. Go to Step 1 with

$$v_l^k = (1 - \rho) v_l^k + \rho[\hat{r}_l^k + v_l^k(1 - \gamma^k)],$$

where $0 \le \rho \le 1$ is a properly chosen stepsize increasing the value $\sum_l \theta_l(v_l)$. The value of ρ is then also transferred to the divisions.

Note that the aggregation origin of this decomposition scheme is "hidden" in Steps 2 and 4, where the local information is averaged/aggregated over different divisions resulting in an aggregated representation of local prices and local resource demands. Moreover, the quantities u^k, γ^k transferred down to the divisions are just the same for all divisions. We see that the control policy of the center is very "democratic" since the same control influence is used for all divisions. The task of the center is only to pass information between subdivisions and construct the mean values. Since the center is almost completely removed, the process can be viewed as an "adhocracy" (in some sense), where the divisions exchange information on an equal basis.

In [11, 12] mean value cross decomposition method was considered which can also be viewed as a combination of resource directive strategy and pricing control policy. In each iteration of this method control influences of the center are the mean values of all previously obtained prices and resource allocations.

So in this case averaging over iteration is used which can be motivated by a wish to even out the fluctuations from iteration to iteration.

Let us now illustrate our decomposition/aggregation method on a small numerical example used in [16] to compare their (heuristic) aggregation approach with classical decomposition techniques. A similar example was studied in [13] by mean value cross decomposition. There are two divisions and one common resource:

$$
\begin{aligned}
f^* = \max\ & 2x_1 + 2x_2 + 4x_3 + 2x_4, \\
\text{s.t.}\ & 2x_1 + x_2 + 2x_3 + x_4 \le 24, \\
& x_1 + 2x_2 \le 12, \\
& 3x_1 + x_2 \le 18, \\
& x_3 + x_4 \le 9, \\
& x_3 \le 6, \\
& x_4 \le 6, \\
& x_1,\ x_2,\ x_3,\ x_4 \ge 0.
\end{aligned}
$$

The problem has the unique optimal solution $x^* = \{2, 5, 6, 3\}$ with $f^* = 44$. The optimal allocation of the common resource $b = 24$ is 9 units for the first division and 15 units for the second. We will use our second iterative scheme refined from aggregation terminology.

Iteration 1.

Let $\delta = 0$. Since the center has no preference among divisions, start with resource allocation $v_1 = v_2 = 12$. This information is transferred to the lower level and the divisions react by solving (3.5):

Division 1

$$
\begin{aligned}
\theta_1(v_1) = \max\ & 2x_1 + 2x_2, \\
\text{s.t.}\ & 2x_1 + x_2 \le 12, \\
& x_1 + 2x_2 \le 12, \\
& 3x_1 + x_2 \le 18, \\
& x_1,\ x_2 \ge 0,
\end{aligned}
$$

Division 2

$$
\begin{aligned}
\theta_2(v_2) = \max \quad & 4x_3 + 2x_4, \\
\text{s.t.} \quad & 2x_3 + x_4 \le 12, \\
& x_3 + x_4 \le 9, \\
& x_3 \le 6, \\
& x_4 \le 6, \\
& x_3, \; x_4 \ge 0
\end{aligned}
$$

with resulting unique solutions $(x_1, x_2) = (4, 4)$, $\lambda_1 = 2/3$, $\theta_1(v_1) = 16$, and $(x_3, x_4) = (6, 0)$, $\lambda_2 = 2$, $\theta_2(v_2) = 24$.

The values $p_1 = \lambda_1 v_1 = 8$, $p_2 = \lambda_2 v_2 = 24$, and $\theta_1(v_1)$, $\theta_2(v_2)$ are transferred to the center yielding the total profit $\theta_1(v_1) + \theta_2(v_2) = 40$ and the mean price $u = (p_1 + p_2)/b = 4/3$. The mean price is transferred to the divisions, which react by solving (3.6):

Division 1

$$
\begin{aligned}
\hat{f}_1 = \max \; & 2x_1 + 2x_2 - (4/3)r_1, \\
\text{s.t.} \; & 2x_1 + x_2 - r_1 \le 0, \\
& x_1 + 2x_2 \le 12, \\
& 3x_1 + x_2 \le 18, \\
& x_1, \; x_2, \; r_1 \ge 0,
\end{aligned}
$$

Division 2

$$
\begin{aligned}
\hat{f}_2 = \max \; & 4x_3 + 2x_4 - (4/3)r_2, \\
\text{s.t.} \; & 2x_3 + x_4 - r_2 \le 0, \\
& x_3 + x_4 \le 9, \\
& x_3 \le 6, \\
& x_4 \le 6, \\
& x_3, \; x_4, \; r_2 \ge 0
\end{aligned}
$$

with resulting unique solutions $(\hat{x}_1, \hat{x}_2, \hat{r}_1) = (0, 6, 6)$, $\hat{f}_1 = 4$, and $(\hat{x}_3, \hat{x}_4, \hat{r}_2) = (6, 3, 15)$, $\hat{f}_2 = 10$. This yields $\pi_1(u) = \hat{f}_1 - [\theta_1(v_1) - p_1] = -4$, $\pi_2(u) = \hat{f}_2 - [\theta_2(v_2) - p_2] = 10$.

The values $\pi_1(u), \pi_2(u)$, and $\hat{r}_1, \hat{r}_2$ are transferred to the upper level. The center calculates $\pi(u) = \pi_1(u) + \pi_2(u) = 6$ receiving an estimation $f^* - [\theta_1(v_1) + \theta_2(v_2)] \leq 6$ of the actual difference $f^* - [\theta_1(v_1) + \theta_2(v_2)] = 4$. Since $\pi(u) > 0$, the stopping criterion is not fulfilled. The center computes $\hat{R} = \hat{r}_1 + \hat{r}_2 = 21$ and $\gamma = \hat{R}/b = 7/8$. This γ is transferred to the divisions and the new resource allocation is considered in the form

$$v_1 = 12(1-\rho) + \rho[6 + 12(1 - 7/8)] = 12 - \rho 9/2,$$
$$v_2 = 12(1-\rho) + \rho[15 + 12(1 - 7/8)] = 12 + \rho 9/2.$$

The best $0 \leq \rho \leq 1$, maximizing $\theta_1(v_1) + \theta_2(v_2)$, is computed by the center yielding $\rho = 2/3$. This stepsize is transferred down to the divisions yielding $v_1 = 9, v_2 = 15$ for the second iteration.

Iteration 2.

For $v_1 = 9, v_2 = 15$ the divisions solve (3.5):

Division 1

$$\begin{aligned} \theta_1(v_1) = \max \quad & 2x_1 + 2x_2, \\ \text{s.t.} \quad & 2x_1 + x_2 \leq 9, \\ & x_1 + 2x_2 \leq 12, \\ & 3x_1 + x_2 \leq 18, \\ & x_1, \quad x_2 \geq 0, \end{aligned}$$

Division 2

$$\begin{aligned} \theta_2(v_2) = \max \quad & 4x_3 + 2x_4, \\ \text{s.t.} \quad & 2x_3 + x_4 \leq 15, \\ & x_3 + x_4 \leq 9, \\ & x_3 \leq 6, \\ & x_4 \leq 6, \\ & x_3, \quad x_4 \geq 0 \end{aligned}$$

The first problem has the unique solution $(x_1, x_2) = (2, 5)$, $\lambda_1 = 2/3$, $\theta_1(v_1) = 14$. The unique primal solution of the second problem is $(x_3, x_4) = (6, 3)$, $\theta_2(v_2) = 30$. The dual solution is nonunique: $\lambda_2 \in D_2 = \{\lambda_2 \mid 0 \leq \lambda_2 \leq 2\}$, such that $p_2 \in \tilde{D}_2 = \{p_2 \mid 0 \leq p_2 \leq 30\}$. It is far from obvious which p_2 (or dual price λ_2) should be transferred to the center in the case of degeneracy. We assume that all $p_2 \in \tilde{D}_2$ are equally preferable for the second division, and that the center has the ability to choose any $p_2 \in \tilde{D}_2$. So the values $p_1 = 6$, $\theta_1(v_1)$, $\theta_2(v_2)$, and $\tilde{D}_2 = \{p_2 \mid 0 \leq p_2 \leq 30\}$ are transferred up to the center, yielding the total profit $\theta_1(v_1) + \theta_2(v_2) = 44$.

The mean price, depending on the choice of $p_2 \in \tilde{D}_2$, has the form $u = 1/4 + p_2/24$. This results in the following bounds for u: $1/4 \leq u \leq 3/2$. For any fixed mean price u the divisions subproblems (3.6) have the form:

Division 1

$$\begin{aligned} \hat{f}_1 = \max\ & 2x_1 + 2x_2 - ur_1, \\ \text{s.t. } & 2x_1 + x_2 - r_1 \leq 0, \\ & x_1 + 2x_2 \leq 12, \\ & 3x_1 + x_2 \leq 18, \\ & x_1, \quad x_2, \quad r_1 \geq 0, \end{aligned}$$

Division 2

$$\begin{aligned} \hat{f}_2 = \max\ & 4x_3 + 2x_4 - ur_2, \\ \text{s.t. } & 2x_3 + x_4 - r_2 \leq 0, \\ & x_3 + x_4 \leq 9, \\ & x_3 \leq 6, \\ & x_4 \leq 6, \\ & x_3, \quad x_4, \quad r_2 \geq 0. \end{aligned}$$

The optimal objective $\hat{f}_1$ has the form

$$\hat{f}_1 = \begin{cases} 16.8 - 13.2u & \text{for } 1/4 \leq u \leq 2/3, \\ 8 & \text{for } u = 2/3, \\ 12 - 6u & \text{for } 2/3 \leq u \leq 3/2. \end{cases}$$

The optimal objective $\hat{f}_2 = 30 - 15u$ for all $1/4 \leq u \leq 3/2$. Then $\pi_1 = \hat{f}_1 - 8$, $\pi_2 = p_2 - 15u$. Remembering that $u = 1/4 + p_2/24$ we get $\pi(u) = \pi_1 + \pi_2 = = \hat{f}_1 - 14 + 9u$. Then $\min\{\pi(u) \mid 1/4 \leq u \leq 3/2\} = 0$ is attained for $u = 2/3$. Hence, the optimality criterion is fulfilled and resource allocation $v_1 = 9$, $v_2 = 15$ is optimal.

Note that the optimal mean price $u = 2/3$ corresponds to $\lambda_2 = 2/3$ and in this case $u = \lambda_1 = \lambda_2$ and $\pi_1 = \pi_2 = 0$. Moreover, the subproblem (3.6) associated with the second division has the unique solution $(\hat{x}_3, \hat{x}_4) = (6, 3)$, $\hat{r}_2 = 15$, such that the optimal demand of the resource $\hat{r}_2$, calculated by the division, coincides with the resource v_2 allocated by the center. For the first division the subproblem (3.6) has a nonunique optimal primal solution for $u = 2/3$. Any optimal solution has the form

$$\begin{pmatrix} \hat{x}_1 \\ \hat{x}_2 \\ \hat{r}_1 \end{pmatrix} = \begin{pmatrix} 4.8 \\ 3.6 \\ 13.2 \end{pmatrix} - \tau \begin{pmatrix} 4.8 \\ -2.4 \\ 7.2 \end{pmatrix}, \quad 0 \leq \tau \leq 1$$

and for $\tau = 7/12$ we get $\hat{r}_1 = v_1 = 9$, such that the optimal demand coincides with the resource allocated by the center.

§4. Adaptive Clustering in Variable Aggregation

In most of the approaches on iterative aggregation, the clusters of variables (or constraints) are kept fixed during the iterative procedure. Another idea is to keep the number of clusters constant, but reclassify the variables from one iteration to the next [1, 19]. Various criteria for reclassifying the variables can be constructed using dual or/and primal information. After reclustering, a weight transformation procedure can be performed to guarantee a non-decreasing sequence of optimal objective values of the aggregated problem. In this section a convergent iterative approach [19] is presented where we allow not only weight adjustment, but also adjustment of variable assignment to clusters.

Consider the linear programming problem in the form

$$z^* = \max\left\{\sum_{j=1}^{n} c_j x_j \mid \sum_{j=1}^{n} a_j x_j \leq b,\ x_j \geq 0,\ j = 1, ..., n\right\}, \tag{4.1}$$

where b and a_j are real m-vectors and n is assumed to be (much) larger than m.

A variable aggregated problem corresponding to (4.1) is constructed by a partition Q of the set of variable indices $\{1, ..., n\}$ into a set of clusters S_k, $k = 1, ..., K$, such that

$$\cup_k S_k = \{1, ..., n\} \text{ and } S_i \cap S_j = \varnothing, \; i \neq j.$$

The variables in each cluster S_k are aggregated into one single variable, X_k, using a vector of nonnegative weights, g_k, where the components in each g_k are normalized such that $\sum_{j \in S_k} g_j^k = 1$. The aggregate problem corresponding to (4.1) is then

$$Z^* = \max \left\{ \sum_{k=1}^{K} C_k X_k \mid \sum_{k=1}^{K} A_k X_k \leq b, \; X_k \geq 0, \; k = 1, ..., K \right\}, \qquad (4.2)$$

where $C_k = \sum_{j \in S_k} c_j g_j^k$ and $A_k = \sum_{j \in S_k} a_j g_j^k$. It is assumed that Q and g have been chosen so that a feasible solution X to (4.2) exists.

¿From Proposition 1.1 of Chapter 1 it follows that when two or more optimal basic variables of (4.1) are aggregated into the same variable of (4.2) using optimal weights, the solution of (4.2) is degenerate. Motivated by this result, we now aggregate the n variables into $m + 1$ clusters. One of these clusters will contain $n - m$ variables while the m others are single-variable clusters. Our goal is to iteratively move each of the optimal basic variables of (4.1) into a single-variable cluster and collect the optimal nonbasic variables in the $(m + 1)$st cluster.

More precisely, assume that $S_k = \{j\}$, $k = 1, ..., m$, $j \in \{1, ..., n\}$, and S_{m+1} contains the rest of variables, i.e., $|S_{m+1}| = n - m$. Thus if we assume that the optimal basic variables of (4.1) are the m first variables, the final content of the clusters will be $S_k = \{k\}$, $k = 1, ..., m$, and the final aggregate problem is

$$\max \left\{ \sum_{j=1}^{m} c_j X_j + C_{m+1} X_{m+1} \mid \sum_{j=1}^{m} a_j X_j + A_{m+1} X_{m+1} \leq b, \; X_k \geq 0, \; \forall k \right\}, \qquad (4.3)$$

where A_{m+1} and C_{m+1} are convex combinations of the optimal nonbasic column vectors and cost coefficients of (4.1), respectively. Problem (4.3) has the same optimal solution as (4.1).

In the following we will describe a procedure for iteratively constructing (4.3) (or its equivalent after rearranging variables). Initially, the m single variable clusters may be set to contain any of the n variables. A lot of work can be saved, however, by using some initial heuristic to find m "good" variables which may be potential candidates for being basic variables in the optimal solution of (4.1). We use equal weights initially for the $n-m$ variables in S_{m+1}, i.e., $g_j^{m+1} = 1/(n-m)$, $\forall j \in S_{m+1}$.

The fundamental reclustering operation is to exchange one of the variables in S_{m+1} with the variable in one of the single-variable clusters. Suppose Q and g are given and X is a feasible (but not necessary optimal) solution of (4.2). It was demonstrated in [1] that for any reclustering of variables, a weighting $\bar{g}$ can be found such that the solution obtained by fixed-weight disaggregation is the same as obtained before reclustering, and the aggregate objective value also remains the same. Suppose a reclustering is performed by moving exactly one variable, x_{j_1} say, from cluster S_{k_2} to cluster S_{k_1}. The new partition $\bar{Q}$ consists of exactly the same subsets as Q except for the two subsets

$$\bar{S}_{k_1} = S_{k_1} \cup \{j_1\} \text{ and } \bar{S}_{k_2} = S_{k_2} \backslash \{j_1\}.$$

The new weights and the corresponding aggregate solution, are then:

For cluster $\bar{S}_{k_2}$:

$$\begin{aligned} \bar{g}_j &= \frac{g_j}{1-g_{j_1}}, \; j \in \bar{S}_{k_2}, \\ \bar{X}_{k_2} &= X_{k_2}(1-g_{j_1}), \end{aligned} \tag{4.4}$$

For cluster $\bar{S}_{k_1}$:

$$\begin{aligned} \bar{g}_j &= \frac{g_j X_{k_1}}{g_{j_1} X_{k_2} + X_{k_1}}, \quad j \in \bar{S}_{k_1}, \quad j \neq j_1, \\ \bar{g}_{j_1} &= \frac{g_{j_1} X_{k_2}}{g_{j_1} X_{k_2} + X_{k_1}}, \\ \bar{X}_{k_1} &= g_{j_1} X_{k_2} + X_{k_1}. \end{aligned} \tag{4.5}$$

This weight transformation updates the fixed-weight disaggregated value of the variable to be transferred from cluster S_{k_2} to cluster S_{k_1}, and updates the weights as the relative values of the fixed-weight disaggregated variables in each cluster after reclustering. For clusters k, $k \neq k_1$ and $k \neq k_2$, the respective aggregated variables values and the weights remain unchanged. Since the new solution $\bar{X}$ is feasible to (4.2), the process may be repeated.

Of particular interest in the present context is the exchange of one variable in cluster S_{m+1}, x_k say, by the variable x_r in a single-variable cluster S_q. this can be done by performing first

$$\bar{S}_{m+1} = S_{m+1} \backslash \{k\} \text{ and } \bar{S}_q = S_q \cup \{k\} = \{r, k\}$$

followed by

$$\hat{S}_q = \bar{S}_q \backslash \{r\} = \{k\} \text{ and } \hat{S}_{m+1} = \bar{S}_{m+1} \cup \{r\}.$$

Suppose that X is a nondegenerate basic feasible solution to (4.2) and that X_{m+1} is basic and X_q nonbasic, i.e., $X_{m+1} > 0$, $X_q = 0$. ¿From (4.4) and (4.5) we have

For cluster $\bar{S}_{m+1}$:

$$\bar{g}_j = \tfrac{g_j}{1-g_k}, \; j \in \bar{S}_{m+1},$$
$$\bar{X}_{m+1} = X_{m+1}(1 - g_k),$$

For cluster $\bar{S}_q$:

$$\bar{g}_r = \frac{g_r X_r}{g_k X_{m+1} + X_q} = 0,$$

$$\bar{g}_k = \frac{g_k X_{m+1}}{g_k X_{m+1} + X_q} = 1,$$

$$\bar{X}_q = g_k X_{m+1} + X_q = g_k X_{m+1}.$$

Then it follows:

For cluster $\hat{S}_q$:

$$\begin{aligned} \hat{g}_k &= \frac{\bar{g}_k}{1 - \bar{g}_r} = 1, \\ \hat{X}_q &= \bar{X}_q(1 - \bar{g}_r) = \bar{X}_q = g_k X_{m+1}, \end{aligned} \tag{4.6}$$

For cluster $\hat{S}_{m+1}$:

$$\begin{aligned} \hat{g}_j &= \frac{\bar{g}_j \bar{X}_{m+1}}{\bar{g}_r \bar{X}_q + \bar{X}_{m+1}} = \bar{g}_j = \frac{g_j}{1 - g_k}, \quad j \in \hat{S}_{m+1}, \quad j \neq r, \\ \hat{g}_r &= \frac{\bar{g}_r \bar{X}_q}{\bar{g}_r \bar{X}_q + \bar{X}_{m+1}} = 0, \\ \hat{X}_{m+1} &= \bar{g}_r \bar{X}_q + \bar{X}_{m+1} = \bar{X}_{m+1} = X_{m+1}(1 - g_k). \end{aligned} \tag{4.7}$$

Note that the new feasible solution $\hat{X}$ is nonbasic since both $\hat{X}_q$ and $\hat{X}_{m+1}$ have positive values for $0 < g_k < 1$ (the special cases $g_k = 1$ and $g_k = 0$ are considered below in the convergence proof). Moreover, the variable x_r, which is moved to cluster $\hat{S}_{m+1}$, gets weight $\hat{g}_r = 0$ while the remaining nonzero weights in $\hat{S}_{m+1}$ increase to sum up to 1.

An algorithm which solves (4.1) by iteratively constructing and solving (4.3) can be stated as follows.

Algorithm **Adapt.**

Step 0: Initialize (4.2).

Step 1: Solve (4.2) and get the optimal primal and dual solutions X and U. We always assume nondegenerate and unique optimal solutions for every (4.2) appearing. Compute $\delta_j := c_j - Ua_j$ for all $j \in S_{m+1}$. If $\delta_j \leq 0$ for all $j \in S_{m+1}$ then stop (optimal solution found).

Step 2: Find k such that $\delta_k = \max_{j \in S_{m=1}} \delta_j > 0$.

Consideration of different cases:

2.1. X_{m+1} is nonbasic. Pivot a_k into the present basis and let X_q (where $S_q = \{r\}$) be the leaving basic variable. If the leaving variable is a slack variable, then at least one of the m first variables is nonbasic: add the column index of one of these to S_{m+1}. Set $S_{m+1} := S_{m+1} \backslash \{k\} \cup \{r\}$; $S_q := \{k\}$. For all $j \in S_{m+1}$ set $g_j := 1/(n-m)$. Go to Step 1.

2.2. X_{m+1} is basic (then at least one X_q (where $S_q = \{r\}$) is nonbasic):

2.2.1. If $g_k > 0$ (a_k is a part of A_{m+1}) then exchange aggregation of $k \in S_{m+1}$ and $r \in S_q$ using (4.6) and (4.7). Go to Step 1.

2.2.2. If $g_k = 0$ (a_k is not a part of A_{m+1}) then pivot a_k into the present basis. Set $S_{m+1} := S_{m+1} \backslash \{k\} \cup \{r\}$; $S_q := \{k\}$; $g_r = 0$. Go to Step 1.

end **Adapt**.

Theorem 4.1. *Assume (4.1) has a nondegenerate optimal solution, and assume that each (4.2) occurring during the process has a nondegenerate and unique optimal solution. Then* **Adapt** *will find the optimal solution of (4.1) in a finite number of steps.*

Proof. The termination criterion $\delta_j \leq 0$ for all $j \in S_{m+1}$ is obvious when X_{m+1} is nonbasic. If X_{m+1} is basic, $\delta_j \leq 0$ implies that U is dual feasible for (4.1). Since X may be fixed-weight disaggregated to x, which is primal feasible to (4.1), and $CX = cx = Ub$, strong duality is fulfilled, i.e., x is optimal to (4.1).

Now prove that the objective value increases monotonously. Three cases may occur:

- X_{m+1} is nonbasic. Then a_k can be pivoted into the present basis and a strict improvement in the objective value will occur. The resetting of all the weights in S_{m+1} to be equal may then give an even better aggregated problem (4.2).
- X_{m+1} is basic and $g_k > 0$ (a_k is a part of the aggregated column A_{m+1}). Then use (4.6) and (4.7) to find a new aggregated problem (4.2) which has a nonbasic feasible solution with the same objective value as the previous aggregated problem. Since the optimal solution of the new aggregated problem is assumed to be unique, a strictly better optimal (basic feasible) solution exists. Note that $g_k = 1$ cannot occur when X_{m+1} is basic and X is optimal.
- X_{m+1} is basic but $g_k = 0$. Then since a_k is not a part of the aggregated column A_{m+1}, it can be pivoted into the present basis immediately. A strict increase in the objective value then occur when nondegeneracy is assumed and the column vector a_r replacing a_k in S_{m+1} is given the weight $g_r = 0$ (A_{m+1} is unchanged).

Since the objective value increases monotonously from one aggregated problem to the next, and since only a finite number of aggregation combinations exist, the optimal solution of (4.1) is reached in a finite number of iterations.

The above Theorem together with the algorithm **Adapt** give a strategy on how an adaptive clustering procedure in iterative aggregation may be designed. Of course, the assumption of nondegeneracy in each iteration is rather restrictive. Nevertheless, combining **Adapt** with standard weight updating schemes may help to avoid running into a deadlock of suboptimality, which often occurs in iterative aggregation with sole weight updating techniques.

To illustrate the use of the proposed approach, consider the following example:

$$
\begin{aligned}
z^* = \max\ & x_1 + 2x_2 - x_3 + x_4 + 4x_5 - 2x_6, \\
\text{s.t.}\ & x_1 + x_2 + x_3 + x_4 + x_5 + x_6 \le 6, \\
& 2x_1 - x_2 - 2x_3 + x_4 \le 4, \\
& x_2 + x_3 + x_4 + 2x_5 + x_6 \le 4, \\
& x_j \ge 0 \quad \forall j.
\end{aligned}
$$

The optimal primal solution is $x_1^* = 3$, $x_2^* = 2$, $x_5^* = 1$, $x_3^* = x_4^* = x_6^* = 0$ and $z^* = 11$. The optimal dual solution is $u_1^* = 1/2$, $u_2^* = 1/4$, $u_3^* = 7/4$.

Let the initial clustering be $S_1 = \{1\}$, $S_2 = \{3\}$, $S_3 = \{4\}$ and $S_4 = \{2, 5, 6\}$. Using equal weights in cluster S_4 we get the following aggregated problem:

$$
\begin{aligned}
Z^* = \max X_1 & - X_2 + X_3 + 4/3X_4, \\
\text{s.t.}\quad X_1 & + X_2 + X_3 + X_4 \leq 6, \\
2X_1 & - 2X_2 + X_3 - 1/3X_4 \leq 4, \\
& X_2 + X_3 + 4/3X_4 \leq 4, \\
X_k & \geq 0 \quad \forall k.
\end{aligned}
$$

Introducing slack variables X_5, X_6 and X_7, respectively, and solving this problem, we obtain

$$
\begin{aligned}
& X_1 = 5/2,\ X_4 = 3,\ X_5 = 1/2,\ X_2 = X_3 = X_6 = X_7 = 0 \text{ and } Z^* = 13/2, \\
& U_1 = 0,\ U_2 = 1/2,\ U_3 = 9/8.
\end{aligned}
$$

We then compute the "relative costs" δ_j for $j \in S_4$:

$$
\begin{aligned}
\delta_2 &= c_2 - Ua_2 = 11/8 > 0, \\
\delta_5 &= c_5 - Ua_5 = 7/4 > 0, \\
\delta_6 &= c_6 - Ua_6 = -25/8 < 0.
\end{aligned}
$$

Since X_4 is basic (case 2.2.1), we exchange $5 \in S_4$ (the most positive) and $3 \in S_2$ (X_2 is nonbasic). The new clustering is then $S_1 = \{1\}$, $S_2 = \{5\}$, $S_3 = \{4\}$ and $S_4 = \{2, 3, 6\}$. The weights in S_4 are now $g_2 = 1/2$, $g_3 = 0$, $g_6 = 1/2$.

So the second aggregated problem is

$$
\begin{aligned}
Z^* = \max X_1 & + 4X_2 + X_3, \\
\text{s.t.}\quad X_1 & + X_2 + X_3 + X_4 \leq 6, \\
2X_1 & \qquad + X_3 - 1/2X_4 \leq 4, \\
& 2X_2 + X_3 + X_4 \leq 4, \\
X_k & \geq 0 \quad \forall k.
\end{aligned}
$$

The solution of this problem, easily found from the previous solution by introducing an artificial variable, is

$$X_1 = 2,\ X_2 = 2,\ X_3 = 2,\ X_3 = X_4 = X_6 = X_7 = 0 \text{ and } Z^* = 10,$$
$$U_1 = 0,\ U_2 = 1/2,\ U_3 = 2.$$

Again we compute the relative costs:

$$c_2 - Ua_2 = 1/2 > 0,$$
$$c_3 - Ua_3 = -2 < 0,$$
$$c_6 - Ua_6 = -4 < 0.$$

Since X_4 is nonbasic (case 2.1), we pivot a_2 into the present basis. The leaving basic variable is a slack variable, so we replace $2 \in S_4$ by $4 \in S_3$ since X_3 is nonbasic. the third clustering is then $S_1 = \{1\}$, $S_2 = \{5\}$, $S_3 = \{2\}$ and $S_4 = \{3,4,6\}$. The weights in S_4 are now $g_3 = g_4 = g_6 = 1/3$, so the third aggregated problem is

$$\begin{aligned} Z^* = \max\ & X_1 + 4X_2 + 2X_3 - 2/3X_4, \\ \text{s.t.}\ & X_1 + X_2 + X_3 + X_4 \le 6, \\ & 2X_1 - X_3 - 1/3X_4 \le 4, \\ & 2X_2 + X_3 + X_4 \le 4, \\ & X_k \ge 0 \quad \forall k. \end{aligned}$$

The basic feasible solution found after pivoting a_2 into the basis is also the optimal solution of the third aggregated problem and is as follows:

$$X_1 = 3,\ X_2 = 1,\ X_3 = 2,\ X_4 = 0 \text{ and } Z^* = 11,$$
$$U_1 = 1/2,\ U_2 = 1/4,\ U_3 = 7/4.$$

The relative costs are now:

$$c_3 - Ua_3 = -11/4 < 0,$$
$$c_4 - Ua_4 = -3/2 < 0,$$
$$c_6 - Ua_6 = -17/4 < 0.$$

The optimal solution is found. The third aggregated problem is (4.3) and the optimal solution of (4.1) is

$$x_1^* = X_1 = 3,\ x_2^* = X_3 = 2,\ x_5^* = X_2 = 1,$$
$$x_3^* = x_4^* = x_6^* = 1/3X_4 = 0 \text{ and } Z^* = 11.$$

§5. Aggregation of Constraints

So far, we considered approaches related to aggregation of variables. Now consider aggregation of constraints.

Suppose that, in the original problem (2.1), the set of constraints indices $k = 1, ..., K$ is divided into P disjoint subsets K_p, $p = 1, ..., P$, such that $|K_p| = L_p$ and $\sum_p L_p = K$. Then we can rewrite (2.1) in the form

$$\max\{f(x) \mid g_l^p(x) \le 0, \quad p = 1, ..., P, \quad l = 1, ..., L_p, \quad x \ge 0\}. \tag{5.1}$$

The problem dual to (5.1) is

$$\begin{aligned}
&f(x) - \sum_l \sum_p v_l^p g_l^p(x) \to \min,\\
&\partial f(x)/\partial x_i - \sum_l \sum_p v_l^p \partial g_l^p(x)/\partial x_i \le 0, \quad x_i = 0,\\
&\partial f(x)/\partial x_i - \sum_l \sum_p v_l^p \partial g_l^p(x)/\partial x_i = 0, \quad x_i > 0, \quad i = 1, 2, ..., n,\\
&v_l^p \ge 0, \quad p = 1, ..., P, \quad l = 1, ..., L_p.
\end{aligned}$$

Let us introduce the set

$$B = \left\{ \beta_l^p : \sum_l \beta_l^p = 1, \ \beta_l^p \ge 0, \quad p = 1, ..., P, \quad l = 1, ..., L_p \right\}$$

of aggregation parameters (weights) and consider the problem with aggregated constraints

$$\max\{f(x) \mid \sum_l \beta_l^p g_l^p(x) \le 0, \quad p = 1, ..., P, \quad x \ge 0\}, \tag{5.2}$$

for a fixed $\beta \in B$. The problem (5.2) satisfies the regularity condition for any $\beta \in B$, provided the original problem (5.1) satisfies the Slater conditions.

The problem dual to (5.2) has the form

$$\begin{aligned}
&f(x) - \sum_p V^p \sum_l \beta_l^p g_l^p(x) \to \min,\\
&\partial f(x)/\partial x_i - \sum_p V^p \sum_l \beta_l^p \partial g_l^p(x)/\partial x_i \le 0, \quad x_i = 0,\\
&\partial f(x)/\partial x_i - \sum_p V^p \sum_l \beta_l^p \partial g_l^p(x)/\partial x_i = 0, \quad x_i > 0, \quad i = 1, 2, ..., n,\\
&V^p \ge 0, \quad p = 1, ..., P.
\end{aligned} \tag{5.3}$$

As expected, if in the original problem the constraints are aggregated, then in the dual problem the variables are. Moreover, the dual variables in the original problem and in the macroproblem are related by

$$v_l^p = \beta_l^p V^p, \quad p = 1, ..., P, \quad l = 1, ..., L_p.$$

Let us denote the objective value function of (5.3) by $\theta(\beta)$ and consider the problem

$$\theta(\beta) \to \min, \quad \beta \in B. \tag{5.4}$$

Note that, by the duality theorem, $\theta(\beta)$ equals the optimal value of the objective function in (5.2). The problem (5.3) is an analogue of the macroproblem (2.4), while the problem (5.4) corresponds to the problem (2.5) of selecting the aggregation weights. These considerations allow us to extend the technique for aggregating variables developed earlier to the macroproblems of type (5.3). By virtue of the mentioned duality relation, the iterative aggregation of the dual variables is equivalent to the iterative aggregation of the constraints in the original problem. Also, the variables and constraints of the original problem can be aggregated jointly in a similar manner. This approach is implemented in Section 7 for discrete linear-quadratic optimal control problems.

Consider now a more general procedure to adjust the aggregation weights, which is presented in [6].

Let the original problem has the form

$$\min f(x), \tag{5.5}$$

$$Ax = b, \tag{5.6}$$

$$x \in X. \tag{5.7}$$

It is assumed that the function $f : \mathbf{R}^n \to \mathbf{R}$ is convex and $X \subset \mathbf{R}^n$ is convex and compact. We also assume that the feasible set of (5.5)–(5.7) is non-empty which guarantees that the problem has an optimal solution.

Suppose that the structure of X is simple and the main difficulty comes from the large number of constraints. To overcome this, we shall replace the original problem by a sequence of simplified problems in which the complicated constraints are represented by one surrogate constraint. The basic iterative procedure can be written as

$$x^{k+1} = x^k + \tau_k(u^k - x^k), \quad k = 0, 1, 2, ..., \tag{5.8}$$

where u^k is the solution of the subproblem

$$f^* = \min f(u), \tag{5.9}$$

$$\left(Ax^k - b,\ Au - b\right) \le 0, \tag{5.10}$$

$$u \in X. \tag{5.11}$$

Procedure (5.8)–(5.11) can be viewed as an iterative constraint aggregation method. The initial equality constraints (5.6) are replaced by a sequence of non-stationary scalar inequalities (5.10). Obviously, (5.10) is a relaxation of (5.5), so u^k exists and $f(u^k) \le f^*$. It is rather clear that the sequence $\{x^k\}$ can converge to the optimal solution set X^* only if $\left|Ax^k - b\right|^2 \to 0$ as $k \to \infty$. This can be analyzed by the following key inequality:

$$\begin{aligned}\left|Ax^{k+1} - b\right|^2 &= \left|(1-\tau_k)(Ax^k - b) + \tau_k(Au^k - b)\right|^2 = \\ &= (1-\tau_k)^2\left|Ax^k - b\right|^2 + 2\tau_k(1-\tau_k)\left(Ax^k - b,\ Au^k - b\right) + \\ &\quad + \tau_k^2\left|Au^k - b\right|^2 \le (1 - 2\tau_k)\left|Ax^k - b\right|^2 + 2K_A\tau_k^2,\end{aligned} \tag{5.12}$$

where K_A is an upper bound on $|Ax - b|^2$ in X.

The convergence of various iterative algorithms analyzed below can be proved by using the following simple lemma.

Lemma 5.1. *Let the sequences* $\{\beta_k\}$, $\{\tau_k\}, \{\delta_k\}$ *and* $\{\gamma_k\}$ *satisfy the inequality*

$$0 \le \beta_{k+1} \le \beta_k - \tau_k\delta_k + \gamma_k. \tag{5.13}$$

If
(i) $\liminf \delta_k \ge 0$;
(ii) *for every subsequence* $\{k_i\}$ *one has*

$$[\liminf \beta_{k_i} > 0] \Longrightarrow [\liminf \delta_{k_i} > 0];$$

(iii) $\tau_k \ge 0$, $\lim \tau_k = 0$, $\sum\limits_{k=0}^{\infty} \tau_k = \infty$;
(iv) $\lim \gamma_k/\tau_k = 0$,
then $\lim_{k\to\infty} \beta_k = 0$.

Proof. Suppose that $\liminf \delta_k = \delta > 0$. Then (5.13) for large k yields $\beta_{k+1} \le \beta_k - \tau_k\delta/2 + \gamma_k \le \beta_k - \tau_k\delta/4$. This contradicts (iii). Therefore there is a subsequence $\{k_i\}$ such that $\beta_{k_i} \to 0$. Suppose that there is another subsequence $\{s_j\}$ such that $\beta_{s_j} \geq \beta > 0$ for $j = 0, 1, \ldots$ With no loss of generality we may assume that $k_1 < s_1 < k_2 < s_2, \ldots$ By (i), (iii) and (iv), for all sufficiently large j there must exist indices $r_j \in [k_j, s_j]$ such that $\beta_{r_j} > \beta/2$ and $\beta_{r_{j+1}} > \beta_{r_j}$. But the, by (ii), $\liminf \delta_{r_j} = \delta > 0$ and we obtain a contradiction with (5.13) for large j.

We are now ready to prove the convergence of the method.

Theorem 5.1. *Consider the algorithm* (5.8)–(5.11) *and assume that* $\tau_k \in [0,1]$, $\lim_{k\to\infty} \tau_k = 0$, $\sum\limits_{k=0}^{\infty} \tau_k = \infty$ *and* $x^0 \in X$. *Then every accumulation point of the sequence* $\{x^k\}$ *is a solution of* (5.5)–(5.7).

Proof. By (5.12) and Lemma 5.1, taking $\beta_k = \left|Ax^k - b\right|^2$, $\delta_k = 2\beta_k$ and $\gamma_k = 2K_A\tau_k^2$, we obtain $\left|Ax^k - b\right| \to 0$ as $k \to \infty$. Since X is compact, we must have $\liminf f(x^k) \geq f^*$. Next, by the convexity of f,

$$f(x^{k+1}) \leq (1-\tau_k)f(x^k) + \tau_k f(u^k) \leq (1-\tau_k)f(x^k) + \tau_k f^*.$$

Thus,

$$\max\left(f(x^{k+1}) - f^*,\ 0\right) \leq (1-\tau_k)\max\left(f(x^k) - f^*,\ 0\right).$$

Using Lemma 5.1 again, with $\beta_k = \delta_k = \max(0,\ f(x^k) - f^*)$, we obtain $f(x^k) \to f^*$. Our assertion follows then from the compactness of X.

Clearly, in view of the compactness of X we also have $\operatorname{dist}(x^k, X^*) \to 0$, where X^* is the set of optimal solutions of (5.5)–(5.4). The stepsize in the basic procedure can also be generated in a more systematic way:

$$\tau_k = \operatorname{Arg} \min_{0\leq\tau\leq 1} \left|(1-\tau)(Ax^k - b) + \tau(Au^k - b)\right|^2. \tag{5.14}$$

Theorem 5.2. *Consider the algorithm* (5.8)–(5.11) *and assume that* $f(x^0) \leq f^*$ *and the stepsizes* τ_k *are defined by* (5.14). *Then every accumulation point of the sequence* $\{x^k\}$ *is a solution of* (5.5)–(5.7).

Proof. From the key inequality (5.12) we obtain

$$\left|Ax^{k+1} - b\right|^2 \leq \min_{0\leq\tau\leq 1}\left((1-2\tau)\left|Ax^k - b\right|^2 + 2K_A\tau^2\right) \leq$$

$$\leq \left(1 - \frac{\left|Ax^k - b\right|^2}{2K_A}\right) \left|Ax^k - b\right|^2.$$

Therefore $\left|Ax^k - b\right|^2 \to 0$. Since $f(x^k) \leq f^*$ for all k, the required result follows.

Clear, that it is not necessary to generate exactly one aggregated constraint. We can aggregate the constraints in groups as follows. Let

$$A_l x = b_l, \quad l = 1, ..., L$$

be (possibly overlapping) subgroups of constraints (5.6), such that each row of (5.6) is represented at least once. Then we can replace (5.10) by L aggregated constraints:

$$\left(A_l x^k - b_l,\ A_l u - b_l\right) \leq 0, \quad l = 1, ..., L.$$

Indeed, the key inequality (5.12) remains valid for each subgroup $\left|A_l x^k - b_l\right|^2$ and all the results follow. The optimizing stepsize (5.14) should be replaced by

$$\tau_k = \text{Arg} \min_{0 \leq \tau \leq 1} \sum_{l=1}^{L} \left|(1-\tau)(A_l x^k - b_l) + \tau(A_l u^k - b_l)\right|^2.$$

Moreover, we can keep in the auxiliary subproblem as many previous aggregated constraints as we wish. They are all relaxations of the original constraints and do not cut off points of the true feasible set. Therefore $f(u^k) \leq f^*$ for all k, so the results remain true.

The basic procedure requires the solution of the auxiliary subproblems (5.9)–(5.11). We shall show now that it is sufficient to use individual subgradients of f within the constraint aggregation scheme to achieve convergence.

Consider the same procedure as before:

$$x^{k+1} = ((1 - \tau_k)x^k + \tau_k u^k, \tag{5.15}$$

but with u^k generated in a simpler fashion:

$$u^k = \Pi_k(x^k - \lambda_k g^k). \tag{5.16}$$

Here a subgradient $g^k \in \partial f(x^k)$ and $\Pi_k(\cdot)$ denotes the orthogonal projection onto the set

$$X_k = \left\{x \in X \mid \left(Ax^k - b,\ Ax - b_l\right) \leq 0\right\}.$$

Theorem 5.3 Assume $\tau_k \in [0,1]$, $\lambda_k \geq 0$, $\tau_k \to 0$, $\lambda_k \to 0$, $\sum_{k=0}^{\infty} \tau_k = \infty$, $\sum_{k=0}^{\infty} \lambda_k = \infty$. Then every accumulation point of the sequence generated by the method (5.15), (5.16) is a solution of (5.5)–(5.7).

Proof. For every optimal solution x^* we have

$$\left|x^{k+1} - x^*\right|^2 \leq \left((1-\tau_k)\left|x^k - x^*\right| + \tau_k\left|u^k - x^*\right|\right)^2 =$$
$$= (1-\tau_k)^2\left|x^k - x^*\right|^2 + 2\tau_k(1-\tau_k)\left|x^k - x^*\right|\left|u^k - x^*\right| + \tau_k^2\left|u^k - x^*\right|^2. \qquad (5.17)$$

Since $x^* \in X_k$, from the non-expansiveness of $\Pi_k(\cdot)$ we get

$$\left|u^k - x^*\right|^2 = \left|\Pi_k(x^k - \lambda_k g^k) - \Pi_k(x^*)\right|^2$$
$$\leq \left|x^k - x^*\right|^2 - 2\lambda_k(g^k,\ x^k - x^*) + C\lambda_k^2,$$

where C is an upper bound on $\left|g^k\right|^2$. Thus, by definition of the subgradient,

$$\left|u^k - x^*\right|^2 \leq \left|x^k - x^*\right|^2 - 2\lambda_k\delta_k + C\lambda_k^2 \qquad (5.18)$$

with $\delta_k = f(x^k) - f^*$. Taking the square root of both sides and using the concavity of $\sqrt{\cdot}$ we get

$$\left|u^k - x^*\right| \leq \left|x^k - x^*\right|\left(1 - \frac{2\lambda_k\delta_k - C\lambda_k^2}{|x^k - x^*|^2}\right)^{1/2} \leq \left|x^k - x^*\right| - \frac{2\lambda_k\delta_k - C\lambda_k^2}{2\,|x^k - x^*|}.$$

Multiplication by $\left|x^k - x^*\right|$ yields

$$\left|x^k - x^*\right|\left|u^k - x^*\right| \leq \left|x^k - x^*\right|^2 - \lambda_k\delta_k + C\lambda_k^2/2. \qquad (5.19)$$

Combining (5.17), (5.18) and (5.19), after elementary simplifications, for every $x^* \in X^*$ one has

$$\left|x^{k+1} - x^*\right|^2 \leq \left|x^k - x^*\right|^2 - 2\tau_k\lambda_k\delta_k + C\tau_k\lambda_k^2.$$

After minimizing both sides with respect to x^* we get

$$\operatorname{dist}(x^{k+1}, X^*)^2 \leq \operatorname{dist}(x^k, X^*)^2 - 2\tau_k\lambda_k\delta_k + C\tau_k\lambda_k^2.$$

Similar to Theorem 5.1, $\left|Ax^k - b\right| \to 0$. Therefore the sequence $\{\delta_k\}$ satisfies conditions (i) and (ii) of Lemma 5.1 with $\beta_k =$dist$(x^{k+1}, X^*)^2$. By virtue of Lemma 5.1 (with $\tau_k \lambda_k$ playing the role of τ_k), dist$(x^k, X^*) \to 0$ as $k \to \infty$.

Now we briefly discuss how to extend the aggregation principle to nonlinear constraints. Consider the problem

$$\min\{f(x) \mid h_j(x) \leq 0, \quad j = 1, ..., m, \quad x \in X\}, \tag{5.20}$$

where it is assumed that the functions $f, h_j : \mathbf{R}^n \to \mathbf{R}$ are convex and $X \subset \mathbf{R}^n$ is convex and compact. Denote $h_j^+(x) = \max(0,\ h_j(x))$, $h(x) = (h_1(x), ..., h_m(x))$ and $h^+(x) = (h_1^+(x), ..., h_m^+(x))$.

The basic iterative procedure (5.8)–(5.11) can be adapted to the nonlinear constraints as follows:

$$x^{k+1} = ((1 - \tau_k)x^k + \tau_k u^k, \tag{5.21}$$

where u^k is the solution of the subproblem

$$\min\{f(u) \mid \left(h^+(x^k),\ h(u)\right) \leq 0, \quad u \in X\}. \tag{5.22}$$

Note that if linear constraints $Ax = b$ are written as inequalities $Ax - b \leq 0$ and $Ax - b \geq 0$, then the aggregated constraint in (5.22) is identical to (5.10). The convergence of the algorithm (5.21), (5.22) is proved similar to Theorem 5.1. As before, the stepsizes in the basic procedure (5.21) can be generated in a more systematic way:

$$\tau_k = \text{Arg} \min_{0 \leq \tau \leq 1} \left| h^+ \left((1 - \tau)x^k + \tau u^k \right) \right|^2.$$

For continuously differentiable constraint functions $h(\cdot)$ the subproblems can be further simplified by replacing the aggregated constraint in (5.22) with the inequality

$$\left(h^+(x^k),\ h(x^k) + \nabla h(x^k)(u - x^k)\right) \leq 0,$$

where $\nabla h(\cdot)$ denotes the Jacobian of $h(\cdot)$. Note that by the convexity of $h(\cdot)$, this is still a relaxation of the nonlinear constraints in (5.20).

Similar results can be obtained for the subgradient projection method with

$$u^k = \Pi_k(x^k - \lambda_k g^k),$$

$g^k \in \partial f(x^k)$ and $\Pi_k(\cdot)$ denotes the orthogonal projection onto the set

$$X_k = \left\{ x \in X \mid \left(h^+(x^k),\ h(x) \right) \le 0 \right\},$$

or onto the set

$$X_k = \left\{ x \in X \mid \left(h^+(x^k),\ h(x^k) + \nabla h(x^k)(x - x^k) \right) \le 0 \right\}$$

if linearization of smooth constraints is employed.

Assume now that both f and h_j are convex and continuously differentiable. Then the idea of linearization can be extended to the objective function, yielding the subproblem to define u^k :

$$\min \left(\nabla f(x^k),\ u - x^k \right),$$

$$\left(h^+(x^k),\ h(x^k) + \nabla h(x^k)(u - x^k) \right) \le 0,$$

$$u \in X.$$

This approach can be treated as a modification of the classical Frank-Wolf method of nonlinear programming, but with constraint linearization and aggregation. Aggregation and linearization in subgroups is also possible here.

So far we did not assume any constraint qualification conditions. Let us now look closer at the properties of the basic procedure (5.8)–(5.11) for the problem (5.5)–(5.7) under the following condition: the set X is polyhedral, or at some feasible point $\tilde{x}$ one has $(X \backslash \{\tilde{x}\}) \cap \{d \mid Ad = 0\} \ne \varnothing$.

Consider the Lagrangian

$$L(x, \pi) = f(x) + (\pi,\ Ax - b) \tag{5.23}$$

and the dual function

$$D(\pi) = \min_{x \in X} L(x, \pi).$$

Under the above constraint qualification condition, there exists π^* such that

$$D(\pi^*) = \max_{\pi \in \mathbf{R}^m} D(\pi) = f^*.$$

The constraint qualification condition also guarantees the existence of an optimal Lagrange multiplier μ^k corresponding to the aggregated constraint (5.10) at each iteration of the method. Let us define

$$s^k = Ax^k - b,$$

$$\pi^k = \mu^k s^k.$$

It follows that u^k minimizes the Lagrangian for (5.9)–(5.11), which turns out to be the Lagrangian (5.23) at π^k :

$$u^k \in \text{Arg} \min_{u \in X} \left\{ f(u) + \mu^k \left(Ax^k - b,\ Au - b \right) \right\} = \text{Arg} \min_{u \in X} L(u, \pi^k).$$

Therefore, the basic algorithm in the dual space can be interpreted as follows:

$$\pi^k = \text{Arg } \max\{D(\pi) \mid \pi = \mu s^k,\ \mu \geq 0\},$$

$$w^k = Au^k - b \in \partial D(\pi^k),$$

$$s^{k+1} = (1 - \tau_k) s^k + \tau_k w^k.$$

We also have

$$f(u^k) = \min_{u \in X} L(u, \pi^k) = D(\pi^k).$$

In [6] this constraint aggregation technique was applied to the dual of the transportation problem. The normalized weights were used and two types of aggregation were considered. In the case of full aggregation, all constraints were combined in one surrogate constraint, while for partial aggregation the constraints corresponding to individual nodes were combined in one aggregated constraint. To guarantee the existence of solutions of the auxiliary subproblems, the box constraints for the dual variables were used. Numerical tests demonstrate that the partial aggregation performs much better, especially with optimizing stepsize rule. In [22] the constraint aggregation approach was generalized to control, games and stochastic programming, interpreted as convex problems in infinite-dimensional spaces.

We see, that if the Lagrangian duality works, this technique is very close to the approach considered in the beginning of this section. The principal difference is in the weight updating scheme, which is based here on the subgradient methods. An interesting question arises in connection with the possibilities of applying this approach to variable aggregation. The primal-dual correspondence between aggregation of constraints and aggregation of variables, mentioned in the beginning of this section, can be used here.

§6. Aggregation of Controls in Dynamic Problems with Mixed Constraints

Examination of complex dynamic systems with cross connections involves optimal control problems with a large number of state variables or controls. In this section, we consider an approach aimed at lowering the dimension with respect to controls. We use the iterative aggregation technique and construct macroproblems by aggregating controls. In the cases where the initial problem has block-separable structure with respect to the state variables and controls, the suggested approach can be regarded as a generalization of the method of aggregating controls from different blocks [53, 54]. We also show that the bounds of cross connections can be used in constructing the decomposition algorithm.

Let $L_1[0,T]$ be the Banach space of summable functions on the interval $[0,T]$ with the norm

$$\|f\|_{L_1[0,T]} = \int_0^T |f(t)|\, dt.$$

We also consider the corresponding space $L_1^n[0,T]$ of n-vector functions with the norm

$$\|f\|_{L_1^n[0,T]} = \sum_{r=1}^{N} \|f_r\|_{L_1[0,T]}.$$

Let $L_\infty[0,T]$ be the Banach space of functions summable to all powers on $[0,T]$. It coincides with the space of measurable functions bounded almost everywhere on $[0,T]$. Its norm has the form

$$\|f\|_{L_\infty[0,T]} = \operatorname{vrai} \max_{0 \le t \le T} |f(t)|,$$

where vrai max is the so-called true maximum. The space $L_\infty^n[0,T]$ of vector functions is introduced similarly.

Consider the optimal control problem

$$J(u(\cdot)) = Q(x(T)) + \int_0^T f(x(t), u(t), t)dt \to \max, \tag{6.1}$$

$$dx(t)/dt = A(t)x(t) + b(u(t), t), \quad x(0) = x_0, \tag{6.2}$$

$$\nu(x(T)) \le 0, \tag{6.3}$$

$$p(x(t), u(t), t) \leq 0, \quad u(t) \geq 0. \tag{6.4}$$

Here, $x(\cdot) \in \mathbf{R}^N$, $u(\cdot) \in \mathbf{R}^n$; the dimensions of the vector functions $b(\cdot\,,\cdot)$, $\nu(\cdot)$, and $p(\cdot\,,\cdot,\cdot)$ equal N, V, and P, respectively; and $A(t)$ is a square matrix of order N.

Problem (6.1)–(6.4) is a Bolz-type optimal control problem with mixed constraints. It consists in finding controls $u(t) \in L^n_\infty[0,T]$ and state variables $x(t)$ related to these controls by (6.2) such that they maximize functional (6.1) and satisfy constraints (6.3)–(6.4).

Suppose that the input functions satisfy the Ter-Krikorov conditions [49], namely, the functions $-p(x,u,t)$, $-b(u,t)$, and $f(x,u,t)$ are continuously differentiable in the entire space of variables, concave with respect to x and u, and monotonically increasing with respect to x, and the functions $-\nu(\gamma)$ and $Q(\gamma)$ are continuously differentiable, concave, and monotonically increasing with respect to γ. We also assume that the elements of the matrix $A(t)$ are bounded and measurable, and they are nonnegative (almost everywhere) on the interval $[0,T]$.

In addition, we suppose that the original problem and the intermediate problems obtained below during the execution of the iterative aggregation algorithm satisfy the Ter-Krikorov conditions ensuring the reduction of these problems to convex programming problems for which certain duality principles hold, namely, the Kuhn–Tucker theorem and the saddle point theorem for the Lagrange function [14, 49]. For problem (6.1)–(6.4), these assumptions reduce to the fulfillment of the Slater conditions on its constraints and of the inequalities

$$x(t) \geq \varepsilon > 0, \quad p(0,0,t) \leq 0.$$

Note that these conditions ensure the fulfillment of the Pontryagin maximum principle [45] for the class of problems (6.1)–(6.4) under consideration.

We divide the control vector $u(t)$ into I $(1 \leq I \leq n)$ disjoint subvectors, each with J_i components, so that

$$u(t) = \big(u_1(t), \dots, u_i(t), \dots, u_I(t)\big),$$

$$u_i(t) = \big(u^i_1(t), \dots, u^i_j(t), \dots, u^i_{J_i}(t)\big), \quad \sum_i J_i = n.$$

Let us introduce a vector of aggregated controls

$$U(t) = \big(U^1(t), \dots, U^i(t), \dots, U^I(t)\big)$$

and a linear disaggregation $u = \alpha U$, or, componentwise,

$$u_j^i(t) = \alpha_j^i(t) U^i(t), \quad i = 1, 2, \ldots, I, \quad j = 1, 2, \ldots, J_i, \tag{6.5}$$

where $\alpha_j^i(t)$ are bounded measurable functions satisfying the normalization condition

$$\alpha \in A = \left\{ \alpha_j^i(t) \colon \sum_j \alpha_j^i(t) = 1, \quad \alpha_j^i(t) \geq 0, \right.$$
$$\left. i = 1, 2, \ldots, I, \quad j = 1, 2, \ldots, J_i \right\}. \tag{6.6}$$

almost everywhere on $[0, T]$.

Fixing an $\alpha \in A$ and substituting (6.5) in the original problem (6.1)–(6.4), we obtain the following problem with respect to the aggregated controls:

$$\begin{aligned} &\tilde{J}(U(\cdot)) = Q(x(T)) + \int_0^T F(x(t), U(t), t)\, dt \to \max, \\ &dx(t)/dt = A(t)x(t) + B(U(t), t), \quad x(0) = x_0, \\ &\nu(x(T)) \leq 0, \quad P(x(t),\, U(t),\, t) \leq 0, \quad U(t) \geq 0, \end{aligned} \tag{6.7}$$

where $F(x,\, U,\, t) = f(x,\, \alpha U,\, t)$, $B(U, t) = b(\alpha U,\, t)$, and $P(x,\, U,\, t) =$ $= P(x,\, \alpha U,\, t)$.

Macroproblem (6.7) consists in finding a control vector $U(t) \in L_\infty^i[0, T]$ that maximizes the functional $\tilde{J}(U(\cdot))$ subject to the given constraints. In what follows, we assume that, for all $\mathring{a}(\cdot) \in A$, the components of the optimal solution $\mathring{U}(t)$ to the problem in aggregated variables satisfy the strict inequalities $\mathring{U}^i(t) > 0$ $(i = 1, 2, \ldots, I)$ almost everywhere on $[0, T]$.

Consider the Wolfe dual to problem (6.7):

$$\begin{aligned} &Q(x)(T)) + \int_0^T \Big[F(x,\, U,\, t) + \big(B(U,\, t) + A(t)x(t) - dx/dt\big)\lambda(t) - \\ &\qquad - p(x,\, U,\, t)\eta(t)\Big] dt - \nu(x(T)\mu \to \min, \\ &-d\lambda/dt = A^{\mathrm{T}}(t)\lambda(t) - \eta(t)\partial P(x,\, U,\, t)/\partial x + \partial F(x,\, U,\, t)/\partial x, \\ &\big(\lambda(t),\, \partial B(U,\, t)/\partial U\big) - \big(\eta(t),\, \partial P(x,\, U,\, t)/\partial U\big) + \\ &\qquad + \partial F(x,\, U,\, t)/\partial U = 0, \\ &\lambda(T) = -\big(\partial\nu(x(T))/\partial x,\, \mu\big) + \partial Q(x(T))/\partial x, \\ &\eta(t) \geq 0, \quad \mu \geq 0. \end{aligned} \tag{6.8}$$

Here, the vector functions $\lambda(t)$ are the dual impulses corresponding to the differential constraints in (6.7), and $\eta(t)$ and μ are the Lagrange multiplier vectors.

A criterion for the optimality of a disaggregated solution $\{\overset{\circ}{x}(t),\, \overset{\circ}{u}(t) = \overset{\circ}{\alpha}(t)\overset{\circ}{U}(t)\}$ obtained at some $\overset{\circ}{\alpha} \in A$ is similar to that given in Section 2. A disaggregated solution is optimal for the initial problem (6.1)–(6.4) if and only if there exists an optimal solution $\overset{\circ}{\lambda}(t)$, $\overset{\circ}{\eta}(t)$ to the dual problem (6.8) such that

$$\psi^i_j(t) = \left[\left(\overset{\circ}{\lambda}(t),\, \frac{\partial b(u,\, t)}{\partial u^i_j}\right) - \left(\overset{\circ}{\eta}(t),\, \frac{\partial p(x,\, u,\, t)}{\partial u^i_j}\right) + \frac{\partial f(x,\, u,\, t)}{\partial u^i_j}\right]\Bigg|_{\overset{\circ}{x},\overset{\circ}{u}} \leq 0, \tag{6.9}$$

$$\left(\psi^i_j(t),\, \overset{\circ}{u}{}^i_j(t)\right) = 0, \quad t \in [0, T], \quad i = 1, 2, \ldots, I, \quad j = 1, 2, \ldots, J_j.$$

The proof of this assertion uses the Kuhn–Tucker theorem, whose validity in the case under consideration is established in [49].

Now, suppose that macroproblem (6.7) has a finite optimal solution for each $\alpha \in A$. Let us denote the extremal value of the objective functional of (6.7) by $\theta(\alpha(\cdot))$ and consider the problem

$$\theta(\alpha) \to \max, \quad \alpha \in A. \tag{6.10}$$

As in the preceding sections, to examine the differential properties of $\theta(\alpha)$, we use the marginal-value theorem [27]. We assume that the dual problems (6.7) and (6.8) have unique solutions for all $\alpha \in A$. In the case under consideration, the uniform convergence condition of the marginal-value theorem is fulfilled if the partial derivatives of $f(x,\, u,\, t)$, $b(u,\, t)$, and $p(x,\, u,\, t)$ with respect to x and u satisfy the Lipschitz condition at every $t \in [0, T]$ and are equicontinuous on any bounded subset of $\mathbf{R}^N \times \mathbf{R}^n$. Since the solutions to (6.7) and (6.8) are unique, the derivative $\xi^{ij}_{\overset{\circ}{\alpha}}(t)$ of the functional $\theta(\alpha(\cdot))$ with respect to the parameter $\alpha^i_j(\cdot)$ at a point $\overset{\circ}{\alpha} \in A$ coincides with the derivative of the Lagrange functional with respect to α^i_j at the optimal solution to (6.7), or, equivalently, with the derivative of the functional of the dual problem (6.8). We have

$$\xi^{ij}_{\overset{\circ}{\alpha}}(t) = \overset{\circ}{U}{}^i(t)\left[\frac{\partial f(\overset{\circ}{x},\, \overset{\circ}{u},\, t)}{\partial u^i_j} + \left(\overset{\circ}{\lambda}(T),\, \frac{\partial b(\overset{\circ}{u},\, t)}{\partial u^i_j}\right) - \left(\overset{\circ}{\eta}(t),\, \frac{\partial p(\overset{\circ}{x},\, \overset{\circ}{u},\, t)}{\partial u^i_j}\right)\right], \tag{6.11}$$

where $\overset{\circ}{\lambda}(t)$, $\overset{\circ}{\eta}(t)$ and $\overset{\circ}{x}(t)$, $\overset{\circ}{U}(t)$ are the unique optimal solutions to (6.8) and (6.7), respectively, at $\alpha = \overset{\circ}{\alpha}$, and $\overset{\circ}{u} = \overset{\circ}{\alpha}\overset{\circ}{U}$ is the corresponding disaggregated solution. Note that, if the solutions to the dual problems (6.7) and (6.8) are jointly bounded, then, under the assumptions made above, the gradient of $\theta(\alpha(\cdot))$ satisfies the Lipschitz condition by virtue of (6.11).

The following assertion shows that the necessary optimality conditions for problem (6.10) are also sufficient.

Theorem 6.1. *Suppose that $\overset{\circ}{\alpha} \in A$ satisfies the necessary optimality conditions $\langle \nabla\theta(\overset{\circ}{\alpha}(\cdot)), \alpha - \overset{\circ}{\alpha} \rangle \leq 0$, where the components of the Frechet gradient $\nabla\theta(\overset{\circ}{\alpha}(\cdot))$ are defined according to (6.11), for all $\alpha \in A$. Then the disaggregated solution $\{\overset{\circ}{x}(t), \overset{\circ}{u}(t)\}$ is optimal for the original problem (6.1)–(6.4).*

Proof. Rewriting the necessary optimality conditions with taking into account the criterion stated in [56], we obtain

$$S(\overset{\circ}{\alpha}) = \max_{\alpha} \left\{ \sum_i \sum_j \int_0^T \overset{\circ}{U}^i(t)\psi^i_j(t)\alpha^i_j \, dt \mid \sum_j \alpha^i_j(t) = 1, \quad \alpha^i_j(t) \geq 0, \right.$$

$$\left. t \in [0,T], \quad i = 1,2,\dots,I, \quad j = 1,2,\dots,J_i \right\} = 0, \tag{6.12}$$

where $\psi^i_j(t)$ is the expression in the brackets in (6.11). As mentioned, conditions (6.9) are necessary and sufficient for the disaggregated solution to be optimal. Note that the conditions of problem (6.8) and the inequalities $\overset{\circ}{U}^i(t) > 0$ $(i = 1,2,\dots,I)$ imply $0 = \overset{\circ}{U}^i(t)\sum_j \overset{\circ}{\alpha}^i_j(t)\psi^i_j(t) = \sum_j \overset{\circ}{u}^i_j(t)\psi^i_j(t)$ for all $t \in [0,T]$. Therefore, if $\psi^i_j(t) \leq 0$ for $i = 1,2,\dots,I$ and $j = 1,2,\dots,J_i$, then the second part of conditions (6.9) are fulfilled. Let us show that (6.12) implies the nonpositiveness of all $\psi^i_j(t)$. Suppose that, on the contrary, (6.12) holds but there exist indices i_0 and j_0 and an interval $[\tau_1, \tau_2] \subset [0,T]$ such that $\psi^{i_0}_{j_0}(t) > 0$ for all $t \in [\tau_1, \tau_2]$. The condition $\sum_j \overset{\circ}{\alpha}^i_j(t) = 1$ and the relation $\sum_j \overset{\circ}{\alpha}^i_j(t)\psi^i_j(t) = 0$, which is implied by the constraints of the dual problem (6.8), give $\max_j \psi^i_j(t) \geq 0$ for each $i = 1,2,\dots,I$ and all $t \in [0,T]$; moreover, by assumption, we have $\max_j \psi^{i_0}_j(t) > 0$ for all $t \in [\tau_1, \tau_2]$. Let $j_i(t)$ denote the subscript j for which the value $\psi^i_j(t)$ is maximum. If the maximum value is attained at several j's for some $t \in [0,T]$, then we define $j_i(t)$ to be the smallest one. Consider $\tilde{\alpha}^i_j(t)$ defined as

$$\tilde{\alpha}^i_j(t) = \begin{cases} 1 \text{ if } j = j_i(t), \\ 0 \text{ if } j \neq j_i(t), \end{cases} \qquad i = 1, 2, \ldots, I.$$

Clearly, $\tilde{\alpha} \in A$. Let us evaluate the objective functional of the problem in the right-hand side of (6.12) at $\alpha = \tilde{\alpha}$. We have

$$\sum_i \sum_j \int_0^T \overset{\circ}{U}{}^i(t)\psi^i_j(t)\tilde{\alpha}^i_j(t)\,dt = \sum_i \int_0^T \overset{\circ}{U}{}^i(t)\psi^i_{j_i(t)}(t)\,dt =$$

$$= \sum_i \int_0^{\tau_1} \overset{\circ}{U}{}^i(t)\psi^i_{j_i(t)}(t)\,dt + \sum_i \int_{\tau_1}^{\tau_2} \overset{\circ}{U}{}^i(t)\psi^i_{j_i(t)}(t)\,dt + \sum_i \int_{\tau_2}^{T} \overset{\circ}{U}{}^i(t)\psi^i_{j_i(t)}(t)\,dt.$$

By the definition of the subscript $j_i(t)$, the integrands in all the three terms are nonnegative, and $\overset{\circ}{U}{}^i(t)\psi^i_{j_i(t)}(t) > 0$ for $i = i_0$ and $t \in [\tau_1, \tau_2]$. Thus, at $\alpha = \tilde{\alpha}$, the objective functional of the maximization problem in (6.12) takes a strictly positive value, and condition (6.12) is violated, which contradicts our assumption.

Therefore, if (6.12) holds, then $\psi^i_j(t) \le 0$ $(i = 1, 2, \ldots, I;\ j = 1, 2, \ldots, J_i)$ almost everywhere on $[0, T]$; this and criterion (6.9) imply the optimality of the disaggregated solution to the original problem (6.1)–(6.4), which completes the proof of Theorem 6.1.

Theorem 6.10 makes it possible to apply the first-order methods, such as the method of conditional gradient [56], to problem (6.10). The auxiliary ascent direction problem involved in such methods decomposes into I independent subproblems, because $A = A_1 \times A_2 \times \cdots \times A_I$, where

$$A_I = \left\{ \alpha^i_j(t) \colon \sum_j \alpha^i_j(t) = 1, \quad \alpha^i_j(t) \ge 0, \quad j = 1, 2, \ldots, J_i \right\}.$$

The convergence of the conditional gradient method is proved as in [56] with the use of the Lipschitz continuity of the Frechet gradient $\nabla\theta(\alpha(\cdot))$.

Now, let us show how the structural properties of the original problem can be taken into account in selecting an ascent direction for the functional $\theta(\alpha)$ in problem (6.10).

Take a closed convex subset Ω in the space $L^n_\infty[0, T]$ of controls with the following property. Let $\bar{u}(t) \in L^n_\infty[0, T]$ be admissible for (6.1)–(6.4), i.e., the pair $\bar{x}(t)$, $\bar{u}(t)$ obeys the conditions (6.3) and (6.4) whenever $\bar{x}(t)$ is a solution to (6.2) with $u(t) = \bar{u}(t)$. Then $\bar{u}(t) \in \Omega$. As in the preceding section, consider the following auxiliary problem at a fixed $\mathring{\alpha}(\cdot) \in A$:

$$\delta(\mathring{\alpha}) = \max_{u(t)} \left\{ \int_0^T \sum_i \sum_j u_j^i(t)\psi_j^i(t)\, dt \mid u(t) \in \Omega, \quad u(t) \geq 0, \quad t \in [0,T] \right\}, \tag{6.13}$$

where $\psi_j^i(t)$, as above, equals the expression in the brackets in (6.11). It is assumed that problem (6.13) has a bounded optimal solution $\hat{u}(t)$ such that $\sum_j \hat{u}_j^i(t) = \hat{U}^i(t) > 0$ for each $i = 1, 2, \ldots, I$ and any $t \in [0,T]$. Put

$$\hat{\alpha}_j^i(t) = \hat{u}_j^i(t)/\hat{U}^i(t), \quad \hat{s}_j^i(t) = \hat{U}^i(t)\left[\hat{\alpha}_j^i(t) - \mathring{\alpha}_j^i(t)\right] / \mathring{U}^i(t),$$
$$i = 1, 2, \ldots, I, \quad j = 1, 2, \ldots, J_i.$$

Theorem 6.2. *Suppose that macroproblems (6.7) and (6.8) and problem (6.13) satisfy all assumptions made above and* $\mathring{U}^i(t) \geq \varepsilon > 0$ *for* $i = 1, 2, \ldots, I$ *and* $t \in [0,T]$. *Then the following assertions are valid.*

(a) If $\delta(\mathring{\alpha}) > 0$, *then* $\hat{s}(\cdot)$ *is an ascent direction for* $\theta(\alpha)$ *in problem (6.10);*

(b) If $\delta(\mathring{\alpha}) = 0$, *then the disaggregated control* $\mathring{u} = \mathring{\alpha}\mathring{U}$ *is optimal for the original problem (6.1)–(6.4);*

(c) For all $\mathring{\alpha}(\cdot) \in A$, *there exists an* $M > 0$ *such that* $S(\mathring{\alpha}) \geq M\delta(\mathring{\alpha})$, *where* $S(\mathring{\alpha})$ *is the value defined by (6.12).*

Theorem 6.2 is proved similarly to Theorem 2.5 from Section 2. The point $\tilde{\alpha} \in A$ used in the proof of assertion (b) of Theorem 2.5 is now constructed as in the proof of Theorem 6.1 in this section.

By virtue of Theorem 6.2, instead of problem (6.12), which is used to determine an ascent direction for the functional $\theta(\alpha(\cdot))$ in the conditional gradient method, we can solve problem (6.13) and define the direction $\hat{s}(\cdot)$ as specified. The convergence of such a modified method of conditional gradient is proved similarly (see [44]).

We can exploit the freedom in the choice of the set Ω in such a way that problem (6.13) decomposes into a number of independent subproblems. The ways to do this are thoroughly analyzed in Section 2. In particular, an analogue of the block-separable problem (2.21), (2.29) is

$$\sum_l^L Q_l(x_l(T)) + \int_0^T \sum_l f_l(x_l(t)),\, u_l(t),\, t)\, dt \to \max, \tag{6.14}$$

$$dx_l(t)/dt = A_l(t)x_l(t) + b_l\big(u_l(t),\, t\big), \quad x_l(0) = x_{l0}, \tag{6.15}$$

$$p_l\big(x_l(t),\, u_l(t),\, t\big) \le 0, \quad \nu_l(x_l(T)) \le 0, \quad u_l(t) \ge 0, \quad l = 1,2,\dots,L, \tag{6.16}$$

$$\sum_l d_l\big(x_l(t),\, u_l(t),\, t\big) \le 0, \quad \sum_l q_l(x_l(T)) \le 0, \tag{6.17}$$

where $x_l(\cdot) \in \mathbf{R}^{N_l}$, $u_l(\cdot) \in \mathbf{R}^{n_l}$, $\sum_l N_l = N$, and $\sum_l n_l = n$; it is supposed that the assumptions made about the original problem (6.1)–(6.4) hold also for problem (6.14)–(6.17).

Suppose that we have obtained an aggregated problem for (6.14)–(6.17) at a fixed $\mathring{\alpha}(\cdot) \in A$, and $\mathring{U}(t)$, $\mathring{x}(t)$ is its unique optimal solution; suppose also that $\mathring{\eta}_l(t)$, $\mathring{\mu}_l$ and $\mathring{\eta}(t)$, $\mathring{\mu}$ are the unique Lagrange multipliers for constraints (6.16) and (6.17), respectively. Let Ω be the set of controls satisfying the block constraints (6.16) and (6.17). Consider L independent subproblems similar to (2.30):

$$Q_l(x(T)) + \int_0^T f_l(x_l(t),\, u_l(t), t)\, dt - \int_0^T \Big[d_l\big(x_l,\, u_l,\, t\big),\, \mathring{\nu}(t)\Big]\, dt - \\ - \Big[\mathring{\mu},\, q_l(x_l(T))\Big] \to \max, \tag{6.18}$$

$$dx_l(t)/dt = A_l(t)x_l(t) + b_l(u_l,\, t), \quad x_l(0) = x_{l0},$$

$$p_l\big(x_l(t),\, u_l(t),\, t\big) \le 0, \quad \nu_l(x_l(T)) \le 0, \quad u_l(t) \ge 0.$$

Let $\delta_l(\mathring{\alpha})$ denote an optimal value of the objective functional in problem (6.18). Using the convexity of the functions in (6.14)–(6.17) and the inequalities $\mathring{\eta}_l(t)p_l(x_l, u_l, t) \le 0$ and $\mathring{\mu}_l\nu_l(x_l(T)) \le 0$, which hold for all x_l and u_l satisfying (6.15) and (6.16), we finally obtain the following relation similar to (2.26):

$$\delta(\mathring{\alpha}) \ge \sum_l \delta_l(\mathring{\alpha}) - \sum_l Q_l(\mathring{x}_l(T)) - \int_0^T \sum_l f_l(\mathring{x}_l, \mathring{u}_l, t)\, dt, \tag{6.19}$$

here, $\delta(\mathring{\alpha})$ has the same sense as in (6.13). The saddle point theorem for the Lagrange functional of problem (6.14)–(6.17) is valid [314, 215]; according to this theorem, the right-hand side of (6.19) is strictly positive if $\mathring{u}$ is nonoptimal for (6.14)–(317) and vanishes otherwise. If $\hat{u}(t)$ is an optimal solution to the local problems (6.18) such that $\hat{U}^i(t) = \sum_j \hat{u}^i_j(t) > 0$ for

$i = 1, 2, \ldots, I$ and $t \in [0, T]$, and if $\overset{\circ}{u}$ is nonoptimal, then

$$\hat{s}_j^i(t) = \hat{U}^i(t)\left[\hat{\alpha}_j^i(t) - \overset{\circ}{\alpha}{}_j^i(t)\right] \Big/ \overset{\circ}{U}{}^i(t), \quad \hat{\alpha}_j^i(t) = \hat{u}_j^i(t)/\hat{U}^i(t),$$

$$i = 1, 2, \ldots, I, \quad j = 1, 2, \ldots, J_i$$

determines the ascent direction of $\theta(\alpha(\cdot))$ at the point $\overset{\circ}{\alpha}(\cdot) \in A$.

The case where constraints (6.16), (6.17) in problem (6.14)–(6.17) are replaced by the conditions

$$p_l(x_l,\, u_l,\, t) + r_l(x,\, u,\, t) \le 0, \quad \nu_l(x_l(T)) + c_l(x(T)) \le 0,$$

$$x = (x_1 \ldots x_L), \quad u = (u_1 \ldots u_L) \ge 0,$$

and the cross terms $r_l(\cdot\,,\cdot\,,\cdot)$ and $c_l(\cdot)$ have the lower bounds

$$r_l(x,\, u,\, t) \ge \xi_l, \quad c_l(x(T)) \ge \gamma_l$$

for all $x(t)$ and $u(t)$ satisfying the constraints (6.15) and (6.17) is handled similarly. Here, ξ_l and γ_l are given constant vectors of the same dimensions as $r_l(\cdot\,,\cdot\,,\cdot)$ and $c_l(\cdot)$, respectively.

§7. Aggregation of State Variables in Linear-Quadratic Problems

In the preceding section, the macroproblem is constructed by aggregating controls, and the optimal aggregation parameters $\alpha^* \in A$ are determined with the use of the extremal properties of α^* and the first-order optimization methods. In this section, we consider aggregation of state variables in linear-quadratic optimal control problems. For recalculating aggregation parameters, we suggest a procedure based on the method of successive approximations, which is applied to solve a system formed by necessary optimality conditions for the aggregation parameters.

Consider the discrete linear-quadratic optimal control problem

$$\begin{aligned}
&(d^{\mathrm{T}},\, x) + (p^{\mathrm{T}},\, u) + 0,5x^{\mathrm{T}}Qx + 0,5u^{\mathrm{T}}Ru \to \max,\\
&x = Ax + Bu + a,\\
&-Dx + Cu = b,\\
&u \ge 0.
\end{aligned} \tag{7.1}$$

Here, x is the full column vector of state variables, u is the full column vector of discretized controls, Q is a symmetric nonnegative definite square matrix, and R is a symmetric negative definite square matrix. The dimensions of the column vectors a, b, d, and p and of the matrices A, B, C, and D are consistent with those of the vectors x and u. Under these assumptions, the functional of problem (7.1) is strictly concave. We also assume that the equalities in the mixed constraints of problem (7.1) are obtained by adding weak variables to inequalities and extending the control vector so that the corresponding Lagrange multiplier vector is nonnegative.

In statement (7.1), the vector x includes both values of the same variable at different points of the mesh and values of different state variables at the same points of the discrete time mesh. This also refers to the vector u. Such a statement is convenient because it allows us to uniformly describe aggregation with respect to both discrete mesh points and state variables. Static problems of type (7.1) arise in problems of mathematical economics related to analyzing models of input–output balance [5, 55].

Suppose that the vector x is divided into I $(1 \leq I \leq \dim x)$ subvectors, i.e., $x = (x_1, \dots x_i \dots x_I)$, where $\dim x = \sum_i \dim x_i$. We can assume without loss of generality that $\dim x_i = n$ for $i = 1, 2, \dots, I$.

Taking into account the partition of the vector x, we can rewrite the original problem in the form

$$
\begin{aligned}
&\sum_{i=1}^{I}(d_i^{\mathrm{T}}, x_i) + (p^{\mathrm{T}}, u) + 0,5\sum_{i=1}^{I}\sum_{j=1}^{I} x_i^{\mathrm{T}} Q_{ij} x_j + 0,5 u^{\mathrm{T}} R u \to \max,\\
&x_i = \sum_j A_j^i x_j + B_i u + a_i, \quad i = 1, 2, \dots, I, \qquad (7.2)\\
&-\sum_j D_j x_j + C u = b, \quad u \geq 0.
\end{aligned}
$$

We assume that problem (7.2) satisfies conditions similar to those stated in [49]; namely, for a solution to be optimal, it is necessary that $x^* \geq \varepsilon > 0$ componentwise; the elements of the matrices A, B, and D and of the vector a are nonnegative; and the objective function of (7.1) is monotonically nondecreasing with respect to x at $x \geq 0$, i.e., $Qx + d \geq 0$ componentwise. In addition, we assume that the above partition of the vector x is such that each row and each column of the matrix A_j^i contains at least one nonzero element.

The necessary and sufficient optimality conditions for problem (7.2) are

$$x_i = \sum_j A^i_j x_j + B_i u + a_i, \tag{7.3}$$

$$\lambda_i = \sum_j \lambda_j^{\mathrm{T}} A_i^j + \sum_j Q_{ij} x_j + d_i + \nu^{\mathrm{T}} D_i, \quad i = 1, 2, \ldots, I, \tag{7.4}$$

$$-\sum_j D_j x_j + Cu = b, \tag{7.5}$$

$$p + Ru + \sum_i \lambda_i^{\mathrm{T}} B_i - \nu^{\mathrm{T}} C \geq 0, \quad \nu \geq 0, \tag{7.6}$$

$$\left(u^{\mathrm{T}}, p + Ru + \sum_i \lambda_i^{\mathrm{T}} B_i - \nu^{\mathrm{T}} C\right) = 0, \quad u \geq 0. \tag{7.7}$$

We seek a solution to system (7.3)–(7.7) in the form

$$x_i = \alpha_i X_i, \quad \lambda_i = \beta_i \Lambda_i, \quad i = 1, 2, \ldots, I, \tag{7.8}$$

where α_i and β_i are the column n-vectors of aggregation parameters, and X_i and Λ_i are the aggregated scalar variables. It is easy to see that, if x_i^*, λ_i^* is a solution to system (7.3)–(7.7), $(e, x_i^*) \neq 0$, and $(e, \lambda_i^*) \neq 0$, where e is the unit n-vector, then a solution of the form (7.8) exists. We can put, for example,

$$X_i^* = (e, x_i^*), \quad \Lambda_i = (e, \lambda_i^*), \quad \alpha_i^* = (X_i^*)^{-1} x_i^*, \quad \beta_i^* = (\Lambda_i^*)^{-1} \lambda_i^*.$$

It follows from (7.8) that α_i, X_i and β_i, Λ_i are determined to a constant scalar factor.

Our motivation in seeking a solution in the form (7.8) is the desire to construct a two-level scheme for solving problem (7.3)–(7.7) such that, at one level, the X and Λ from the aggregated problem be determined for fixed α and β, and at the other level, the aggregation parameters α and β be adjusted.

As the aggregated problem for (7.3)–(7.7), consider the system

$$(\alpha_j^{\mathrm{T}}, \beta_i \Lambda_i) = \left(\alpha_i^{\mathrm{T}}, \sum_j \Lambda_j \beta_j^{\mathrm{T}} A_i^j + \sum_j X_j Q_{ij} \alpha_j + d_i + \nu^{\mathrm{T}} D_i\right),$$

$$(\beta_i^{\mathrm{T}}, \alpha_i X_i) = \left(\beta_i^{\mathrm{T}}, \sum_j X_j A_j^i \alpha_j + B_i u_i + a_i\right), \quad i = 1, 2, \ldots, I,$$

$$
\begin{gathered}
-\sum_j X_j D_j \alpha_j + Cu = b, \\
p + Ru + \sum_i \Lambda_i \beta_i^{\mathrm{T}} B_i - \nu^{\mathrm{T}} C \geq 0, \quad \nu \geq 0, \\
\left(u^{\mathrm{T}}, p + Ru + \sum_i \Lambda_i \beta_i^{\mathrm{T}} B_i - \nu^{\mathrm{T}} C \right) = 0, \quad u \geq 0,
\end{gathered}
\tag{7.9}
$$

which is a consequence of (7.3)–(7.7); indeed, instead of equations (7.3) and (7.4), (7.9) includes their linear combinations with coefficients β_i and α_i, respectively. A direct verification shows that (7.9) is a system of optimality conditions for the problem

$$
\begin{gathered}
\sum_i X_i (d_i^{\mathrm{T}}, \alpha_i) + (p^{\mathrm{T}}, u) + 0,5 \sum_i \sum_j X_i X_j \alpha_i^{\mathrm{T}} Q_{ij} \alpha_j + \\
+ 0,5 u^{\mathrm{T}} R u \to \max, \\
X_i = (\beta_i^{\mathrm{T}}, \alpha_j)^{-1} \left[\sum_j X_j (\beta_i^{\mathrm{T}}, A_j^i \alpha_j) + (\beta_i^{\mathrm{T}}, B_i u) + (\beta_i^{\mathrm{T}}, a_i) \right], \\
-\sum_j X_j D_j \alpha_j + Cu = b, \quad u \geq 0, \quad i = 1, 2, \ldots, I
\end{gathered}
\tag{7.10}
$$

for fixed aggregation parameters α and β. Here, we assume that $(\beta_i^T, \alpha_i) \neq 0$ for $i = 1, 2, \ldots, I$. In what follows, we suppose that problem (7.10) and its dual have unique optimal solutions at all α and β such that $(\alpha_i^T, \beta_i) \neq 0$ for $i = 1, 2, \ldots, I$.

The aggregated problem (7.10) is of the same form as the initial problem (i.e., it is a discrete linear-quadratic optimal control problem), but the number of the state variables in (7.10) is n times smaller.

Suppose, we have obtained a solution $\mathring{X}_i$, $\mathring{\Lambda}_i$, $\mathring{u}$, $\mathring{\nu}$ to system (7.9) (or, which is the same thing, optimal solutions to problem (7.10) and its dual) for fixed $\mathring{\alpha}_i$ and $\mathring{\beta}_i$. Note that $\mathring{x}_i = \mathring{\alpha}_i \mathring{X}_i$ and $\mathring{\lambda}_i = \mathring{\beta}_i \mathring{\Lambda}_i$ may not satisfy constraints (7.3) and (7.4). The fulfillment of these constraints serves as a criterion for the optimality of the solution $\mathring{x}_i$, $\mathring{\lambda}_i$, $\mathring{u}$, $\mathring{\nu}$ to the original problem (7.1). We could obtain the aggregated problem (7.10) in the way described in Section 2, by substituting the disaggregated solutions $x_i = \alpha_i X_i$ in (7.1) and summing the dynamical equations with coefficients β_i; the optimality criterion would then be implied by Theorem 2.2.

Taking into account the specific dynamical features of equations (7.3) and (7.4), we suggest the following iterative scheme for solving problem (7.1). At each iteration, two steps are performed.

1. Let the current values of the aggregation parameters be $\overset{\circ}{\alpha}_i$ and $\overset{\circ}{\beta}_i$, where $(\overset{\circ}{\beta}{}_i^T, \overset{\circ}{\alpha}_i) \neq 0$ $(i = 1, 2, \dots, I)$. Macroproblem (7.10) and its dual are solved at $\alpha = \overset{\circ}{\alpha}$ and $\beta = \overset{\circ}{\beta}$. Let $\overset{\circ}{X}_i \neq 0$, $\overset{\circ}{u}$ and $\overset{\circ}{\Lambda}_i \neq 0$, $\overset{\circ}{\nu}$ be the unique solutions to the respective problems, and let $\overset{\circ}{x}_i = \overset{\circ}{\alpha}_i \overset{\circ}{X}_i$ and $\overset{\circ}{\lambda}_i = \overset{\circ}{\beta}_i \overset{\circ}{\Lambda}_i$ be the disaggregated solution.

2. The fulfillment of constraints (7.3) and (7.4) for $\overset{\circ}{x}$, $\overset{\circ}{\lambda}$, $\overset{\circ}{u}$, and $\overset{\circ}{\nu}$ is verified. If these constraints are fulfilled (with given accuracy), then $\overset{\circ}{x}$ and $\overset{\circ}{u}$ are optimal for the initial problem (7.1). Otherwise, new $\bar{\alpha}_i$ and $\bar{\beta}_i$ are defined by setting

$$
\begin{aligned}
\bar{\alpha}_i &= (e, \bar{x}_i)^{-1} \bar{x}_i, \quad \bar{\beta}_i = (e, \bar{\lambda}_i)^{-1} \bar{\lambda}_i, \\
\bar{x}_i &= \sum_j A_j^i \overset{\circ}{x}_j + B_i \overset{\circ}{u} + a_i, \\
\bar{\lambda}_i &= \sum_j \overset{\circ}{\lambda}{}_j^{\mathrm{T}} A_i^j + \sum_j Q_{ij} \overset{\circ}{x}_j + d_i + \overset{\circ}{\nu}{}^{\mathrm{T}} D_i, \quad i = 1, 2, \dots, I.
\end{aligned}
$$

The scheme suggested above is an implementation of the method of successive approximations aimed at solving the system of equations (7.3), (7.4), where the values u and v are found from the macroproblem (7.10) or from the corresponding system (7.9) of optimality conditions. Note that, in the case under consideration, optimization is only involved in solving macroproblem (7.10), which contains a small number of aggregated state variables.

To determine conditions for the convergence of the iterative procedure, we have to find out how the components u and v of the optimal solution to macroproblem (7.10) depend on the parameters α and β.

Suppose that an optimal solution u^* to the original problem is nondegenerate in the sense that the equality in (7.6) is only attained for the positive components of the vector u^*. Let δ be the set of the numbers of the positive components of u^*. We denote the vector formed by the positive components of u^* by u^δ and the matrix formed by those columns of C whose numbers belong to δ by C_δ; the other vectors and matrices labeled by δ are formed similarly.

According to (7.6), we have

$$u^\delta = R_\delta^{-1}\left(-\sum_i \Lambda_i \beta_i^{\mathrm{T}} B_{i\delta} - p_\delta + \nu^{\mathrm{T}} C_\delta\right).$$

The substitution of u^δ in the mixed constraints of problem (7.10) yields

$$C_\delta R_\delta^{-1} C_\delta^{\mathrm{T}} \nu = b + \sum_j X_j D_j \alpha_j + C_\delta R_\delta^{-1} \sum_i \Lambda_i \beta_i^{\mathrm{T}} B_{i\delta} + C_\delta R_\delta^{-1} p_\delta.$$

If $P = \left(C_\delta R_\delta^{-1} C_\delta^{\mathrm{T}}\right)^{-1}$ is defined, then

$$\nu = P\left(b + \sum_j X_j D_j \alpha_j + C_\delta R_\delta^{-1} \sum_i \Lambda_i \beta_i^{\mathrm{T}} B_{i\delta} + C_\delta R_\delta^{-1} p_\delta\right).$$

Substituting the expressions for u^δ and ν in (7.3) and (7.4), we obtain the system

$$\begin{aligned} \alpha_i X_i &= \sum_j \left(A_j^i + B_{i\delta} R_\delta^{-1} C_\delta^{\mathrm{T}} P D_j\right) \alpha_j X_j - \\ &\quad - B_{i\delta} R_\delta^{-1} \sum_j \left(B_{j\delta}^{\mathrm{T}} - C_\delta^{\mathrm{T}} P C_\delta R_\delta^{-1} B_{j\delta}^{\mathrm{T}}\right) \Lambda_j B_j + \xi_1^i, \\ \Lambda_i \beta_i &= \sum_j \left[(A_i^j)^{\mathrm{T}} + D_i^{\mathrm{T}} P C_\delta R_\delta^{-1} B_{j\delta}^{\mathrm{T}}\right] \Lambda_j \beta_j + \\ &\quad + \sum_j \left(Q_{ij} + D_i^{\mathrm{T}} P D_j\right) \alpha_j X_j + \xi_2^i, \end{aligned} \tag{7.11}$$

where ξ_1^i and ξ_2^i do not depend on α and β.

Thus, equations (7.3) and (7.4) are reduced to the system

$$z = \Phi_\delta z + r,$$

where z is the vector formed by $\Lambda_i \beta_i$ and $X_i \alpha_i$ and Φ_δ is a matrix, which is determined from (7.11) in an obvious way and does not depend on the numbers of the positive components of the vector u^* of optimal controls in the original problem. The iterative scheme described above is an implementation of the method of successive approximations for system (7.11). It locally converges if

$$\|\Phi_\delta\| \le \Delta < 1.$$

If the initial values of $\overset{\circ}{\alpha}_i$ and $\overset{\circ}{\beta}_i$ $(i = 1, 2, \ldots, I)$ are componentwise strictly positive, then all succeeding values of $\overset{\circ}{\alpha}_i$ and $\overset{\circ}{\beta}_i$ $(i = 1, 2, \ldots, I)$ are also strictly positive (this is implied by the formulas for recalculating the aggregation parameters at step 2), and $(\beta_i^T, \alpha_i) \neq 0$ at each iteration. Since the aggregation parameters are determined to a constant factor, we can assume without loss of generality that $(\beta_i^T, \alpha_i) = 1$ for $i = 1, 2, \ldots, I$.

It is easy to show with the use of a similar argument that, if the original problem is linear, i.e., $Q = 0$ and $R = 0$, then system (7.11) has the form

$$\alpha_i X_i = \sum_j \Big(A_j^i + B_{i\delta} C_\delta^{-1} D_j\Big)\alpha_j X_j + \theta_1^i,$$

$$\beta_i \Lambda_i = \sum_j \Big[(A_i^j)^{\mathrm{T}} + D_i^{\mathrm{T}} (C_\delta^{-1})^{\mathrm{T}} B_{j\delta}^{\mathrm{T}}\Big] B_j \Lambda_j + \theta_2^i,$$

where θ_1^i and θ_2^i are constants independent of α_i and β_i; it is assumed that C_δ^{-1} exists.

For $I = 1$, when the aggregated problem has only one state variable, we obtain

$$\Lambda\beta = \Big[A^{\mathrm{T}} + D^{\mathrm{T}} \big(C_\delta^{-1}\big)^{\mathrm{T}} B_\delta^{\mathrm{T}}\Big] \Lambda\beta + \theta_1,$$

$$X\alpha = \big(A + B_\delta C_\delta^{-1} D\big) X\alpha + \theta_2.$$

The local convergence condition then has the form $\|A + B_\delta C_\delta^{-1} D\| < 1$, since $\big(B_\delta C_\delta^{-1} D\big)^{\mathrm{T}} = D^{\mathrm{T}} \big(C_\delta^{-1}\big)^{\mathrm{T}} B_\delta^{\mathrm{T}}$. If $D = 0$, i.e., only controls are subject to constraints, we obtain the condition $\|A\| < 1$, which is traditional for problems of mathematical economics.

A similar approach to linear problems is considered in [55] in analyzing models of input–output balance, where one macrovariable is introduced and the aggregation is determined by the economical meaning of the problem. In the notation of problem (7.1), the local convergence condition obtained in [55] has the form $\|A + B_\delta C_\delta^{-1} D\| < 0,5$.

§8. Distributed Parameter Systems

In this section, we examine statements of optimal control problems in mathematical physics and apply the iterative aggregation method to such problems. The systems can be described by various partial differential equations of classical types.

Consider the process in the domain $\Omega = (t \in (0,T), x \in (0,L))$ described by the polylinear system of the first-order hyperbolic partial differential equations [58] written in terms of the Riemann invariants [46] as

$$\partial z(t,x)/dt + \xi(t,x)\partial z(t,x)/dt = A(t,x)z(t,x) + b(t,\, x,\, u(t,x)), \tag{8.1}$$

where x is a one-dimensional spatial coordinate varying over the interval $[0, L]$, $z = z(t,x)$ is the N-vector function of the state, $u(x,t)$ is the n-vector of controls, and $A(t,x)$ is an $N \times N$ matrix with nonnegative elements. All components $\xi_{lk}(t,x)$ $(l,k = 1,2,\dots,N)$ of the matrix of characteristic directions $\xi(t,x)$ are assumed to be positive and monotonically increasing with respect to x. The controls $u(t,x)$ are sought in the class of functions measurable and almost everywhere bounded along any characteristic curve corresponding to equations (8.1) and belonging to the rectangle $[0,L] \times [0,T]$.

The system of initial and boundary conditions is given in the form

$$z(0,x) = z^0(x), \quad x \in [0,L], \quad z(t,0) = z^0(t), \quad t \in [0,T]. \tag{8.2}$$

There are also the mixed constraints

$$d(t,\, x,\, z(t,x),\, u(t,x)) \le 0, \quad u(t,x) \ge 0, \tag{8.3}$$

where $d(\cdot)$ is a k-vector function.

The objective functional has the form

$$J(u(\cdot)) = \int_0^T \int_0^L f_1(t,\, x,\, z(t,x),\, u(t,x))\, dt\, dx \, +$$

$$+ \int_0^T f_2(t,\, z(t,L))\, dt + \int_0^L f_3(x,\, z(T,x))\, dx \to \max. \tag{8.4}$$

We assume that the functions b, f_1, f_2, and f_3 are twice continuously differentiable with respect to their arguments, have bounded second derivatives, are concave with respect to z and u, and monotonically increase with respect to z. The elements of the matrix $A(t,x)$ are twice continuously differentiable, and the components of $\xi(t,x)$ are continuously differentiable with respect to both arguments.

Suppose that the initial problem (8.1)–(8.4) and the auxiliary problems obtained below satisfy the Slater condition [48] (with respect to constraints (8.3)) and the Ter-Krikorov conditions [49]

$$z(t,x) \ge \varepsilon > 0, \quad d(t,\, x,\, 0,\, 0) \le 0.$$

As in [51], these conditions make it possible to reduce (8.1)–(8.4) to a concave programming problem in a Banach space so that the duality theorems are valid and the complementary slackness conditions (an analogue of the Kuhn–Tucker theorem) hold. In this case, the Pontryagin maximum principle also holds.

Let us aggregate the controls. As above, we divide the n-vector $u(t,x)$ into I subvectors, each with J_i components ($i = 1,2,\dots,I$), so that $u = (u_1 \dots u_i \dots u_I)$, $u_i = \left(u_1^i \dots u_j^i \dots u_{J_i}^i\right)$, and $\sum\limits_i J_i = n$, and seek the controls in the form

$$u_j^i(t,x) = \alpha_j^i(t,x)U^i(t,x), \quad i = 1,2,\dots,I, \quad j = 1,2,\dots J_i, \tag{8.5}$$

where the aggregation parameters $\alpha_j^i(t,x)$ are bounded and measurable on Ω and satisfy the condition

$$\alpha \in A = \left\{\alpha_j^i(t,x)\colon \sum_j \alpha_j^i(t,x) = 1, \quad \alpha_j^i(t,x) \geq 0, \quad i = 1,2,\dots,I, \quad j = 1,2,\dots,J_i\right\} \tag{8.6}$$

almost everywhere on Ω.

Fixing an $a \in A$ and substituting (8.5) in the original problem, we come to the following macroproblem with respect to the aggregated controls $U^i(t,x)$:

$$\begin{aligned} \tilde{J}(U(\cdot)) = \int_0^T \int_0^L F_1(t,x,z,U)\,dt\,dx + \int_0^T f_2\big(t,\, z(t,L)\big)\,dt \,+ \\ + \int_0^L f_3\big(x,\, z(T,x)\big)\,dx \to \max, \end{aligned}$$

$$\begin{aligned} &\partial z(t)/\partial t + \xi(t,x)\partial z(t,x)/\partial x = A(t,x)z(t,x) + B\big(t,\, x,\, U(t,x)\big), \\ &z(0,x) = z^0(x), \quad x \in [0,L], \\ &z(t,0) = z^0(t), \quad t \in [0,T], \\ &D\big(t,\, x,\, z(t,x),\, U(t,x)\big) \leq 0, \quad U(t,x) = \big(U^1(t,x),\dots,U^I(t,x)\big) \geq 0, \end{aligned} \tag{8.7}$$

where

$$F_1(t,\, x,\, z,\, U) = f_1\Big(t,\, x,\, z,\, \alpha_1^1 U^1,\dots,\alpha_{J_I}^I U^I\Big),$$

$$B(t,\,x,\,U) = b\Big(t,\,x,\,\alpha_1^1 U^1,\dots,\alpha_{J_I}^I U^I\Big),$$

$$D(t,\,x,\,z,\,U) = d\Big(t,\,x,\,z,\,\alpha_1^1 U^1,\dots,\alpha_{J_I}^I U^I\Big),$$

and F_1, B, and D obviously meet the requirements of concavity and monotonicity with respect to z and U mentioned above.

The problem dual to the aggregated problem is stated with the use of the results of [48] and has the form

$$\begin{aligned}
&L\big(U(t,x),\, z(t,x),\, y(t,x),\, \delta(t,x)\big) \to \min,\\
&\partial y(t,x)/\partial t = \xi(t,x)\partial y(t,x)/\partial x - \big[\partial\xi(t,x)/\partial x\big]y(t,x) -\\
&- A^{\mathrm{T}}(t,x)y(t,x) - \partial\big(F_1(t,\,x,\,z,\,U) - D(t,\,x,\,z,\,U)\delta(t,x)\big)/\partial z,\\
&y(T,x) = -\partial f_3\big(x,\,z(T,x)\big)/\partial z(T,x), \quad x \in [0,L],\\
&\xi(t,L)y(t,L) = -\partial f_2\big(t,\,z(t,L)\big)/\partial z(t,L), \quad t \in [0,T],\\
&\frac{\partial}{\partial U(t,x)}\Big[F_1(t,\,x,\,z,\,U) - B(t,\,x,\,U)y(t,x) -\\
&\qquad\qquad\qquad - D(t,\,x,\,z,\,U)\delta(t,x)\Big] \le 0,\\
&U(t,x) \ge 0, \quad \delta(t,x) \ge 0,
\end{aligned} \tag{8.8}$$

where $y(t,x)$ are the dual functions, $\delta(t,x)$ are the Lagrange multipliers for the mixed constraints in (8.7), and $L(U,z,y,\delta)$ is the Lagrange functional defined by

$$\begin{aligned}
L = \int_0^T\!\!\int_0^L \Big\{F_1(t,\,x,\,z,\,U) + \Big[A(t,x)z + B(t,\,x,\,U) - \partial z/\partial t -\\
- \xi(t,x)\partial z/\partial x\Big]y(t,x) - D(t,\,x,\,z,\,U)\delta(t,x)\Big\}\,dt\,dx +\\
+ \int_0^T f_2\big(t,\,z(t,L)\big)\,dt + \int_0^L f_3\big(t,\,z(T,x)\big)\,dx.
\end{aligned}$$

Suppose that, at a given $\overset{\circ}{\alpha} \in A$, we have found the unique optimal solution $\overset{\circ}{U}(t,x)$ to the macroproblem, and this solution is componentwise positive on Ω. Suppose also that the corresponding state functions $\overset{\circ}{z}(t,x)$ and the unique solutions $\overset{\circ}{y}(t,x)$ and $\overset{\circ}{\delta}(t,x)$ to the dual problem (8.8) are found, too. A criterion for the disaggregated control $\overset{\circ}{u}(t,x) = \overset{\circ}{\alpha}(t,x)\overset{\circ}{U}(t,x)$ to be optimal for the original problem is the fulfillment of the condition

$$\frac{\partial}{\partial u}\Big[f_1(t,\, x,\, \overset{\circ}{z},\, \overset{\circ}{u}) - b(t,\, x,\, \overset{\circ}{u})\overset{\circ}{y} - d(t,\, x,\, \overset{\circ}{z}, \overset{\circ}{u})\overset{\circ}{\delta}\Big] \le 0, \qquad (8.9)$$

where the derivatives are calculated for the disaggregated solution $\overset{\circ}{u}(t,x)$, $\overset{\circ}{z}(t,x)$. This is proved similarly to the corresponding assertion in Section 2. In what follows, we assume that the set of solutions to macroproblem (8.7) and its dual (8.8) is bounded uniformly with respect to $\alpha \in A$.

Let us denote the optimal value of the objective functional of macroproblem (8.7) in the dependence on the aggregation parameter $\alpha \in A$ by $\theta(\alpha(\cdot))$ and consider the problem

$$\theta(\alpha(\cdot)) \to \max, \quad \alpha(\cdot) \in A. \qquad (8.10)$$

To analyze the differential properties of $\theta(\alpha(\cdot))$, consider the parametric family of macroproblems (8.7) depending on the functional parameter $\alpha(t,x)$. Since the second derivatives of the functions involved in the objective functional and in constraints (8.1)–(8.4) are bounded and the solutions to the dual problems (8.7) and (8.8) are uniformly bounded, the basic assumption of the marginal-value theorem [27], that the input functions must be uniformly bounded with respect to the parameter $\alpha(\cdot)$, holds. Applying this theorem with taking into account the supposed uniqueness of the solutions to problems (8.7) and (8.8), we conclude that the functional $\theta(\alpha(\cdot))$ is Frechet differentiable. Its derivative $\nu^{ij}_{\overset{\circ}{\alpha}}(t,x)$ with respect to the parameter $\alpha^i_j(\cdot)$ at a point $\overset{\circ}{\alpha} \in A$ is evaluated as the derivative with respect to α^i_j of the Lagrange functional at the optimal solution to (8.7). We have

$$\begin{aligned}\nu^{ij}_{\overset{\circ}{\alpha}}(t,x) = \overset{\circ}{U}{}^i(t,x)\Big[\partial f_1(t,\, x,\, \overset{\circ}{z},\, \overset{\circ}{u})/\partial u^i_j\, - \\ - \Big(\partial b(t,\, x,\, \overset{\circ}{u})/\partial u^i_j,\, \overset{\circ}{y}(t,x)\Big) - \Big(\partial d(t,\, x,\, \overset{\circ}{z},\, \overset{\circ}{u})/\partial u^i_j,\, \overset{\circ}{\delta}(t,x)\Big)\Big].\end{aligned} \qquad (8.11)$$

Theorem 8.1. *Let $\overset{\circ}{\alpha} \in A$ be such that $\langle \nabla\theta(\overset{\circ}{\alpha}),\, \alpha - \overset{\circ}{\alpha}\rangle \le 0$ for all $\alpha \in A$, and let $\overset{\circ}{U}(t,x) > 0$ be the extremal solution to macroproblem (8.7) at $\alpha = \overset{\circ}{\alpha}$. Then the disaggregated solution $\overset{\circ}{u} = \overset{\circ}{\alpha}\overset{\circ}{U}$ is optimal for the original problem. Here, $\nabla\theta(\overset{\circ}{\alpha})$ is the operator corresponding to the Frechet gradient, whose components are defined by (8.11), and $\langle \nabla\theta(\overset{\circ}{\alpha}),\, \alpha - \overset{\circ}{\alpha}\rangle$ is the value of this operator at the element $\alpha - \overset{\circ}{\alpha}$.*

Theorem 8.1 is proved similarly to the corresponding assertion from Section 2 with the use of the necessary and sufficient conditions (8.9) and the complementary slackness theorem.

Theorem 8.1 implies that an optimal solution to problem (8.1)–(8.4) can be constructed by applying the first-order methods (such as the methods of conditional gradient, feasible directions, etc. [56]) to problem (8.10). The auxiliary problem of selecting an ascent direction for $\theta(\alpha(\cdot))$ involved in such methods decomposes into I independent subproblems, because $A = A_1 \times A_2 \times \cdots \times A_I$.

If the original problem has block constraints of type (8.1)–(8.3) while the binding constraints (8.3), as well as the functional of (8.1), are additively separable with respect to the blocks, then we can take into account the block structure of the constraints in determining the ascent direction for $\theta(\alpha(\cdot))$ the same way we did in Section 2. The ascent direction problem then decomposes into independent subproblems, whose number equals the number of blocks in the original statement. Such a decomposition does not depend on the aggregation determined by (8.6); it only depends on the block-separable structure of the original problem.

Next, consider the problem of controlling a second-order hyperbolic system. We use the following notation [28, 58]. Let $x = (x_1, \ldots, x_n)$ be a spatial variable that varies in a bounded open set $\Omega \subseteq \mathbf{R}^n$ with boundary Γ. We assume that Γ is an infinitely differentiable manifold of dimension $n-1$, and Ω lies on one side of Γ. Let $t \in [0,T]$ be a time variable. By $Q = \Omega \times (0,T)$, we denote the cylinder with lateral surface $\Sigma = \Gamma \times (0,T)$. We use the notation $L^2(S)$ for the space of functions square-summable in a domain S and $H^1(\Omega)$ for the first-order Sobolev space of functions $\phi(x)$ such that $\partial\phi/\partial x_r \in L^2(\Omega)$ for $r = 1, 2, \ldots, n$. By $L^2(0,T;E)$, we denote the space of functions $t \to f(t)$ from the interval $(0,T)$ to a Hilbert space E such that $\int_0^T \|f(t)\|_E \, dt < \infty$.

Given functions $a^l_{pr}(x,t)$ $(p, r = 1, 2, \ldots, n)$ such that (i) they are measurable and bounded almost everywhere on Ω for any fixed $t \in [0,T]$, (ii) they have continuous first partial derivatives with respect to $t \in [0,T]$ at any fixed $x \in \Omega$, (iii) $a^l_{pr}(x,t) = a^l_{rp}(x,t)$ for $r, p = 1, 2, \ldots, n$, and (iv) $\sum_{p,r}^{n} a^l_{pr}(x,t)\xi_p\xi_r \geq \beta \sum_r \xi_r^2$ for $\beta > 0$ and ξ_p, $\xi_r \in \mathbf{R}$, we introduce the operator A_l defined by

$$A_l\varphi = -\sum_{p,r}^{n} (\partial/\partial x_p)\Big(a^l_{pr}(x,t)\partial\varphi/\partial x_r\Big)$$

and the normal derivative

$$\partial\varphi/\partial\nu_{A_l} = \sum_{p,r}^{n} a_{pr}^{l}\,(\partial\varphi/\partial x_r)\big(\cos(N,\, x_p)\big)$$

related to the operator A_l; here, $\cos(N,\, x_p)$ is the p-th direction cosine of the outer normal N to the boundary Σ.

Consider the optimal control problem for the system whose state coordinates satisfy the hyperbolic equations

$$\partial^2 z_l/\partial t^2 + A_l z_l(x,t) = f_l(x,t) \in L^2(Q), \quad l = 1, 2, \ldots, L \tag{8.12}$$

with initial conditions

$$z_l(x,0) = \varphi_l^1(x) \in H^1(\Omega), \quad \partial z_l(x,0)/\partial t = \varphi_l^2(x) \in L^2(\Omega), \tag{8.13}$$

where $f_l(x,t)$, $\varphi_l^1(x)$, and $\varphi_l^2(x)$ are given functions.

The boundary conditions are specified in the form

$$\partial z_l(s,t)/\partial\nu_{A_l} = b_l\big((s,\, t,\, u(s,t)\big), \quad s \in \Gamma, \quad t \in [0,T], \tag{8.14}$$

where $u(s,t) = \{u_1(s,t) \ldots u_N(s,t)\}$ is the N-vector of controls $u(s,t) \in L_N^2(\Omega)$.

On the state variables and controls, the constraints

$$d\big(s,\, t,\, u(s,t),\, z(s,t)\big) \le 0, \quad q\big(s,\, z(s,T)\big) \le 0, \quad u(s,t) \ge 0, \quad s,t \in \Sigma. \tag{8.15}$$

are imposed. The vector functions d and q have dimensions P and M, respectively, and the inequalities are assumed to hold almost everywhere on Σ and Γ; $z(s,t) = \{z_1(s,t) \ldots z_L(s,t)\}$.

The control objective is to maximize the functional

$$\Phi = \int_\Gamma \int_0^T c\big(s,\, \tau,\, u(s,\tau),\, z(s,\tau)\big)\, ds\, d\tau \,+$$

$$+ \int_\Omega \psi\big(x,\, z(x,T),\, z_x(x,T),\, z_t(x,t)\big)\, dx + \int_\Gamma w\big(s,\, z(s,T)\big)\, ds, \tag{8.16}$$

where c, ψ, and w are given functions and z_x and z_t are the vectors of partial derivatives with respect to x and t.

We assume that the first-order partial derivatives of the functions d, q, b, c, w, and ψ satisfy the Lipschitz condition with constants bounded in the finite balls in the spaces of their variables, and the partial derivatives of d, q, c, and w with respect to z are bounded in the entire spaces of their variables. Problem (8.12)–(8.16) consists in finding a vector of controls $u(s,t) \in L^2_N(\Sigma)$ and the corresponding weak (in the sense of [116]) solutions $z(x,t) \in L^2(Q)$ to problems (8.12)–(8.14) that satisfy constraints (8.15) and maximize the functional (8.16). The dimension of the initial problem with respect to controls is N. Below, we suggest a method of iterative aggregation of controls for this problem.

We assume that the functions c, b_l, and d are everywhere concave with respect to u and z and monotonically increase with respect to z; ψ is concave with respect to z, z_x, and z_t; and w and $-q$ are concave with respect to z.

An argument similar to that used in [51, 53] reduces problem (8.12)–(8.16) to a concave programming problem in a Banach space. If the Ter-Krikorov conditions [48] $z(x,t) \geq \varepsilon > 0$ and $d(s,t,0,0) \leq 0$ hold and the constraints satisfy the Slater condition, then the maximum principle, the duality theorems about the saddle point of the Lagrange function, and the complementary slackness relations are valid. In what follows, it is assumed that the original problem and all intermediate problems arising in the iterative aggregation method satisfy the Ter-Krikorov and Slater conditions. Note that $z(x,t)$ is strictly positive if, e.g., $\varphi_l^1(x), \varphi_l^2(x), f_l, b_l > 0$ and the Green function of the mixed problem for the hyperbolic equation is positive [50, 58].

Let us divide the N-vector $u(s,t)$ of controls into I subvectors, each with J_i components ($i = 1, 2, \dots, I$), so that

$$u = (u_1 \dots u_i \dots u_I), \quad u_i = \left(u_1^i \dots u_j^i \dots u_J^i\right), \quad \sum_i J_i = N.$$

We seek the controls in the form

$$u_j^i(s,t) = \alpha_j^i(s,t) U^i(s,t), \quad i = 1, 2, \dots I, \ j = 1, 2, \dots J_i, \quad s, t \in \Sigma \quad (8.17)$$

(or, briefly, $u = \alpha U$), where the parameters $\alpha_j^i(t,x)$ satisfy the condition

$$\alpha \in A = \Bigg\{ \alpha_j^i(s,t) \colon \sum_j \alpha_j^i(s,t) = 1, \quad \alpha_j^i(s,t) \geq 0,$$

$$i = 1, 2, \dots, I, \quad j = 1, 2, \dots, J_i, \quad s, t \in \Sigma \Bigg\}. \quad (8.18)$$

Fixing an $a \in A$ and substituting (8.17) in the original problem, we obtain a problem with aggregated controls. Let us introduce the notation

$$B_l(s,\, t,\, U) = b_l(s,\, t,\, \alpha U), \quad D(s,\, t,\, U) = d(s,\, t,\, \alpha U),$$

$$C(s,\, t,\, U) = c(s,\, t,\, \alpha U).$$

In the aggregated problem, the constraints involving controls are replaced by

$$U(s,t) \geq 0, \quad \partial z_l(s,t)/\partial \nu_{A_l} = B_l(s,\, t,\, U), \quad D(s,\, t,\, U,\, z) \leq 0;$$

functional (8.16) changes similarly.

If $(\overset{\circ}{z}, \overset{\circ}{U})$ is an optimal solution to the aggregated problem at a fixed $\overset{\circ}{\alpha} \in A$, then the disaggregated solution $(\overset{\circ}{z},\, \overset{\circ}{u})$, where $\overset{\circ}{u}$ is defined by (8.17), is obviously feasible in the original problem, but it may be nonoptimal.

Consider the dual to the problem with aggregated controls; using the results of [28, 48], we can state it as

$$\begin{aligned}
&\partial^2 y_l(t,x)/\partial t^2 + A_l y_l(x,t) = 0, \quad x, t \in Q,\\
&y_l(T,x) = \partial\psi/\partial z_{lx},\\
&\partial y_l(T,x)/\partial t = -\partial\psi/\partial z_l + \sum_r (\partial/\partial x_r)(\partial\psi/\partial z_{lx_r}), \quad x \in \Omega,\\
&\partial y_l/\partial\nu_{A_l} = (\partial/\partial z_l)\left[C - D\delta(s,t)\right], \quad s, t \in \Sigma.\\
&(\partial/\partial U)\left[C + \sum_l B_l y_l(s,t) - D\delta(s,t)\right] \leq 0, \quad s, t \in \Sigma,\\
&(\partial/\partial z_l)\left[w - \omega(s)q\right] - \sum_r (\partial\psi/\partial z_{lx_r}) \cos(N, x_r) = 0, \quad s \in \Gamma,\\
&\delta(s,t) \geq 0, \quad \omega(s) \geq 0, \quad U(s,t) \geq 0,\\
&L(U,\, z,\, y,\, \delta,\, \omega) \to \min,
\end{aligned} \tag{8.19}$$

where $\delta(s,t) \in L^2_M(\Sigma)$ and $\omega(s) \in L^2_p(\Gamma)$ are vector functions of dual variables and $y(t,x)$ are the adjoint variables being weak solutions to the corresponding problems. The functional L is the Lagrangian of the problem with aggregated controls.

Suppose that, at a given $\overset{\circ}{\alpha} \in A$, we have found the unique optimal solution $\overset{\circ}{U} > 0$ to the macroproblem and the corresponding $\overset{\circ}{z}$. The disaggregated solution $(\overset{\circ}{z},\, \overset{\circ}{u},\, \overset{\circ}{\alpha}\overset{\circ}{U})$ is optimal for the original problem (8.12)–(8.16) if and only if the unique optimal solution $\overset{\circ}{y}$, $\overset{\circ}{\delta}$ to the dual problem (8.19) is such that

$$(\partial/\partial u)\Big[c(s,\, t,\, \overset{\circ}{u},\, \overset{\circ}{z}) + \sum_l b_l(s,\, t,\, \overset{\circ}{u})\overset{\circ}{y}_l(s,t) - \\ - d(s,\, t,\, \overset{\circ}{u},\, \overset{\circ}{z})\overset{\circ}{\delta}(s,t)\Big] \le 0, \quad s,t \in \Sigma. \tag{8.20}$$

This is proved similarly to the corresponding assertion in Section 2.

In what follows, we assume that the solutions to the macroproblem and its dual are unique at each $\alpha \in A$ and bounded uniformly with respect to $\alpha \in A$; we also assume that $\overset{\circ}{U} > 0$.

Let us denote the optimal value of the objective functional of the macroproblem by $\theta(\alpha(\cdot))$ and consider the problem

$$\theta(\alpha(\cdot)) \to \max, \quad \alpha(\cdot) \in A. \tag{8.21}$$

By virtue of the assumptions made above about the Lipschitz continuity of the derivatives and convexity of the input functions, the marginal-value theorem [27] is valid; according to this theorem, the functional $\theta(\alpha(\cdot))$ is Fréchet differentiable. Its derivative $\nu^{ij}_{\overset{\circ}{\alpha}}(s,t)$ with respect to the parameter $\alpha^i_j(s,t)$ at a point $\overset{\circ}{\alpha} \in A$ is obtained by differentiating the Lagrange functional, which gives

$$\nu^{ij}_{\overset{\circ}{\alpha}}(s,t) = \overset{\circ}{U}{}^i(s,t)(\partial/\partial u^i_j)\Big[c(s,\, t,\, \overset{\circ}{u},\, \overset{\circ}{z}) + \\ + \sum_l b_l(s,\, t,\, \overset{\circ}{u})\overset{\circ}{y}_l - d(s,\, t,\, \overset{\circ}{u},\, \overset{\circ}{z})\overset{\circ}{\delta}\Big], \quad s,t \in \Sigma. \tag{8.22}$$

Using expression (8.22) for the derivative and the optimality criterion for the disaggregated solution in the same way as in Section 4, we conclude that, if $\overset{\circ}{\alpha}$ is a stationary point of problem (8.21) and $\overset{\circ}{U}$ is the corresponding macrocontrol, then $\overset{\circ}{u} = \overset{\circ}{\alpha}\overset{\circ}{U}$ is an optimal control in the original problem. Therefore, as above, to iteratively recalculate α, we can apply the first-order methods [56], and the problem of determining an ascent direction for $\theta(\alpha(\cdot))$ decomposes into I independent subproblems.

The example given below can be used to analytically examine all constructions involved in the iterative aggregation method.

Consider the problem of optimally controlling the longitudinal vibrations of a system of two one-sided rods of length L with left ends fastened and right ends subjected to forces acting in one direction. On the sum of the forces, a constraint is imposed. The initial deviations and velocities are zero. The problem is to maximize the integral sum of the rod deviations with certain coefficients at a given time T. We have

$$\begin{aligned}
&\partial^2 z_1/\partial t^2 - \partial^2 z_1/\partial x^2 = 0, \quad \partial^2 z_2/\partial t^2 - \partial^2 z_2/\partial x^2 = 0,\\
&z_1(0,x) = z_{1t}(0,x) = 0, \quad z_2(0,x) = z_{2t}(0,x) = 0, \quad x \in [0,L],\\
&z_1(t,0) = 0, \quad z_{1x}(t,L) = a_{11}u_1 + a_{12}u_2,\\
&z_2(t,0) = 0, \quad z_{2x}(t,L) = a_{21}u_1(t) + a_{22}u_2(t), \quad t \in [0,T],\\
&u_1(t) \geq 0, \quad u_2(t) \geq 0, \quad u_1(t) + u_2(t) \leq m,\\
&F = \gamma_1 \int_0^L z_1(T,\xi)d\xi + \gamma_2 \int_0^L z_2(T,\xi)d\xi \to \max;
\end{aligned} \tag{8.23}$$

the constants a_{ij}, γ_i, and m are strictly positive.

The problem with aggregated controls differs from (8.23) in the relations

$$\begin{aligned}
&z_{1x}(t,L) = U(t)\big[\alpha_1(t)a_{11} + \alpha_2(t)a_{12}\big],\\
&z_{2x}(t,L) = U(t)\big[\alpha_1(t)a_{21} + \alpha_2(t)a_{22}\big], \quad 0 \leq U(t) \leq m.
\end{aligned}$$

The dual to the aggregated problem has the form

$$\begin{aligned}
&\partial^2 y_1/\partial t^2 - \partial^2 y_1/\partial x^2 = 0, \quad \partial^2 y_2/\partial t^2 - \partial^2 y_2/\partial x^2 = 0,\\
&y_1(T,x) = 0, \quad y_{1t}(T,x) = -\gamma_1, \quad y_2(T,x) = 0,\\
&y_{2t}(T,x) = -\gamma_2, \quad x \in [0,L],\\
&y_1(t,0) = 0, \quad y_{1x}(t,L) = 0, \quad y_2(t,0) = 0,\\
&y_{2x}(t,L) = 0, \quad t \in [0,T],\\
&(\partial/\partial U)\Big\{y_1(t,L)U(\alpha_1 a_{11} + \alpha_2 a_{12}) +\\
&\qquad + y_2(t,L)U(\alpha_1 a_{21} + \alpha_2 a_{22}) - \mathring{\delta} U\Big\} \leq 0.\\
&\int_0^T m\delta(t)\,dt \to \min.
\end{aligned} \tag{8.24}$$

Solving the mixed problems for the first five relations, we obtain $\mathring{y}_1(t,L) = \gamma_1(T-t)$ and $\mathring{y}_2(t,L) = \gamma_2(T-t)$ for each pair in (8.24). Let us specify aggregation parameters $\overset{\circ}{\alpha}_1 \geq 0$ and $\overset{\circ}{\alpha} \geq 0$ such that they are constant on the interval $[0, T]$ and $\overset{\circ}{\alpha}_1 + \overset{\circ}{\alpha}_2 = 1$. The solution to the aggregated problem is $\overset{\circ}{U}(t) = m$. Since $\overset{\circ}{U}(t) > 0$, the last relation in the dual problem (8.24) becomes an equality, whence

$$\delta(t) = (T - t)\big[\gamma_1(\overset{\circ}{\alpha}_1 a_{11} + \overset{\circ}{\alpha}_2 a_{12}) + \gamma_2(\overset{\circ}{\alpha}_1 a_{21} + \overset{\circ}{\alpha}_2 a_{22})\big]. \tag{8.25}$$

The components of the gradient $\theta'(\overset{\circ}{\alpha})$ are evaluated by (8.20) as

$$\begin{aligned}\frac{\partial \theta(\overset{\circ}{\alpha})}{\partial \alpha_1} &= m\frac{\partial}{\partial u_1}\Big[(a_{11}u_1 + a_{12}u_2)\overset{\circ}{y}_1(t, L) + (a_{21}u_1 + a_{22}u_2)\overset{\circ}{y}_2(t, L) - \\ &\quad - (u_1 + u_2)\overset{\circ}{\delta}(t)\Big] = m(T - t)(A_1 - \overset{\circ}{\alpha}_1 A_1 - \overset{\circ}{\alpha}_2 A_2), \\ \frac{\partial \theta(\overset{\circ}{\alpha})}{\partial \alpha_2} &= m(T - t)(A_2 - \overset{\circ}{\alpha}_1 A_1 - \overset{\circ}{\alpha}_2 A_2),\end{aligned}$$

where $A_1 = \gamma_1 a_{11} + \gamma_2 a_{21}$ and $A_2 = \gamma_1 a_{12} + \gamma_2 a_{22}$.

To solve the problem of type (8.21), we apply the conditional gradient method [56]. In the case under consideration, the auxiliary problem of determining an ascent direction l has the form

$$\begin{gathered} m\int_0^T (T - t)\Big[A_1 - \overset{\circ}{\alpha}_1 A_1 - \overset{\circ}{\alpha}_2 A_2)l_1 + \\ +(A_2 - \overset{\circ}{\alpha}_1 A_1 - \overset{\circ}{\alpha}_2 A_2)l_2\Big]dt \to \max, \\ l_1(t) \geq 0, \quad l_2(t) \geq 0, \quad l_1(t) + l_2(t) = 1. \end{gathered} \tag{8.26}$$

To be definite, we suppose that $A_2 > A_1$. Then the solution to (8.26) is $\overset{\circ}{l}_1 = 0$, $\overset{\circ}{l}_2 = 0$, because $A_1 - \overset{\circ}{\alpha}_1 A_1 - \overset{\circ}{\alpha}_2 A_2 < 0$ and $A_2 - \overset{\circ}{\alpha}_1 A_1 - \overset{\circ}{\alpha}_2 A_2 > 0$.

The new values of the aggregation parameters are determined by $\alpha_i = \overset{\circ}{\alpha}_i + \rho(\overset{\circ}{l}_i - \overset{\circ}{\alpha}_i)$ for $i = 1, 2$, where $0 \leq \rho \leq 1$ is the stepsize. Substituting the new parameter values in the objective functional of problem (8.24) and taking into account expressions (8.25) for $\delta(t)$, we obtain the dependence

$$\theta(\rho) = \int_0^T m(T - t)\Big[A_1\overset{\circ}{\alpha}_1(1 - \rho) + A_2(\overset{\circ}{\alpha}_2 + \rho(1 - \overset{\circ}{\alpha}_2))\Big]dt.$$

The maximum of $\theta(\rho)$ over $[0, 1]$ is attained at $\rho = 1$, independently of the choice of α_1 and α_2. Thus, $\alpha_1 = 0$ and $\alpha_2 = 1$. It is easy to see that, for these values of the α parameters, the optimality criterion stated above is satisfied. Therefore, $u_1^*(t)$ and $u_2^*(t)$ are optimal controls in the initial problem.

§9. Numerical Indications

This section presents the results of numerical calculations by the iterative aggregation method for some model optimal control problems. They confirm the convergence and efficiency of the iterative algorithm. Several iterations yield good approximations to the optimum.

Consider the problem of monitoring a thermal process in a vertical furnace [4, 23]. The furnace has 12 heating zones equidistant in the vertical direction. It is required to achieve a given temperature distribution for a vertical object placed in the furnace. The heat fluxes supplied to each heating zone serve as controls. The twelve subsystems corresponding to the heating zones have cross connections because of the physical processes inside the furnace, largely, convection [23]. The object temperature is measured at 12 equidistant points.

In [23], the following approach (called feedback with interruption) to real-time control of the heating process is suggested. The control time is divided into intervals of length Δ, during which optimal controls for the period $[k\Delta, k\Delta + T]$ are calculated; here, $k\Delta$ is the current time moment corresponding to the start of the current calculation and T is the length of the interval under control. The initial state $y(k\Delta)$ is determined by the real process for each calculation. The obtained control law is applied in real time during period Δ, after which the procedure is repeated. The scheme of such a control is shown in Fig. 9.1. This approach imposes more severe requirements on the speed of the algorithms used to construct the optimal controls.

The simplest linearized model of the process is considered in [131]. We have

$$\begin{aligned} &dy(t)/dt = Ay(t) + Bu(t), \quad y(0) = y_0, \quad t \in [0, T], \\ &z(t) = Cy(t). \end{aligned} \tag{9.1}$$

Here, $y(t)$ is the state vector, $u(t)$ is the control vector, and $z(t)$ is the output vector (of the temperatures at the control points). Their dimensions are 24, 12, and 12, respectively.

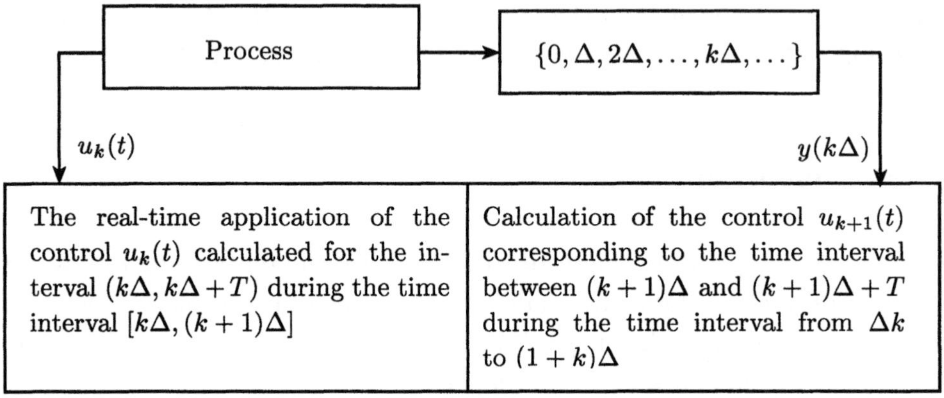

Fig. 9.1.

The matrix A has the form

$$A = \begin{bmatrix} D_1 & S+P & \varepsilon S & \varepsilon^2 S & \dots & \varepsilon^9 S & \varepsilon^{10} S \\ \theta & D & S+P & \varepsilon S & \dots & \varepsilon^8 S & \varepsilon^9 S \\ 0 & \theta & D & S+P & \dots & \varepsilon^7 S & \varepsilon^8 S \\ \dots & \dots & \dots & \dots & \dots & \dots & \dots \\ & & & & \dots & S+P & \varepsilon S \\ & & & & \dots & D & S+P \\ & & & & \dots & \theta & D_{12} \end{bmatrix},$$

where D_1, D_{12}, S, D, and P are 2×2 square matrices and $\varepsilon \in (0,1)$ is a scalar parameter characterizing the cross connections.

The temperatures of the object at the control points at the final moment of time T are subject to the constraint

$$\big(z_i(T) - z_{id}\big)^2 \leq \delta, \quad i = 1, 2, \dots, 12, \tag{9.2}$$

where z_{id} is the preset temperature at the i-th control point at the final time and $\delta > 0$ is a sufficiently small number.

Consider the problem of minimizing the quadratic integral functional

$$J(u(\cdot)) = \frac{\alpha}{\|z_d\|^2} \int_0^T \sum_i (z_i(t) - z_{id})^2 dt + \frac{\beta}{\|u_d\|^2} \int_0^T \sum_i (u_i(t) - u_{id})^2 dt. \tag{9.3}$$

Here, $\alpha \geq 0$ and $\beta \geq 0$ are dimensionless scalar coefficients and $u_d = \{u_{id}\}$ is a control that yields the required temperature $z_d = \{z_{id}\}$ asymptotically as $T \to \infty$. In [23], it is shown that $u_d = -(CA^{-1}B)^{-1}z_d$. The first term in (9.3) serves as the integral measure of the accuracy with which the required temperature of the object is achieved, and the second term characterizes the energy expenditures necessary to achieve the required temperature. The second term in (9.3), which is quadratic with respect to controls, can be replaced, for example, by the expression

$$\int_0^T \sum_i \varphi_i[u_i(t) - u_{id}]dt,$$

where $\varphi_i(\cdot)$ is a sufficiently smooth strictly convex positive definite scalar function.

To problem (9.1)–(9.3), we apply the procedure of aggregation of controls described in Section 6. Because the problem possesses the special feature of having no mixed constraints, we successively apply two types of control aggregation, in space and time. First, we aggregate the controls corresponding to different heating zones, and then, we aggregate the obtained macrocontrol along the time interval.

We divide the interval $[0, T]$ into L equal subintervals $[t_{j-1}, t_j]$ of length $h = T/L$ $(j = 1, 2, \ldots, L)$; $t_0 = 0$, $t_L = T$. Next, we introduce scalar aggregated controls U_j $(j = 1, 2, \ldots, L)$ and aggregation parameters $\alpha_0(t)$, $\alpha_i(t)$ $(i = 1, 2, \ldots, 12)$ satisfying the conditions

$$\int_{t_{j-1}}^{t_j} \alpha_0(t)dt = 0, \quad j = 1, 2, \ldots, L, \quad \sum_{i=1}^{12} \alpha_i(t) = 0, \quad t \in [0, T]. \tag{9.4}$$

The parameters $\alpha_i(t)$ and $\alpha_0(t)$ correspond to the space and time aggregations, respectively.

The disaggregated controls are specified in the form

$$u_i(t) = U_j/(12h) + \alpha_0(t) + \alpha_i(t), \quad t \in [t_{j-1}, t_j]. \tag{9.5}$$

According to (9.4) and (9.5), the aggregated and original controls are related by

$$U_j = \int_{t_{j-1}}^{t_j} \sum_i u_i(t)dt, \quad j = 1, 2, \ldots, L. \tag{9.6}$$

The macrocontrol U_j is the total (over all heating zones) heat supplied to the furnace during time $t_j - t_{j-1}$.

The aggregated problem is obtained by substituting (9.5) in the initial problem (9.1)–(9.3) at fixed values $\mathring{\alpha}_0(t)$ and $\mathring{\alpha}_i(t)$ ($i = 1, 2, \ldots, 12$) satisfying (9.4). The macroproblem obtained in this way contains L scalar controls U_j. The objective functional θ of the macroproblem depends on the chosen values of the aggregation parameters, i.e., $\theta = \theta(\alpha_0, \alpha_1, \ldots, \alpha_{12})$. To minimize the functional $\theta(\cdot)$ over the set A defined by (9.4), we apply the projected gradient methods. Constraints (9.4) split into $L+1$ independent constraints. As in Section 6, the gradient of the functional $\theta(\cdot)$ at a fixed point $(\mathring{\alpha}_0, \mathring{\alpha}_1, \ldots, \mathring{\alpha}_{12})$ is evaluated by solving the macroproblem and its dual at the corresponding aggregation parameter values. Thus, solving the original problem reduces to iteratively solving the aggregated problem with scalar controls U_j and recalculating the aggregation parameters.

Problem (9.1)–(9.3) was numerically solved for the following values of the elements specifying the matrix A [131]:

$$D_1 = \begin{bmatrix} -0.38 & 0.196 \\ 0.0068 & -0.0629 \end{bmatrix}, \quad D = \begin{bmatrix} -0.39 & 0.196 \\ 0.0068 & -0.0682 \end{bmatrix},$$

$$D_{12} = \begin{bmatrix} -0.44 & 0.196 \\ 0.0068 & -0.0559 \end{bmatrix}, \quad S = \begin{bmatrix} 0.0057 & 0.012 \\ 0.00030 & -0.0008 \end{bmatrix},$$

$$P = \begin{bmatrix} 0.0651 & 0 \\ 0 & -0.0031 \end{bmatrix}.$$

The elements b_{lk} and c_{lk} of the matrices B and C were

$$b_{lk} = \begin{cases} 0.00195 & \text{if } l \text{ is even and } k = l/2, \\ 0 & \text{otherwise,} \end{cases}$$

$$c_{lk} = \begin{cases} 1 & \text{if } l = 1, 2, \ldots, 24 \text{ and } k = 2l-1, \\ 0 & \text{otherwise.} \end{cases}$$

All components of the vector z_d were zero; the other values were $y_0 = 0$, $\varepsilon = 0.67$, and $\delta = 0.01$.

Aggregation (9.6) and normalization (9.4) with $L = 12$ were used. The problem with aggregated controls was discretized on a uniform mesh with 25 points. The macroproblem and its dual were solved with the use of modified Lagrange functions.

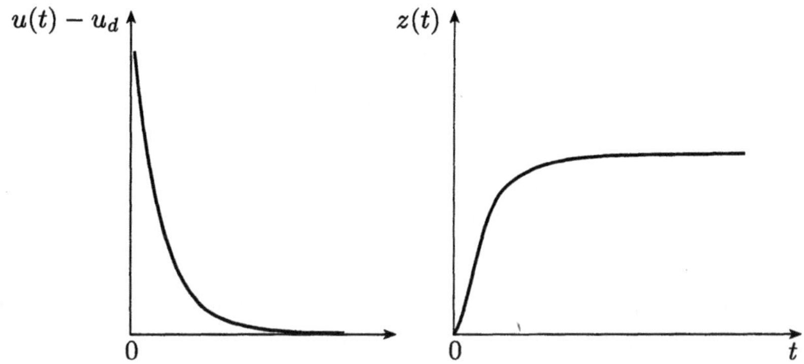

Fig. 9.2.

Calculations with these input data were performed on an IBM PC-XT with the INTEL 8088 processor. The computation terminated when the ratio of the norms of the differences between the aggregation parameters in consequent iterations did not exceed 10^{-3}. The solution was obtained in 3–5 iterations with respect to the aggregation parameters; each iteration step was calculated no longer than 4 min. It should be mentioned that the computation time virtually does not depend on the ratio α/β and is largely determined by δ, i.e., by the accuracy with which the terminal constraints are fulfilled. Figure 9.2 shows typical time dependences of the optimal controls and the temperature of the object.

Another series of experiments was concerned with a problem related to that of hierarchical control of a distributed-parameter system [25, 52].

Consider a system of I homogeneous thin rods of length $l = 1$. The temperature distribution $z_i(x,t)$ of the i-th rod, where x is the coordinate along the rod axis and t is time, satisfies the equation

$$\partial z_i(x,t)/\partial t = \partial^2 z_i(x,t)/\partial x^2, \quad t > 0. \quad 0 < x < 1 \tag{9.7}$$

with the initial conditions

$$z_i(x,0) = 0, \quad 0 \le x \le 1 \tag{9.8}$$

and the boundary conditions

$$\partial z_i(1,t)/\partial x = a_i\left(\sum_j b_{ij}u_j(t) - z_i(1,t)\right), \tag{9.9}$$

$$\partial z_i(0,t)/\partial x = 0, \quad t \ge 0. \tag{9.10}$$

The i-th rod is heated by a separate heater, which provides the temperature $u_i(t)$ of the heating medium at the right end of the rod. According to equation (9.8), there is no heat flux at the left end of the rod, and the heat flux at the right end obeys the Newton law (9.10). In the hierarchical system under consideration, the rods are subsystems. The cross connections between the subsystems arise because the heat supplied to one rod partially transfers to the other rods. The values b_{ij} $(i, j = 1, 2, \dots, I)$ are the cross-coupling coefficients; they depend on the mutual arrangement of the rods and are assumed to be constant in the case under consideration.

We assume that the controls $u_i(t)$ are proportional to the amount of resources, say, fuel, supplied to the i-th heater. In addition to the cross connections specified by (9.9), there is the following constraint on the total consumption of the resources, which combines the rods in one system:

$$\sum_i u_i(t) \leq w, \tag{9.11}$$

where w is a given constant.

Given a finite time interval $[0, T]$, it is required to minimize the functional

$$J(u(\cdot)) = 0.5 \sum_{i=1}^{I} \left[\int_0^1 \sigma_i z_i(x, T)^2 dx + \gamma_1 \int_0^T u_i^2(t) dt \right], \tag{9.12}$$

where σ_i and γ_i are constant coefficients. We seek a solution to problem (9.7)–(9.12) in the form of a finite Fourier series

$$z_i(x, t) = \sum_n z_i^n(t) \cos \lambda_j^n x, \quad i = 1, 2, \dots, I,$$

where λ_i^n is the n-th positive root of the equation $\lambda_i \operatorname{tg} \lambda_i = a_i$. The original problem then reduces to the following problem of optimal control of a lumped-parameter system:

$$\begin{gathered} 0.5 \sum_{i=1}^{I} \left[\gamma_i \int_0^T u_i^2(t) dt + \sum_{n=0}^{N} c_i^n (z_i^n(T))^2 \right] \to \min, \\ \dot{z}_i^n = -d_i^n z_i^n + \sum_{n=1}^{N} b_{ij}^n u_j, \quad z_i^n(0) = 0, \\ n = 0, 1, \dots, N, \quad i = 1, 2, \dots, I, \\ \sum_i u_i(t) \leq w. \end{gathered} \tag{9.13}$$

The constant coefficients $d_i^n > 0$, b_{ij}^n, and c_i^n are obtained by applying the Fourier method to the original problem (see, e.g., [25, 53]).

Problem (9.13) contains I controls and $I(N+1)$ state variables. Directly solving (9.13) by the maximum principle yields a system of linear algebraic equations of order $I(N+1)$ [112], which results in the shortage of computer memory at large I and N. When $a_i = a$ for $i = 1, 2, \ldots, I$ and the matrix $[b_{ij}]$ is diagonal, the dimension of the system is lowered by the reduction method described in [51, 53]. If equation (9.9) has no cross connections, i.e., the matrix $[b_{ij}]$ is diagonal but a_i are different, then problem (9.13) has block-separable structure, and we can apply the decomposition method based on the aggregation of variables from different blocks [53]. If constraint (9.11) is an equality, then the dimension can sometimes be reduced at the expense of introduction of special structural parameters in the system of necessary optimality conditions. The choice of the structural parameters is a separate fairly complicated problem [52].

Problem (9.13) was numerically solved by the method of aggregating controls described in Section 6. One macrocontrol equal to the sum of all controls was introduced. The calculations were performed for $I(N+1) = 180$. The coefficients in the conditions of the problem varied in the following ranges: $d_i^n \in (0,1]$, $b_{ij}^n \in [-1,1]$, $\gamma_i \in [0,1]$, $c_i^n \in [0,1]$, and $w \in [1,24]$.

The parameters were updated by the gradient projection method. In the numerical implementation of the algorithm, the changes in the relative variations $\Delta J_l/\Delta J_{l+1}$ (of the functional) and $||\Delta\alpha_l||/||\Delta\alpha_{l+1}||$ (of the aggregation parameters) from the l-th to the $(l+1)$-th iteration were examined. The rate of change in these characteristics and the approximation of the functional to the optimum were of the same order in virtually all examples. After the first three or four iterations, the relative variations were

$$\Delta J_l/\Delta J_{l+1} \sim 10^{-2} \div 10^{-3}, \quad ||\Delta\alpha_l||/||\Delta\alpha_{l+1}|| \sim 10^{-1} \div 10^{-2}$$

for the optimal values $J_{\text{opt}} \sim 10^4 \div 10^5$ and $||\Delta\alpha_{\text{opt}}|| \sim 10$. The calculations were performed on a BESM-6 computer; each iteration step was calculated no longer than in 1 min.

In the examples considered above, the controls are linear variables in the dynamic equations of the system. Now, consider a problem where the right-hand sides of differential equations depend on controls nonlinearly. The statement of the initial problem is based on the model of noise optimization in dynamical systems in the presence of random perturbations studied in [2]. We have

$$\begin{gathered}
\sum_{j=1}^{n} \sigma_j x_j(T) + \int_0^T \sum_{j=1}^{n} \gamma_j x_j^2(t)dt + \int_0^T \sum_{j=1}^{n} \beta_j u_j^2(t)dt \to \min, \\
dx_j(t)/dt = \left(\sum_{j=1}^{n} a_{ij} u_i(t) \right)^{-1}, \quad x_j(0) = x_{0j}, \quad j = 1, 2, \ldots, n, \\
\sum_j u_j(t) \le w(t), \quad u_j(t) \ge 0, \quad j = 1, 2, \ldots, n.
\end{gathered} \tag{9.14}$$

Here, T is a fixed final time; $w(t)$ is a given function of time, positive on $[0, T]$; σ_j, γ_j, β_j, and a_{ij} are positive constant coefficients; and the initial values are nonnegative.

In solving problem (9.14) by the iterative aggregation method, we introduce one macrocontrol $U(t) = \sum\limits_{j}^{n} u_j(t)$ and specify the set of aggregation parameters by the condition

$$A = \left\{ a_j(t) \colon \sum_j \alpha_j(t) = 1, \ \alpha_j(t) \ge 0, \quad j = 1, 2, \ldots, n \right\}.$$

The macrocontrol in the aggregated problem satisfies the inequalities $0 \le U(t) \le w(t)$; for this reason, we apply the gradient projection method to solve the macroproblem. Since the coefficients γ_j and a_{ij} are positive, the optimal solution $\overset{\circ}{U}(t)$ of the aggregated problem is positive for any $\overset{\circ}{\alpha}(\cdot) \in A$. Therefore, the constraint of the dual macroproblem is fulfilled as an equality and determines the Lagrange multiplier for the condition $U(t) \le w(t)$. The problem of determining a descent direction in the space of aggregation parameters is solved by the gradient projection method with the use of a special projection algorithm, which takes into account that the set under consideration is a simplex with time-independent vertices [52].

Problem (9.14) was numerically solved for $n = 10$, $T = 1$, $x_{0j} = 0$, $\sigma_j = 0.5$ $(j = 1, 2, \ldots, 10)$, and $w(t) = \exp(t)$. The integer coefficients β_j and γ_j and the elements of the matrix were randomly chosen in the interval $[1, 9]$. The macroproblem was discretized on a uniform mesh with 51 points. Computations by the method of iterative aggregation terminated if

$$\max_t \max_j \left| \alpha_i^l(t) - \alpha_i^{l+1} \right| < 10^{-3},$$

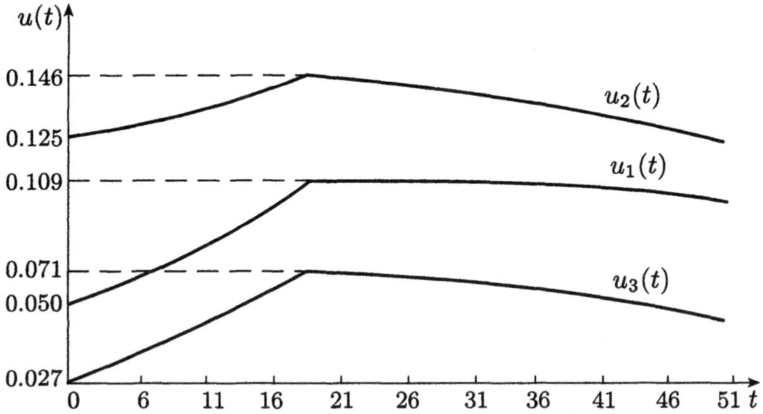

Fig. 9.3.

where l is the number of iteration with respect to the aggregation parameters. Below, the input data for one of the test examples are given:

Matrix $[a_{ij}]$:

$$\begin{bmatrix} 1 & 8 & 4 & 1 & 5 & 5 & 4 & 6 & 5 & 1 \\ 5 & 2 & 5 & 5 & 4 & 8 & 6 & 3 & 5 & 2 \\ 6 & 5 & 4 & 1 & 4 & 8 & 8 & 5 & 8 & 1 \\ 2 & 6 & 2 & 6 & 6 & 2 & 8 & 8 & 8 & 8 \\ 6 & 8 & 2 & 4 & 8 & 8 & 5 & 5 & 3 & 4 \\ 5 & 4 & 4 & 1 & 1 & 6 & 8 & 5 & 3 & 3 \\ 5 & 7 & 4 & 8 & 8 & 8 & 3 & 2 & 8 & 3 \\ 6 & 5 & 2 & 1 & 1 & 5 & 6 & 8 & 6 & 2 \\ 1 & 5 & 5 & 1 & 3 & 8 & 5 & 8 & 3 & 8 \\ 3 & 3 & 1 & 7 & 2 & 6 & 4 & 6 & 3 & 2 \end{bmatrix}$$

Vector β:

$$\begin{bmatrix} 4 & 4 & 6 & 3 & 2 & 4 & 8 & 2 & 6 & 5 \end{bmatrix}$$

Vector γ:

$$\begin{bmatrix} 7 & 7 & 8 & 6 & 3 & 1 & 3 & 2 & 7 & 1 \end{bmatrix}$$

The initial approximations for the aggregation parameters were set constant on the interval $[0, 1]$:

$$\alpha_1(t) = 1, \quad \alpha_j(t) = 0 \quad (j = 2, 3, \dots, 10).$$

The termination condition was fulfilled after nine iteration steps with respect to the aggregation parameters. Figure 9.3 shows the corresponding dependences for some disaggregated controls.

Figure 9.4 illustrates the time dependence of the discrepancy of the constraint $U(t) \leq w(t)$.

In Fig. 9.5, the decrease in the objective functional of the aggregated problem with increasing the number of iterations is shown.

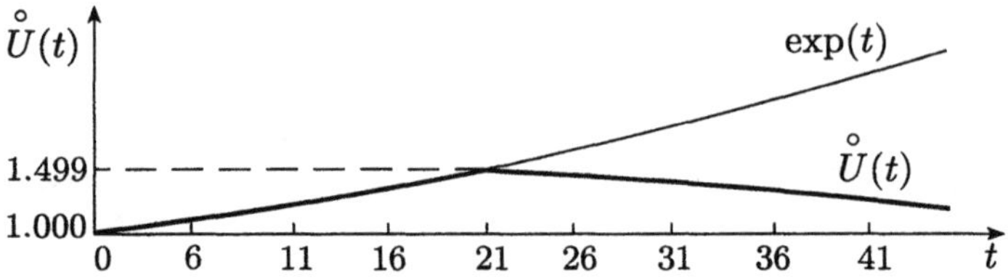

Fig. 9.4.

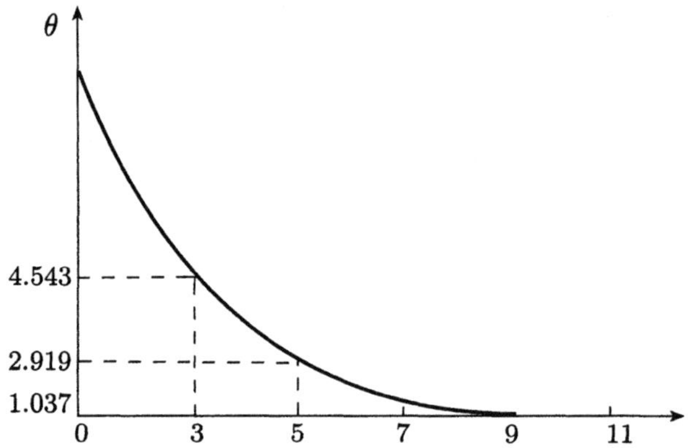

Fig. 9.5.

The numerical calculations were performed on an IBM PC-XT computer; the algorithm was implemented in Pascal, version TURBO-7. In all test examples, the total computation time did not exceed 20 min.

The results presented above are evidence of a fairly high efficiency of the iterative aggregation method.

Comments and References to Chapter 2

The earlier results on iterative aggregation for large-scale programs are summarized in the book by Dudkin L.M., Rabinovich I. and Vakhutinsky I. [5]. They prove a convergence result by showing that the Jacobian matrix of their iteration step describes a contract mapping. Petr C. and Craven B.D. [43] approach iterative aggregation from the standpoint of augmented Lagrangians and the Powell-Hestenes-Rokafellar algorithms for constrained optimization. Various weights updating schemes based on gradient considerations are studied by Garcia L. [8, 9], Jörnsten K., Leisten R. and Storøy S. [17], and Storøy S. [47] for the linear programming problems. Heuristic decomposition-aggregation techniques for linear block problems, their interpretations for hierarchical planning structures, and comparisons with classical decomposition schemes one can find in Jörnsten K. and Leisten R. [15, 16, 8]. The approach of iterative aggregation of variables entering the same blocks is presented in Mednitskii V.G. [42] for linear programs, while aggregation of variables from different blocks is considered in the monographs by Tsurkov V.I. [53, 54] for various classes of extremal problems. Planning aspects of iterative aggregation are broadly covered in the book by Leisten R. [26] and in the cited book by Dudkin L.M., Rabinovich I. and Vakhutinsky I. [5]. The results presented in Sections 1–3 are based on the papers by Litvinchev I.S. [29, 30, 34, 37–39], and Litvinchev I.S., Silva G.N. and Treskov Yu.P. [41]. Iterative clustering approach of Section 4 is considered in the works by Aboudi R., Jörnsten K. and Leisten R. [1], and Jörnsten K., Leisten R. and Storøy S. [17]. Constraint aggregation techniques discussed in Section 5 can be found in the book by Tsurkov V.I. and Litvinchev I.S. [52], and in the paper by Ermoliev Yu.M., Kryazhimskii A.V. and Ruszczyński A. [6]. Applications of iterative aggregation to dynamical and distributed parameter systems, studied in Sections 6-9, as well as various numerical examples, one can find in Litvinchev I.S. [31–33, 35, 36, 38, 40], and in the cited book by Tsurkov V.I. and Litvinchev I.S. [52]. In particular, the linear aggregation presented in Section 1, is generalized to optimal control problems in the paper by Litvinchev I.S. [38].

References to Chapter 2

[1] Aboudi R., Jörnsten K. and Leisten R., Solving Large–Scale Linear Programs with Bounded Variables by Iterative Aggregation, Working Paper, University of Miami, Miami, USA, 1993.

[2] Chernous'ko F.L. and Kolmanovskii V.V., *Optimal'noe upravlenie pri sluchainykh vozmushcheniyakh* (Optimal Control under Random Perturbations), Moscow: Nauka, 1978.

[3] Dantzig G.B. and Wolfe P., Decomposition Principle for Linear Programs, *Oper. Res.*, 1961, vol. 8, no. 1, pp. 101–111.

[4] Drouin M. and Bertrand P., A New Coordination Structure for On-Line Control of Complex Processes, *Large-Scale Systems*, 1982, vol. 3, no. 3, pp. 147–157.

[5] Dudkin L.M., Rabinovich I. and Vakhutinsky I., Iterative Aggregation Theory, New York-Basel: Marcel Dekker, 1987.

[6] Ermoliev Yu.M., Kryazhimskii A.V. and Ruszczyński A., Constraint Aggregation Principle in Convex Optimization, *Math. Progr.*, 1997, vol. 76, pp. 353–372.

[7] Fedorov V.V. *Chislennye metody maxmina* (Numerical Methods of Maximin), Moscow: Nauka, 1979.

[8] Garcia L., A Global Convergence Theorem for Aggregation Algorithms, *Optimization,* 1988, vol. 19, no. 6, pp. 819–828.

[9] Garcia L., An Algorithm of Feasible Directions for the Aggregation Problem, *Optimization,* 1989, vol. 20, no. 2, pp. 193–208.

[10] Gol'shtein E.G., *Teorija dvoistvennosti v matematicheskom programmirovanii i eyo prilozhenija* (Duality Theory in Mathematical Programming and Its Applications), Moscow: Nauka, 1971.

[11] Holmberg K., Linear Mean Value Cross Decomposition: A Generalization of the Kornai-Liptak Method, *Eur. J. Oper. Res.*, 1992, vol. 62, pp. 55–73.

[12] Holmberg K., A Convergence Proof for Linear Mean Value Cross Decomposition, *Zeitschrift fur Oper. Res.*, 1994, vol. 39, pp. 157–186.

[13] Holmberg K., Primal and Dual Decomposition as Organizational Design: Price and/or Resource Directive Decomposition, in Design Models for Hierarchical Organizations: Computation, Information, and Decentralization, eds. Burton R.M. and Obel B., Boston: Kluwer, 1995, pp. 61–92.

[14] Ioffe A.D. and Tikhomirov V.M., *Teorija extremal'nykh zadach* (Extremal Problems Theory), Moscow: Nauka, 1974.

[15] Jörnsten K. and Leisten R., Column Aggregation and Primal Decomposition in Linear Programming: Some Observations, *Optimization*, 1992, vol. 24, no. 2, pp. 193–208.

[16] Jörnsten K. and Leisten R., Aggregation and Decomposition for Multi-Divisional Linear Programs, *Eur. J. Oper. Res.*, 1994, vol. 72, pp. 175–191.

[17] Jörnsten K., Leisten R. and Storøy S., Gradient Schemes in Iterative Aggregation Procedures for Variable-Aggregated LP-Problems, *Optimization*, 1994, vol. 30, pp. 251–268.

[18] Jörnsten K. and Leisten R., Decomposition and Iterative Aggregation in Hierarchical and Decentralized Planning Structures, *Eur. J. Oper. Res.,* 1995, vol. 86, pp. 120–141.

[19] Jörnsten K., Leisten R. and Storøy S., Convergence Aspects of Adaptive Clustering in Variable Aggregation, *Comput. & Oper. Res.*, 1999, vol. 26, no. 10–11, pp. 955–966.

[20] Karmanov V.G., *Matematicheskoe programmirovanie* (Mathematical Programming), Moscow: Nauka, 1986.

[21] Kornai I. and Liptak T., Two Level Planning, *Econometrica*, 1965, vol. 33, no. 1, pp. 141–169.

[22] Kryazhimskii A. and Ruszczyński A., Constraint Aggregation in Infinite-Dimensional Spaces and Applications, *Math. Oper. Res.*, 2001, vol. 26, no. 4, pp. 769–795.

[23] Lang B., El-Awtani S., Spiteri P. and Cheikobeid N.E., Decentralized Calculations in Optimal Control of a Large Thermic Process, *Proc. IFAC Symp. Large-Scale Syst. Theory and Appl.*, Toulouse, France, 1980, London: Pergamon Press, pp. 505–516.

[24] Lasdon L., Optimization Theory for Large Systems, New York: MacMillan, 1970.

[25] Leonov V.Yu. and Tsurkov V.I., *Ierarkhicheskie sistemy s raspredelennymi parametrami i perekrestnymi svyazyami* (Hierarchical Distributed Parameter Interconnected Systems), *Avtom. Telemekh.*, 1988, no. 8, pp. 75–82.

[26] Leisten R., Iterative Aggregation und Mehrstufige Entscheidungsmodelle, Heidelberg: Physica-Verlag, 1995.

[27] Levitin E.S., *O differentsiryemosti po parametry optimal'nogo znachenija parametricheskikh zadach matematicheskogo programmirovaniya* (On Differentiability with Respect to Parameter of the Optimal Value in Parametric Problems of Mathematical Programming), *Kibernetika*, 1976, no. 1, pp. 44–59.

[28] Lions J.-L., *Optimal'noe upravlenie sistemami, opisyvaememi uravneniyami s chastnymi proizvodnymi* (Optimal Control in Systems Described by Partial Differential Equations), Moscow: Mir, 1972 [Russian translation].

[29] Litvinchev I.S., Method of Expansion for Optimization Problems which do not have a Block-Separable Structure, *USSR Comput. Maths. Math. Phys.,* 1987, vol. 27, no. 2, pp. 6–11.

[30] Litvinchev I.S., A Decomposition for Nonseparable Extremal Problems, *Soviet Math. Dokl.*, 1987, vol. 35, no. 1, pp. 25–28.

[31] Litvinchev I.S., An Iterative Decomposition in Dynamic Problems with Cross Couplings, *Soviet Math. Dokl.,* 1989, vol. 39, no. 3, pp. 578–580.

[32] Litvinchev I.S., Investigation of Multidimensional Problems in the Control of Heat Transfer, *Math. Model. Comput. Experiment,* 1993, vol. 1, no. 2, pp. 119–128.

[33] Litvinchev I.S., Decomposition in Control Systems of Hyperbolic Type with Cross Connections, *Soviet J. Comput. Syst. Scis.,* 1990, vol. 28, no. 3, pp. 74–81.

[34] Litvinchev I.S., Decomposition in Extremal Problems of Special Structure, *USSR Comput. Maths. Math. Phys.,* 1990, vol. 30, no. 7, pp. 1008–1016.

[35] Litvinchev I.S., Use of Structure and of Cross Connection Estimates in the Method of Decomposition of Optimal Control Problems, *Soviet J. Comput. Syst. Scis.,* 1991, vol. 29, no. 5, pp. 145–155.

[36] Litvinchev I.S., Iterative Decomposition of Optimization Problems with the First-Order Partial Differential Equations, *Cybernetics*, 1990, no. 6, pp. 869–876.

[37] Litvinchev I.S., Decomposition-Aggregation Method for Convex Programming Problems, *Optimization,* 1991, vol. 39, no. 1, pp. 47–56.

[38] Litvinchev I.S., Iterative Aggregation Method for Constrained Optimal Control Problems, *J. Comput. Appl. Maths.*, 1992, vol. 39, no. 3, pp. 315–328.

[39] Litvinchev I.S., Iterative Aggregation and Decomposition in Block-Separable Extremal Problems, *Comput. Maths. Math. Phys.*, 1992, vol. 32, no. 8, pp. 1179–1181.

[40] Litvinchev I.S., A Two-Level Decomposition-Aggregation Approach for Large-Scale Optimal Control Problems, *J. Comput. Appl. Maths.*, 1995, vol. 61, no. 2, pp. 117–137.

[41] Litvinchev I.S., Silva G.N. and Treskov Yu.P., Aggregation-Based Decomposition for Multi-Divisional Models, *J. Comput. Syst. Scis. Int.,* 1999, vol. 38, no. 2, pp. 244–254.

[42] Mednitskii V.G., *Ob optimal'nosti agregirovanija v blochnoy zadache lineinogo programmirovania* (On Optimality of Aggregation in the Block Problem of Linear Programming), in: Mathematical Methods for Solution of Economical Problems, 1972, vol. 3, pp. 3–17.

[43] Petr C. and Craven B.D., Mathematical Programming with Iterative Aggregation, 1991, *Opsearch,* vol. 28, no 1, pp. 1–17.

[44] Polyak B.T., *Shodimost' metodov vozmojnykh napravlenii v extremal'nykh zadachakh* (Convergence of Feasible Directions Methods in Extremal Problems), *Zh. Vych. Math. Mat. Fiz.*, 1971, vol. 11, no. 4, pp. 855–869.

[45] Pontryagin L.S., Boltyanskii V.G., Gamkrelidze R.V. and Mishchenko E.F., *Matematicheskaya teoriya optimal'nykh protsessov* (Mathematical Theory of Optimal Processes), Moscow: Nauka, 1969. Moscow: Mir, 1972 [English translation].

[46] Rozhdestvenskii B.L. and Yanenko N.N., *Sistemy kvazilineinykh uravnenii* (Systems of Quasilinear Equations), Moscow: Nauka, 1968.

[47] Storøy S. Weights Improvement in Column Aggregation, *Eur. J. Oper. Res.*, 1994, vol. 73, pp. 510–516.

[48] Ter-Krikorov A.M., *Ob odnoi zadache optimal'nogo upravleniya so smeshannymi ogranicheniyami dlya sistem s raspredelennymi parametrami* (On a Problem of Optimal Control with Mixed Constraints for Systems with Distributed Parameters), *Zh. Vych. Mat. Mat. Fiz.*, 1981, vol. 21, no. 3, pp. 561–571.

[49] Ter-Krikorov A.M., *Optimal'noe upravlenie i matematicheskaja economica* (Optimal Control and Mathematical Economics), Moscow: Nauka, 1977.

[50] Tikhonov A.N. and Samarskii A.A., *Uravneniya matematicheskoi fiziki* (Equations of Mathematical Physics), Moscow: Nauka, 1977.

[51] Tsurkov V.I., *Optimizatsiya mnogomernykh sistem s raspredelennymi parametrami* (Optimization of Multidimensional Distributed-Parameter Systems), *Izv. Akad. Nauk SSSR, Ser. Tekh. Kibern.*, 1981, no. 5, pp.13–21.

[52] Tsurkov V.I. and Litvinchev I.S., *Dekompozitsiya v dinamicheskih zadachakh s perekrestnymi svyazyami. Chast' 2* (Decomposition in Dynamical Problems with Cross Connections. Part 2), Moscow: Nauka, 1994.

[53] Tsurkov V.I., Hierarchical Optimization and Mathematical Physics, Dordrecht, Boston, London: Kluwer Acad. Publ., 2000.

[54] Tsurkov V.I., Large-Scale Optimization — Problems and Methods, Dordrecht, Boston, London: Kluwer Acad. Publ., 2001.

[55] Vakhutinskii I.Ya., Dudkin L.M. and Shchennikov B.A., *Iterativnoe agregirovanie v nekotorykh optimal'nykh ekonomicheskykh modelyakh* (Iterative Aggregation in Some Optimal Economic Models), *Ekonomika Mat. Metody*, 1973, vol. 9, no. 3, pp. 420–434.

[56] Vasil'ev F.P., *Metody reshenija extremal'nykh zadach* (Solution Methods for Extremal Problems), Moscow: Nauka, 1981.

[57] Vasil'ev F.P., *Chislennye metody reshenija extremal'nykh zadach* (Numerical Methods for Extremal Problems), Moscow: Nauka, 1988.

[58] Vladimirov V.S., *Uravneniya matematicheskoi fiziki* (Equations of Mathematical Physics), Moscow: Nauka, 1988.

Chapter 3

Consistent Aggregation in Parametric Optimization

Many applied optimization problems require analyzing the effect of changes in problems' data (parameters) on optimal function values. For example, this is the case when parameters vary over a given set, and optimization should be performed not only with respect to the original decision variables but also with respect to the parameters. The situation becomes even more complicated when the set of admissible parameters is corrected in the course of optimization.

Thus, the computational burden is associated not only with optimization itself but also with searching for the parameters that determine more preferable optimization problems, such as problems with better optimal objective values.

The approach used in this chapter is as follows. Suppose that a few parameters are allowed to change. The data associated with the unchanged parameters are aggregated in a special way. Then, the parametric aggregated problem is used to find the parameter changes resulting in an improvement of the original nonaggregated problem. Thus, the parametric analysis of the original problem is, in a certain sense, approximated by the analysis of the parametric aggregated problem.

We stress the difference between this approach and the parametric analysis based on marginal values [28]. The latter provides local characteristics, i.e. is valid for small changes of parameters, while the aggregation approach aims to analyze the original parametric problem globally, allowing large parameters variations.

§1. Consistency of Aggregation with Respect to Parameter Variations

We consider original parametric problem in the form

$$\min\left\{f(a,x) \mid x \in X(a) \subset L\right\} \tag{1.1}$$

over a set $\mho^0$ of feasible parameters a. Here, $X(a)$ is a subset of a linear space L of decision variables x and f is a scalar objective function.

The value function for problem (1.1) is defined by

$$\varphi(a) = \begin{cases} \min\{f(a,x) \mid x \in X(a)\} & \text{if} \quad a \in \mho, \\ +\infty & \text{if} \quad a \in \mho^0 \backslash \mho, \end{cases}$$

where $\mho = \{a \mid X(a) \neq \varnothing\}$. In what follows, we assume that $\mho \neq \varnothing$ and consider $\mho$ as the set of feasible parameters for (1.1).

For a fixed $y \in \mho$, we approximate (1.1) as

$$\min\{f_y(a,p) \mid p \in X_y(a) \subset L_y\}, \tag{1.2}$$

where f_y and X_y are such that (1.2) is "simpler" (say, has fewer variables and/or constraints) than (1.1). We refer to problem (1.2) as the aggregated problem for (1.1) at $y \in \mho$, or merely, the aggregated problem. Later on, we describe procedures for constructing f_y and X_y.

Let us denote the value function for (1.2) by $\varphi_y(a)$ and set

$$\Gamma(y) = \{a \in \mho \mid \varphi(a) < \varphi(y)\},$$

$$\Gamma_A(y) = \{a \in \mho \mid \varphi_y(a) < \varphi_y(y)\}.$$

Definition 1.1. An aggregation of problem (1.1) into problem (1.2) is *consistent* in the sense of parameter variations at a point $y \in \mho$ (or merely consistent) if

$$\begin{aligned} &\Gamma_A(y) \neq \varnothing \Longrightarrow \Gamma_A(y) \subseteq \Gamma(y), \\ &\Gamma_A(y) = \varnothing \Longrightarrow \Gamma(y) = \varnothing. \end{aligned} \tag{1.3}$$

Accordingly, if (1.3) holds, then problem (1.2) is said to be the consistent aggregated problem for (1.1) at $y \in \mho$. The point y is called a base point.

As follows from Definition 1.1, all parameter variations improving the objective of a consistent aggregated problem improve the objective of the original problem as well. At the same time, if the aggregated problem does not have such variations, then the original problem also does not have them.

The consistent aggregated problem can be used to compare the point y with other parameters from $\mho$ based on the value of φ_y. If there are parameters improving φ_y, then all of them decrease the original value-function φ.

Note that, generally, the ordering of parameters by the values of φ_y can differ from the ordering based on φ. Hence, if $\Gamma_A(y) \neq \varnothing$, the aggregated problem cannot be used to choose the best (minimizing φ) parameters. Finally, if $\Gamma_A(y) = \varnothing$, i.e., there are no parameters better than y in the sense of φ_y, then the base point y is the best in the sense of φ as well.

Suppose that only a part of constraints in (1.1) depends on parameters, i.e., $X(a) = Z(a) \cap F$. This situation often occurs in practice, since the analyst/decision-maker can handle only few parameters involved in a small part of restrictions.

The following constructions are used to state the aggregated problem (1.2).

1. Aggregation of variables. Let a point $x_y \in L$ and a linear operator $P_y\colon L_y \to L$ be fixed. We assume that L_y has a lower dimension than L. Here, P_y is an analog of the weight matrix introduced in Section 1.1 of Chapter 1. Considering the disaggregated variables in the form $x = x_y + P_y p$, where $p \in L_y$, we obtain the following variable aggregated problem in the same way as in Remark 1.1 from Chapter 1:

$$\min\{f_y(a,p) \mid p \in Z_y(a) \cap F_y \subset L_y\},$$

where

$$f_y(a,p) = f(a,\, x_y + P_y p),$$
$$Z_y(a) = \{p \mid x_y + P_y p \in Z(a)\},$$
$$F_y = \{p \mid x_y + P_y p \in F\}.$$

Geometrically, the feasible set of the variable aggregated problem is the intersection of $Z(a)$ and F with the affine manifold (hyperplane)

$$\Pi_y = \{x \mid x = x_y + P_y p,\ p \in L_y\}. \tag{1.4}$$

We choose the point x_y and the operator P_y later on so as to ensure the consistency of aggregation.

2. Constraint Reduction. In some cases, aggregation of variables results in redundancy of the constraints determining F_y (see, e.g., Section 2 of Chapter 1). Moreover, sometimes, we may want to use some subsets of F_y rather than the entire F_y to simplify approximations. By reduction of constraints, we mean the construction of a nonempty subset $G_y \subseteq F_y$ having a simpler structure than F_y. In this chapter, we also refer to this processes as constraint aggregation to stress that the reduction is determined by the aggregation.

Thus, the aggregated problem (1.2) is considered in the form

$$\min \{ f_y(a,p) \mid p \in Z_y(a) \cap G_y \}.$$

Remark 1.1. If p is feasible to the aggregated problem for a fixed parameter a, then the disaggregated solution $x = x_y + P_y p$ is feasible to the original problem for the same parameter value. Hence, $\varphi(a) \leq \varphi_y(a)$, and the first condition in (1.3) is fulfilled. It is not so easy to verify the second condition in (1.3). In the following sections, we specify aggregation (1.4) for some classes of parametric optimization problems in order to fulfill the second condition in (1.3).

§2. Parametric Linear Programming

Linear programming problems are best studied from the point of view of sensitivity analysis. The term parametric programming refers largely to studying the behavior of solutions to linear programming problems under a change of parameters. However, the majority of publications on this topic only study the behavior of basic variables in the range where the basis does not change or examine the range of parameter variation where the problem has a finite optimal solution.

A fairly complete survey on this topic is contained in [4]. How can the values of the basic variables in an optimal solution be changed? There are two ways:

1. By changing the right-hand sides;
2. By changing the elements of the matrix A. This can be done

(a) by varying one or several basic columns;

(b) by varying a row;

(c) by combining (a) and (b).

A change in the coefficients of the objective function does not affect the values of the basic variables, but it may lead to a change in the optimal basis.

Examining the consequences of a change in an element of the matrix A involves significant difficulties. Purely theoretically, a variation of elements of A is more convenient for approximating the optimal solution, but in practice, its effect is weak because of computational burden.

When it is sufficient to estimate the variation in the optimal solution by the variation in the optimal objective value, parameter changes of various kinds can be estimated with the use of marginal values [24]. However, the marginal values only apply when the optimal basis does not change, and they cannot be used to estimate large variations in basic columns and arbitrary variations in other columns.

The approach considered in this section is based on constructing an aggregated problem consistent with respect to parameter variations. This approach makes it possible to estimate arbitrary variations, including those leading to a change in the basis of the optimal vertex.

This section deals with constructing partially aggregated problems, i.e., those where a part of variables are aggregated and the other variables retain their original, nonaggregated, form. Such an aggregation method is selected because it is usually inexpedient to aggregate the variables corresponding to columns subjected to variation. The case of complete aggregation is also easy to handle, but the dependence of the aggregated problem on the variable parameters is then obscured.

2.1. The Base Problem

Consider the original parametric linear programming problem in the form

$$\begin{gathered} c_1x_1 + c_2x_2 \to \min, \\ x \in X(A_{11} + \Delta A_{11}) = \\ = \left\{ \begin{pmatrix} x_1 \\ x_2 \end{pmatrix} \middle| \begin{aligned} (A_{11} + \Delta A_{11})x_1 + A_{12}x_2 &= b_1, \\ A_{21}x_1 + A_{22}x_2 &= b_2, \end{aligned} \quad x_1 \geq 0,\ x_2 \geq 0 \right\}, \\ A_{11} + \Delta A_{11} \in \mho, \end{gathered} \tag{2.1}$$

where (c_1, c_2) is the row n-vector of coefficients of the objective function, $x = \begin{pmatrix} x_1 \\ x_2 \end{pmatrix}$ is the n-vector of the variables to be optimized, $A = \begin{pmatrix} A_{11}, & A_{12} \\ A_{21}, & A_{22} \end{pmatrix}$ is the $m \times n$ matrix of constraint coefficients, $b = \begin{pmatrix} b_1 \\ b_2 \end{pmatrix}$ is the m-vector of right-hand sides, ΔA_{11} is the $m_1 \times n_1$ matrix of variations of the constraint coefficients, and $\mho$ is the set of feasible matrices of variable coefficients.

Although the set $\mho$ is specified informally, we assume that it has the property

$$A_{11} + \Delta A_{11} \in \mho \Longrightarrow A_{11} + \Delta A_{11}\alpha \in \mho \tag{2.2}$$

for an arbitrary matrix α of form $\alpha = (\delta_{ij}\alpha_i)_1^{n_1}$, where $0 \le \alpha_i \le 1$, $i = \overline{1, n_1}$, and δ_{ij} is the Kronecker symbol.

To comment this property, suppose that $X(A_{11} + \Delta A_{11})$ describes a Leontief-type system [19] and $A_n + \Delta A_n$ correspond to the choice of technology; then (2.2) says that, in addition to new technologies, the production system can use old ones, and the corresponding parameters are the weighted means of the parameters of the old and new technologies.

We also assume that

$$m_1,\, n_1 \ll \min\{m,\, n\},$$

and the set $X(A_{11} + \Delta A_{11})$ is bounded for all $A_{11} + \Delta A_{11} \in \mho$.

By the base problem, we understand the following problem from family (2.1) corresponding to $\Delta A_{11} = 0$:

$$\begin{aligned} & c_1 x_1 + c_2 x_2 \to \min, \\ & \begin{pmatrix} A_{11}, & A_{12} \\ A_{21}, & A_{22} \end{pmatrix} \begin{pmatrix} x_1 \\ x_2 \end{pmatrix} = \begin{pmatrix} b_1 \\ b_2 \end{pmatrix}, \\ & x_1 \ge 0, \quad x_2 \ge 0. \end{aligned} \tag{2.3}$$

Only the elements of the matrix A are allowed to vary, but this does not cause loss of generality, because, for varying objective function coefficients and right-hand sides, the base problem can be restated as

$$\begin{aligned} & \lambda \to \min, \\ & \lambda - c_1 x_1 - c_2 x_2 = 0, \\ & A_{11} x_1 + A_{12} x_2 - b_1 y = 0, \\ & A_{21} x_1 + A_{22} x_2 - b_2 y = 0, \\ & y = 1, \quad x_1 \ge 0, \quad x_2 \ge 0, \end{aligned}$$

and all variable elements then belong to the matrix of constraint coefficients.

Perhaps, the most severe constraint in this statement is the assumption that all variable elements can be collected into a submatrix A_{11} much smaller than A in size by applying row and column permutations. It is not the point that the variable parameters are small in number, they must be few as it is; the point is that these parameters must refer to certain user activities

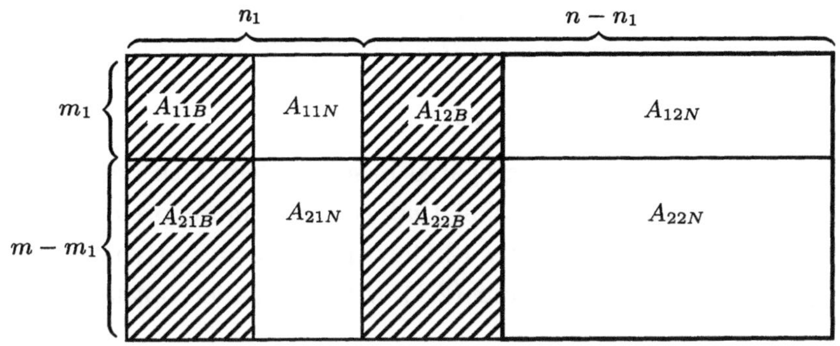

Fig. 2.1.

related to the problem. For instance, if the base problem uses a dynamical model of input–output balance, then the blocks may be the those of major construction work, direct expenses, reproduction, utilization of manpower, etc.

Let x^* be an optimal basic solution to problem (2.3). We use the subscript B to denote the columns forming a basis of this optimal solution; the nonbasic columns have subscript N. Let us reorder the columns of the matrix A as in Fig. 2.1.

Taking into account the decomposition of A into blocks shown in Fig. 2.1, we introduce the following notation for the subvectors and submatrices of the base problem, which can be used in the further constructions:

$$c_1 = (c_{1B},\ c_{1N}), \quad c_2 = (c_{2B},\ c_{2N}),$$

$$x_1 = \begin{pmatrix} x_{1B} \\ x_{1N} \end{pmatrix}, \quad x_2 = \begin{pmatrix} x_{2B} \\ x_{2N} \end{pmatrix},$$

$$\tilde{B} = \begin{pmatrix} A_{11B}, & A_{12B} \\ A_{21B}, & A_{22B} \end{pmatrix}^{-1} = \begin{pmatrix} \tilde{B}_{11}, & \tilde{B}_{12} \\ \tilde{B}_{21}, & \tilde{B}_{22} \end{pmatrix},$$

$$A_{1B} = \begin{pmatrix} A_{11B} \\ A_{21B} \end{pmatrix}, \quad A_{1N} = \begin{pmatrix} A_{11N} \\ A_{21N} \end{pmatrix}, \quad A_{2B} = \begin{pmatrix} A_{12B} \\ A_{22B} \end{pmatrix}, \quad A_{2N} = \begin{pmatrix} A_{12N} \\ A_{22N} \end{pmatrix},$$

$$A_1 = \begin{pmatrix} A_{11} \\ A_{21} \end{pmatrix}, \quad A_2 = \begin{pmatrix} A_{12} \\ A_{22} \end{pmatrix}, \quad \Delta A_1 = \begin{pmatrix} \Delta A_{11} \\ 0 \end{pmatrix}, \quad A_{11} = (a_1^1, \ldots, a_1^{n_1}),$$

$$\tilde{B}_1 = (\tilde{B}_{11},\ \tilde{B}_{12}), \quad \tilde{B}_2 = (\tilde{B}_{21},\ \tilde{B}_{22}), \quad \Delta A_{11} = (\Delta a_1^1, \ldots, \Delta a_1^{n_1}).$$

Here, a_1^s for $1 \le s \le n_1$ are the columns of A_{11}, and the dimensions of the blocks $\tilde{B}_{ij}$ of the inverse basic matrix $\tilde{B}$ coincide with those of the blocks A_{jiB}^T.

2.2. Aggregation of Variables

We start constructing the aggregated base problem with determining an affine variety Π_y of small dimension that contains more preferred feasible points of the problem from family (2.1) with $\Delta A_{11} = \Delta A_{11}^* \alpha$ (if the variation ΔA_{11}^* yields a problem from (2.1) with a smaller optimal objective value).

Suppose that an optimal solution x^0 of problem (2.3) corresponds to a nondegenerate vertex of the polyhedron of feasible solutions.

To describe the variations ΔA_{11} yielding problems with more preferred solutions, we apply the mapping $\Gamma(A_{11})$ constructed by analogy with $\Gamma(y)$ (see Section 1 in this chapter) and defined as

$$\Gamma(A_{11}) = \left\{A_{11}+\Delta A_{11} \in \mho \mid x \in X(A_{11}+\Delta A_{11}),\ c_1x_1+c_2x_2 < c_1x_1^0+c_2x_2^0\right\}.$$

The following lemma shows that, for any improving variation, there exists a one-column variation such that the corresponding problem from parametric family (2.1) has a more preferred optimal solution.

Lemma 2.1. *If* $A_{11} + \Delta A_{11}^* \in \Gamma(A_{11})$, *then there exist numbers* s $(1 \le s \le n_1)$ *and* α_1^* $(0 \le \alpha_1^* \le 1)$ *such that*

$$\left(a_1^1, \dots, a_1^{s-1}, a_1^s + \alpha_1^* \Delta a_1^{s*}, a_1^{s+1}, \dots, a_1^{n_1}\right) \in \Gamma(A_{11}).$$

Proof. It is sufficient to show that the problem from family (2.1) with $\Delta A_{11} = (0, \dots, 0, \alpha_s^* \Delta a_1^{s*}, 0, \dots, 0)$ has a feasible point at which the objective value is smaller than $c_1x_1^2 + c_2x_2^s$.

Let x' be an optimal solution to problem (2.1) with $\Delta A_{11} = \Delta A_{11}^*$.

Consider the extended problem

$$\begin{aligned} & c_1(u_1 + v) + c_2u_2 \to \min, \\ & A_1u_1 + (A_1 + \Delta A_1^s)v + A_2u_2 = b, \\ & u_1 \ge 0, \quad v \ge 0, \quad u_2 \ge 0. \end{aligned} \tag{2.4}$$

At $v = 0$, this problem turns into (2.3), and at $u_1 = 0$, it turns into problem (2.1) with $\Delta A_{11} = \Delta A_{11}^*$. By the condition of the lemma, the corner point $u_1 = x_1^0$, $v = 0$, $u_2 = x_2^0$ is not optimal, because the value of the objective function at this point is strictly larger than its value at $u_1 = 0$, $v = x_1'$, $u_2 = x_2'$. Hence [32], there exists an edge containing a point u_1^*, v^*, u_2^* such that it is feasible for problem (2.4) and the objective value at this point is strictly smaller than that at the corner point. Suppose that this edge corresponds to the inclusion of the s-th column of the matrix $A_1 + \Delta A_1^*$ in the basis. Put

$$x_1^* = u_1^* + v^*, \quad x_2^* = u_2^*;$$

then, at

$$\alpha_s^* = v^{s*}/(u_1^{s*} + v^{s*}),$$

the equations and the nonnegativity constraints on the variables in problem (2.1) with $\Delta A_{11} = (0, \dots, 0, \alpha_s^* \Delta a_1^{s*}, 0, \dots, 0)$ are fulfilled (v^{s*} and u_1^{s*} denote the s-th components of the vectors v^* and u_1^*, respectively). Indeed, for $\alpha^* = \operatorname{diag}(0, \dots, \alpha_s^*, \dots, 0)$, the feasibility of the point u_1^*, v^*, u_2^* in problem (2.4) implies

$$\begin{aligned}(A_1 + \Delta A_1^* \alpha^*) x_1^* + A_{12} x_2^* = (A_1 + \Delta A_1^*)(u_1^* + v^*) + A_{12} u_2^* = \\ = A_1 u_1^* + (A_1 + \Delta A_1^*) v^* + A_{12} u_2^* = b.\end{aligned}$$

In addition, we have

$$c_1 x_1^* + c_2 x_2^* \le c_1 x_1^0 + c_2 x_2^0;$$

this completes the proof of the lemma.

The proof of Lemma 2.1 shows that the preferred points x^* can be chosen on the edges of the polyhedron of (2.4) incident to the vertex $u_1 = x_1^0$, $v = 0$, $u_2 = x_2^0$. Now, let us show that they belong to the affine variety

$$\Pi_y = \{x \mid x_1 = x_1,\ x_{2B} = x_{2B}^0 + \tilde{B}_{21}\eta - \tilde{B}_{22} A_{21N} \nu,\ x_{2N} = 0\},$$

where x_1, η, and ν are independent vector variables of dimensions n_1, m_1, and no larger than n_1, respectively (thus, the dimension of the manifold Π_y does not exceed $m_1 + 2n_1$).

Lemma 2.2. $x \in \Pi_y$.

Proof. In the notation of Lemma 2.1, we have

$$x_{2B}^* = u_{2B}^* = x_{2B}^0 - (\tilde{B}_{21}, \tilde{B}_{22})(a^s + \Delta a^s) v_s^*,$$

$$x_{2N}^* = u_{2N}^* = 0,$$

where a^s and Δa^s are the s-th columns of the matrices A_1 and ΔA_1, respectively.

Let e_s denote the vector whose components are put in correspondence with the columns of the matrix A_{1N} and equal zero, except the component that corresponds to the column a^s (if we find it among the columns of A_{1N}) and equals one.

To obtain the assertion of the lemma, it suffices to set

$$x_1 = x_1^*, \quad \eta = \eta^* = -\Delta a_1^s v_s^*, \quad \nu = \nu^* = 0$$

if a^s is a basic column (i.e., if $\tilde{B}_2 a^s = 0$) and

$$x_1 = x_1^*, \quad \eta = \eta^* = -(a_1^s + \Delta a_1^s) v_s^*, \quad \nu = \nu^* = e_s v_s^*$$

otherwise.

Corollary 2.1. *All preferred points x^* can be expressed in terms of the variables of the variety Π_y with $\nu \geq 0$.*

Corollary 2.2. *By virtue of the linearity of the preceding relations, under the replacement of α^* with*

$$\tilde{\alpha} = \beta v^{s*} / \big(\beta u_1^{s*} + (1-\beta)(x_1^0)^s + \beta v^{s*}\big),$$

where $0 < \beta \leq 1$, the points $\tilde{x} = (1-\beta)x^0 + \beta x^0 + \beta x^*$ retain all properties of the points x^*.

Let us construct a problem whose feasible solution set is formed by the points corresponding to $\nu \geq 0$ in the intersection of the affine variety Π_y with feasible solution set of problem (2.3). The problem is

$$\begin{aligned}
&c_1 x_1 + c_{2B}\big(x_{2B}^0 + \tilde{B}_{21}\eta - \tilde{B}_{22}A_{21N}\nu\big) \to \min,\\
&A_1 x_1 + A_{2B}\big(\tilde{B}_{21}\eta - \tilde{B}_{22}A_{21N}\nu\big) = b - A_2 x_2^0,\\
&x_{2B}^0 + \tilde{B}_{21}\eta - \tilde{B}_{22}A_{21N}\nu \geq 0,\\
&x_1 \geq 0, \quad \nu \geq 0.
\end{aligned} \tag{2.5}$$

The number of variables in this problem is substantially smaller than in the base problem (2.3) (it does not exceed $m_1 + 2n_1$), but the number of equations is the same, and problem (2.5) includes the additional inequalities

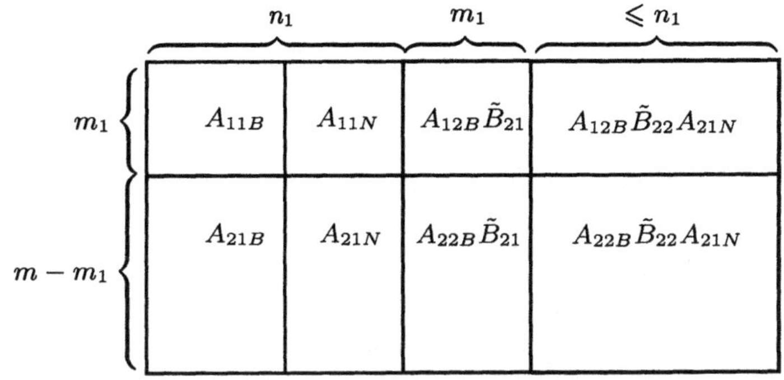

Fig. 2.2.

corresponding to the nonnegativity constraints on the x_{2B} variables in the base problem.

2.3. Constraint Reduction

The matrix of the system of equations of problem (2.5) has the form shown in Fig. 2.2.

Lemma 2.3. *The columns of the matrix* $A_{2B}\tilde{B}_{22}A_{21N}$ *have a linear expression in terms of the columns of the matrix* $(A_1,\ A_{2B}\tilde{B}_{21})$.

Proof. By the definition of $\tilde{B}$, we have

$$(A_{1B},\ A_{2B}\tilde{B}_{21})\begin{pmatrix}\tilde{B}_{11}\\ E\end{pmatrix} = \begin{pmatrix}E\\ 0\end{pmatrix},\quad (A_{21B},\ A_{22B})\begin{pmatrix}\tilde{B}_{12}\\ \tilde{B}_{22}\end{pmatrix} = E. \tag{2.6}$$

Here and in what follows, E denotes the identity matrix of appropriate size.

Multiplying the latter relation by A_{21N} on the right, we obtain the following expression for the last $m - m_1$ elements of the columns of the matrix $A_{2B}\tilde{B}_{22}A_{21N}$ (see Fig. 2.2):

$$A_{22B}\tilde{B}_{22}A_{21N} = A_{21N} - A_{21B}\tilde{B}_{12}A_{21N}.$$

Taking into account that the addition of the linear combination of the columns specified by the first relation in (2.6) does not change the last $m-m_1$ vector components, we obtain

$$A_{2B}\tilde{B}_{22}A_{21N} = A_{1N} - A_{1B}\tilde{B}_{12}A_{21N} + (A_{1B},\ A_{2B}\tilde{B}_{21})\begin{pmatrix}\tilde{B}_{11}\\ E\end{pmatrix}\times$$
$$\times\ (A_{12B}\tilde{B}_{22}A_{21N} - A_{11N} + A_{11B}\tilde{B}_{12}A_{21N}).$$

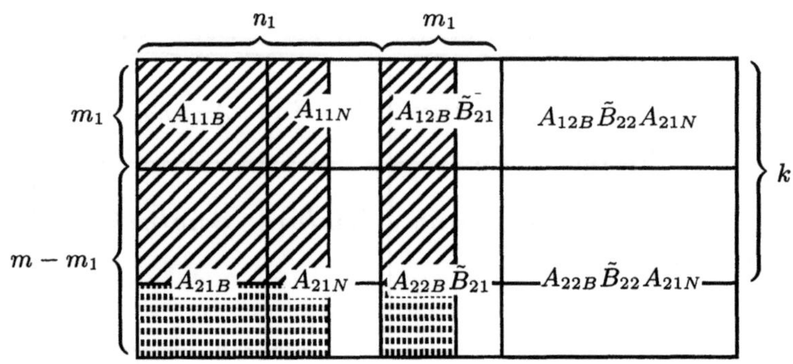

Fig. 2.3.

This completes the proof of the lemma.

Corollary 2.3. *The rank of the matrix* $(A_1,\ A_{2B}\tilde{B}_{21},\ -A_{2B}\tilde{B}_{22}A_{21N})$ *does not exceed* $m_1 + n_1$.

Let $\tilde{\tilde{B}}_{21}$ be a matrix formed by linearly independent columns of $\tilde{B}_{21}$, and let T be the right matrix of the skeleton decomposition

$$\tilde{B}_{21} = \tilde{\tilde{B}}_{21}T \tag{2.7}$$

of $\tilde{B}_{21}$.

Suppose that the rank of the matrix $(A_1,\ A_{2B}\tilde{B}_{21},\ -A_{2B}\tilde{B}_{22}A_{21N})$ equals k.

By Lemma 2.3, the matrix $(A_1,\ A_{2B}\tilde{B}_{21})$ contains a nonsingular submatrix G_1 of rank k. This submatrix G_1 can be chosen so as to include the elements of the matrices A_{11B} and $A_{12B}\tilde{\tilde{B}}_{21}$ [22], because the elements of these matrices are at the intersections of linearly independent rows with linearly independent columns. The columns of A_{1B} and A_{2B} are linearly independent, because they belong to the basis. The multiplication of A_{2B} by $\tilde{\tilde{B}}_{21}$ preserves the linear independence of A_{1B}, and the rank of the product equals the number of columns in $\tilde{\tilde{B}}_{21}$ according to the Sylvester inequality. The first m_1 rows in the matrix $(A_1,\ A_{2B}\tilde{B}_{21})$ are linearly independent (see the first formula in (2.6)).

We keep the same order of rows in G_1 as in $(A_1,\ A_{2B}\tilde{B}_{21})$ and reorder the columns as shown in Fig. 2.3, i.e., move the columns included in G_1 to the beginning of the matrices A_{1N} and $A_{2B}\tilde{B}_{21}$.

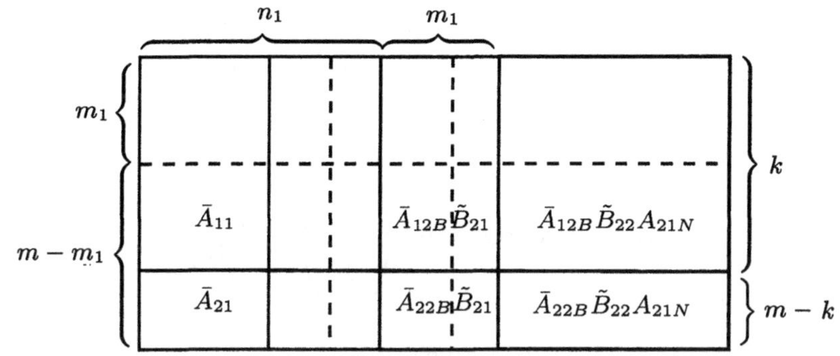

Fig. 2.4.

Let G_2 denote the matrix formed by the elements of $(A_1,\ A_{2B}\tilde{B}_{21})$ in the intersections of the columns included in G_1 and the rows not included in G_1 (G_2 is shown by dots in Fig. 2.3).

Let us divide the matrix A_1 into the parts $\bar{A}_{11}$ and $\bar{A}_{12}$ consisting of the rows included and not included in G_1, respectively. Similar partitions of the matrices A_2 and A_{2B} and of the vector b determine submatrices $\bar{A}_{11}$, $\bar{A}_{22}$, $\bar{A}_{12B}$, and $\bar{A}_{22B}$ and subvectors $\bar{b}_1$ and $\bar{b}_2$. Reordering the rows of $(A_1,\ A_{2B}\tilde{B}_{21},\ A_{2B}\tilde{B}_{22}A_{21N})$, we can represent the resulting decomposition of matrices as shown in Fig. 2.4.

Lemma 2.4. *The last* $m-k$ *rows of the matrix* $(A_1,\ A_{1B}\tilde{B}_{21},$ $-A_{2B}\tilde{B}_{22}A_{21N})$ *have a linear expression in terms of the rows with numbers* $m_1+1,\dots,k$.

Proof. It follows from the construction of the matrices G_1 and G_2 that the columns of $(A_1,\ A_{2B}\tilde{B}_{21},\ A_{2B}\tilde{B}_{22}A_{21N})$ linearly depend on the columns of $\begin{pmatrix} G_1 \\ G_2 \end{pmatrix}$; hence, there exists a matrix F such that

$$(\bar{A}_{11},\ \bar{A}_{12B}\tilde{B}_{21},\ -\bar{A}_{12B}\tilde{B}_{22}A_{21N}) = G_1F,$$

$$(\bar{A}_{21},\ \bar{A}_{22B}\tilde{B}_{21},\ -\bar{A}_{22B}\tilde{B}_{22}A_{21N}) = G_2F.$$

Therefore,

$$(\bar{A}_{21},\ \bar{A}_{22B}\tilde{B}_{21},\ -\bar{A}_{22}\tilde{B}_{22}A_{21N}) = G_2G_1^{-1}(\bar{A}_{11},\ \bar{A}_{12B}\tilde{B}_{21},\ -\bar{A}_{21B}\tilde{B}_{22}A_{21}). \tag{2.8}$$

Let us show that the first m_1 columns of the matrix $G_2G_1^{-1}$ in this decomposition are nonzero. For this purpose, we add the matrix $\tilde{A}_{1N}$ formed by the columns of A_{1N} included in G to the first factor in the left-hand side of (2.6) and augment the second factor with the corresponding number of zero rows; as a result, we obtain

$$(A_{1B},\ \tilde{A}_{1N}, A_{2B}\tilde{B}_{21}) \begin{pmatrix} \tilde{B}_{11} \\ 0 \\ E \end{pmatrix} = \begin{pmatrix} E \\ 0 \end{pmatrix}.$$

The skeleton decomposition (2.7) gives

$$(A_{1B},\ \tilde{A}_{1N}, A_{2B}\tilde{\tilde{B}}_{21}) \begin{pmatrix} \tilde{\tilde{B}}_{11} \\ 0 \\ T \end{pmatrix} = \begin{pmatrix} E \\ 0 \end{pmatrix}. \tag{2.9}$$

By the definition of the matrices G_1 and G_2, we have

$$(A_{1B},\ \tilde{A}_{1N}, A_{2B}\tilde{\tilde{B}}_{21}) \begin{pmatrix} G_1 \\ G_2 \end{pmatrix};$$

so, looking at the right-hand side of (2.9), we conclude that, first, the second factor in (2.9) is the matrix formed by the first m_1 columns of G^{-1} and, secondly,

$$G_2 \begin{pmatrix} \tilde{B}_{11} \\ 0 \\ T \end{pmatrix} = 0.$$

Thus, we have shown that the first m_1 columns of the matrix $G_2G_1^{-1}$ are zero. Therefore, the first m_1 rows of $(\bar{A}_{11}, \bar{A}_{12}\tilde{B}_{21},\ -\bar{A}_{21}\tilde{B}_{22}A_{21N})$ in the right-hand side of (2.9) play no role, and they can be discarded together with the first columns of $G_2G_1^{-1}$. This proves the lemma.

Lemma 2.4 implies that the last $m-k$ rows of the matrix $(A_1A_{1B}\tilde{B}_{21}- -A_{2B}\tilde{B}_{22}A_{21N})$ can be linearly expressed in terms of other rows of this matrix, which contain no variable parameters.

Leaving only linearly independent equations in problem (2.5) and introducing the new independent variables

$$\vartheta = T\eta, \tag{2.10}$$

(T is the right matrix in the skeleton decomposition (2.7)), we obtain the partially aggregated base problem

$$\begin{aligned}
&c_1x_1 + c_{2B}(x_{2B}^0 + \tilde{\tilde{B}}_{21}(\vartheta - \tilde{B}_{22}A_{21N}\nu) \to \min,\\
&\bar{A}_{11}x_1 + \bar{A}_{12B}(\tilde{\tilde{B}}_{21}\vartheta - \tilde{B}_{22}A_{21N}\nu) = \bar{b} - \bar{A}_{12}x_2^0,\\
&x_{2B}^0 + \tilde{\tilde{B}}_{21}\vartheta - \tilde{B}_{22}A_{21N}\nu \geq 0,\\
&x_1 \geq 0, \quad \nu \geq 0,
\end{aligned} \tag{2.11}$$

where the dimension of the new vector ϑ of aggregation variables is no higher than m_1 and the number of equations does not exceed $m_1 + n_1$ (see Corollary 2.3).

The only difference between problems (2.5) (with the change of variables (2.10)) and (2.11) is the presence of the additional equations

$$\bar{A}_{21}x_1 + \bar{A}_{22B}(\tilde{\tilde{B}}_{21}\vartheta - \tilde{B}_{22}A_{21N}\nu) = \bar{b}_2 - \bar{A}_{22}x_2^0. \tag{2.12}$$

It is easy to see that removing equations (2.12) does not enlarge the set of feasible solutions to problem (2.11). Indeed, if the equations of problem (2.5) are solvable, then the right-hand side vector $\bar{b}_2 - \bar{A}_{22}x_2^0$ have the same expression in terms of the components of the right-hand side vector of (2.5) with numbers $m_1 + 1, \ldots, k$ as the expression of the corresponding rows of the matrices in the left-hand sides given by Lemma 2.4. Therefore, all feasible solutions to problem (2.11) satisfy equations (2.12) independently of the variation in the elements of the matrix A_{11} involved in the first m_1 equations of this problem.

2.4. The Aggregated Problem in the Standard Form and the Proof of Its Consistency

This section continues the examination of the case where the point x^0 corresponds to a nondegenerate vertex of the base problem.

We start reducing problem (2.11) to the standard form with replacing the variables ϑ, which are not subject to the nonnegativity constraint, by twice their number of nonnegative variables τ. The problem becomes

$$\begin{aligned}
&c_1x_1 + c_{2B}x_{2B}^0 + c_{2B}(\tilde{\tilde{B}}_{21},\ -\tilde{\tilde{B}}_{21})\tau - c_{2B}\tilde{B}_{22}A_{21N}\nu \to \min,\\
&\bar{A}_{11}x_1 + \bar{A}_{12B}(\tilde{\tilde{B}}_{21},\ -\tilde{\tilde{B}}_{21})\tau - \bar{A}_{12B}\tilde{B}_{22}A_{21N}\nu = \bar{b}_1 - \bar{A}_{12}x_2^0,\\
&x_{2B}^0 + (\tilde{\tilde{B}}_{21},\ -\tilde{\tilde{B}}_{21})\tau - \tilde{B}_{22}A_{21N}\nu \geq 0,\\
&x_1 \geq 0, \quad \tau \geq 0, \quad \nu \geq 0.
\end{aligned}$$

Using the notation

$$p = \begin{pmatrix} \tau \\ \nu \end{pmatrix}, \quad g = \bar{b}_1 - \bar{A}_{12}x_2^0, \quad q^0 = c_{2B}x_{2B}^0,$$

$$R = (\tilde{\tilde{B}}_{21},\ -\tilde{\tilde{B}}_{21},\ -\tilde{B}_{22}A_{21N}), \quad q = c_{2B}R, \quad Q = \bar{A}_{12B}R,$$

we rewrite the constructed partially aggregated problem in the form

$$\begin{aligned}
&c_1x_1 + q^0 + qp \to \min,\\
&\bar{A}_{11}x_1 + Qp = g,\\
&-Rp \leq x_{2B}^0,\\
&x_1 \geq 0, \quad p \geq 0.
\end{aligned} \tag{2.13}$$

An important shortcoming of this problem is the presence of a large number of inequalities corresponding to the nonnegativity constraints on the basic components in the original problem; their number can amount to m (the dimension of the vector of basic components). To remove this shortcoming, we use the observation that the values of the variables do not affect the consistency with respect to parameter variations at the point A_{11}; only the direction in which they change is important.

Since the optimal vertex x^0 is nondegenerate, all the components $(x_{2B}^0)^i$ of the vector x_{2B}^0 are strictly positive. Denoting the elements of the matrix R by r^{ij}, we rewrite the inequalities $-Rp \leq x_{2B}^0$ in (2.13) as

$$\sum_j \Big(-r^{ij}/(x_{2B}^0)^i\Big)p^j \leq 1, \quad i = \overline{1,\theta}, \tag{2.14}$$

where p^j are the components of the vector p and θ is the dimension of x_{2B}^0.

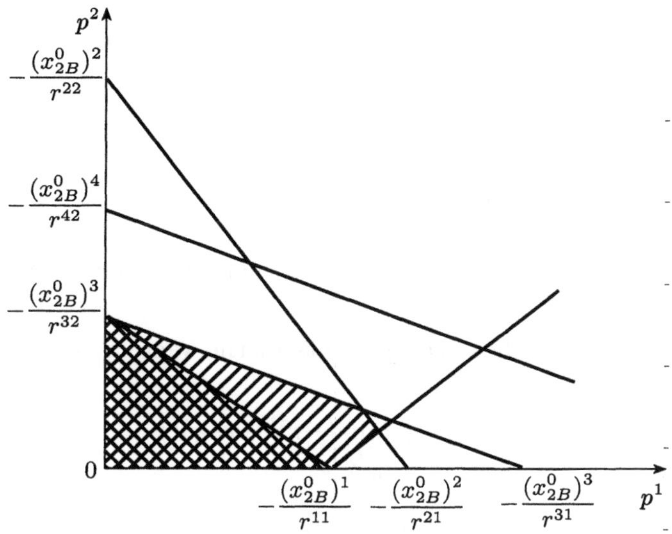

Fig. 2.5.

Let us denote the row vector with the components

$$t^j = \begin{cases} \max\limits_{i \in I_j} \left(-r^{ij}/(x_{2B}^0)^i\right) & \text{if} \quad I_j = \{i \mid r^{ij} < 0\} \neq \varnothing, \\ 0 & \text{otherwise} \end{cases}$$

by t and replace inequalities (2.14) with the inequality

$$tp \leq 1. \tag{2.15}$$

In Fig. 2.5, the region specified by the constraints $-Rp \leq x_{2B}^0$, $p \geq 0$ is shown by hatching, and the region specified by the constraints $tp \leq 1$, $p \geq 0$ is crosshatched. Any point $p \geq 0$ satisfying inequality (2.15) satisfies inequalities (2.14). Indeed,

$$t^j p^j \leq \left(-r^{ij}/(x_{2B}^0)^i\right) p^j \quad \text{for all} \quad i = \overline{1, \theta},$$

and therefore,

$$\sum_j t^j p^j \leq \sum_j \left(-r^{ij}/(x_{2B}^0)^i\right) p^j \quad \text{for all} \quad i = \overline{1, \theta}.$$

At the same time, if $p > 0$ satisfies inequalities (2.14), then there exists a number $\alpha > 0$ such that αp satisfies (2.15). To show this, consider the set of indices

$$J = \{j \mid I_j \neq \varnothing\}.$$

Inequality (2.15) only imposes constraints on the components of p with numbers $j \in J$, and if

$$p^j \leq \underline{p} = 1 \,/\, |J| \max_{j \in J} t^j,$$

where $|J|$ is the number of elements in J, then the corresponding vector p satisfies (2.15).

On the other hand, these components of the vectors p satisfying (2.14) are bounded by

$$p^j \leq \bar{p} = 1 / \min_{\substack{i \in I_j \\ j \in J}} \left(-r^{ij} / (x_{2B}^0)^i\right).$$

Therefore, as α, we can take, for example,

$$\alpha = \underline{p} / \bar{p}.$$

Adding a scalar variable $z \geq 0$ to inequality (2.15) reduces the partially aggregated base problem to the final form

$$\begin{aligned}
&c_1 x_1 + q^0 + qp \to \min, \\
&\bar{A}_{11} x_1 + Qp = g, \\
&tp + z = 1, \\
&x_1 \geq 0, \quad p \geq 0, \quad z \geq 0.
\end{aligned} \tag{2.16}$$

The partially aggregated parametric linear programming problem is

$$\begin{aligned}
&c_1 x_1 + q^0 + qp \to \min, \\
&\begin{pmatrix} x_1 \\ p \end{pmatrix} \in X_y(A_{11} + \Delta A_{11}) = \\
&\qquad = \left\{ \begin{pmatrix} x_1 \\ p \end{pmatrix} \,\middle|\, \begin{array}{l} (\bar{A}_{11} + \Delta \bar{A}_{11}) x_1 + Qp = q, \quad x_1 \geq 0 \\ tp + z = 1, \quad p \geq 0, \quad z \geq 0 \end{array} \right\}, \\
&A_{11} + \Delta A_{11} \in \mho,
\end{aligned} \tag{2.17}$$

where $y = A_{11}$.

Theorem 2.1. *If the point x^0 corresponds to a nondegenerate vertex, then the aggregation of parametric problem (2.1) into problem (2.17) is consistent with respect to parameter variations at the point $y = A_{11}$.*

Proof. According to the definition of aggregation consistency with respect to parameter variations given in Section 1, we must verify the fulfillment of conditions (1.3).

The first condition in (1.3) requires that the substitution of $A_{11}+\Delta\tilde{A}_{11} \in$ $\in \mho$ in problem (2.3) yields a smaller optimal objective value of this problem if its substitution in (2.16) leads to a decrease in the optimal objective value.

Suppose that $\tilde{x}_1$, $\tilde{p}$, $\tilde{z}$ is an optimal solution to problem (2.16) with A_{11} replaced by $A_{11} + \Delta\tilde{A}_{11}$. Let us apply the same parameter variation $\Delta\tilde{A}_{11}$ to problems (2.11), (2.5), and (2.3) and represent $\tilde{p}$ in the form

$$\tilde{p} = \begin{pmatrix} \tilde{p}_1 \\ \tilde{p}_2 \end{pmatrix},$$

where $\tilde{p}_2$ is the vector of the components of $\tilde{p}$ corresponding to the columns $-\tilde{B}_{22}A_{21N}$ in the matrix R. Then the point

$$x_1, \ \tilde{\vartheta} = (E, \ -E)\tilde{p}_1, \quad \tilde{\nu} = \tilde{p}_2$$

is feasible in problem (2.11) and gives the same objective value.

It can be shown with the use of the skeleton decomposition (2.7) that any point $\tilde{x}_1$, $\tilde{\eta}$, $\tilde{\nu}$, where $\tilde{\eta}$ is a solution to the equation

$$T\eta = \tilde{\vartheta}, \tag{2.18}$$

is a feasible solution to (2.5) with the same objective value. Equation (2.18) is solvable for any ϑ, because the rank of the right matrix in the skeleton decomposition coincides with the number of its rows.

Applying the relations determining the linear variety Π_y, we see that the point

$$\tilde{x}_1, \ \tilde{x}_{2B} = x^0_{2B} + \tilde{B}_{21}\tilde{\eta} - \tilde{B}_{22}A_{21N}\tilde{\nu}, \quad x_{2N} = 0$$

is a feasible point of problem (2.3), and the corresponding objective value coincides with the optimal objective value of problem (2.16) to which the variation $\Delta\tilde{A}_{11}$ is applied. Therefore, the optimal value of the objective function of (2.3) also decreases under the variation $\Delta\tilde{A}_{11}$.

Let us prove that the second condition in (1.3) holds. It is sufficient to verify that

$$\Gamma(A_{11}) \neq \varnothing \Longrightarrow \Gamma_A(A_{11}) \neq \varnothing.$$

To prove the fulfillment of this new condition, it suffices to show that, for any A_{11}^* and ΔA_{11}^* such that $\Gamma(A_{11}^*) \neq \varnothing$ and $A_{11}^* + \Delta A_{11}^* \in \Gamma(A_{11}^*)$, there exists a matrix

$$\alpha = (\delta_{ij}\alpha_i)_1^{n_i}, \quad 0 \leq \alpha_i \leq 1, \quad i = \overline{1, n_1},$$

satisfying the condition $A_{11}^* + \Delta A_{11}^* \alpha \in \Gamma_A(A_{11}^*)$.

According to Lemma 2.1, there exist an s $(1 \leq s \leq n_1)$ and an α_s^* $(0 \leq \alpha_s^* \leq 1)$ such that changing the s-th column of A_{11}^* by $\Delta a_1^{s*}\alpha_s^*$ makes the point x^* feasible in problem (2.3) and yields

$$c_1 x_1^* + c_2 x_2^* \leq c_1 x_1^0 + c_2 x_2^0. \tag{2.19}$$

By the construction of problem (2.5), the point x_1^*, η^*, ν^* used in the proof of Lemma 2.2 is feasible in problem (2.5) where the s-th column of A_{11}^* is changed by $\Delta a_1^{s*}\alpha_s^*$.

As is shown in Section 2.3, the point

$$x_1^*, \; \vartheta^* = T\eta^*, \; \nu^*$$

is feasible in problem (2.11) where the s-th column of A_{11}^* is changed by $\Delta a_1^{s*}\alpha_s^*$.

Let us denote the vectors obtained from ϑ^* by replacing the positive and negative components with zeros as ϑ_+^* and ϑ_-^*, respectively, and set

$$p^* = \begin{pmatrix} \vartheta_+^* \\ -\vartheta_-^* \\ -\nu^* \end{pmatrix}.$$

Consider

$$\beta^* = \begin{cases} 1 & \text{if} \quad tp^* \leq 1, \\ 1/tp^* & \text{if} \quad tp^* > 1. \end{cases}$$

According to Corollary 2.2, the points

$$\tilde{x}_1 = (1 - \beta)x_1^0 + \beta x_1^0, \quad \tilde{p} = \beta p^*, \quad \tilde{z} = 1 - \beta t p^*,$$

where $0 < \beta \leq \beta^*$, are feasible in problem (2.16) where the s-th column of the matrix A_{11}^* is changed by

$$\Delta a_1^{s*} \beta \vartheta^{s*} / (\beta u_1^{s*} + (1-\beta)(x_1^0)^s + \beta \vartheta^{s*}).$$

Let us find the change of the objective function of (2.16) caused by the passage from the point $x_1 = x_1^0$, $p = 0$, $z = 1$ to the point $\tilde{x}_1$, $\tilde{p}$, $\tilde{z}$:

$$\begin{aligned} c_1 x_1^0 - c_1 \tilde{x}_1 - q\tilde{p} &= c_1 x_1^0 - c_1 \tilde{x}_1 - \beta q p^* = \\ &= c_1 x_1^0 - c_1 \tilde{x}_1 - \beta c_{2B}(\tilde{\tilde{B}}_{21} \vartheta^* + \tilde{B}_{22} A_{21N} \nu^*) = \\ &= c_1 x_1^0 - c_1 \tilde{x}_1 - \beta c_{2B}(\tilde{B}_{21} \eta^* + \tilde{B}_{22} A_{21N} \nu^*) = \\ &= c_1 x_1^0 - c_1 \tilde{x}_1 - \beta c_{2B}(\tilde{x}_{2B}^* - x_{2B}^0) = \\ &= c_1 x_1^0 - c_1 \tilde{x}_2 - c_{2B}(\tilde{x}_{2B} - x_{2B}^0) = \\ &= \beta \left[c_1 x_1^0 + c_{2B} x_{2B}^0 - (c_1 x_1^* + c_{2B} x_{2B}^*) \right] = \\ &= \beta \left[c_1 x_1^0 + c_2 x_2^0 - (c_1 x_1^* + c_2 x_2^*) \right] > 0. \end{aligned}$$

The last expression is positive by virtue of (2.19).

Thus, we have proved that, if the variation is specified by the matrix $\alpha = \|\delta_{ij} \alpha_i\|_1^{n_1}$, where

$$\alpha_i = \begin{cases} \beta \vartheta^{s*} / \left(\beta u_1 + (1-\beta)(x_1^0)^s + \beta v^{s*} \right) & \text{if} \quad i = s, \\ 0 & \text{if} \quad i \neq s, \end{cases}$$

then

$$A_{11}^* + \Delta A_{11}^* \alpha \in \Gamma_A(A_{11}^*).$$

This completes the proof of the theorem.

We conclude this section by more rigorously describing the algorithm for aggregating the base problem (2.3) into (2.16) and demonstrating its performance for a numerical example.

The aggregation algorithm can be represented as the following sequence of steps.

Step 1. Constructing the matrix $(A_{1B},\ A_{2B}\tilde{B}_{21})$ and determining its nondegenerate submatrix G_1 of maximum rank.

Step 2. Determining the left matrix $\tilde{\tilde{B}}_{21}$ of the skeleton decomposition $\tilde{B}_{21} = \tilde{\tilde{B}}_{21} T$ of $\tilde{B}_{21}$ according to the numbers of the columns included in G_1.

Step 3. Determining the submatrices $\bar{A}_{11}$ and $\bar{A}_{12B}$ and the subvector $\bar{b}_1$ according to the numbers of the rows included in G_1.

Step 4. Constructing the matrix

$$R = (\tilde{\tilde{B}}_{21},\ -\tilde{\tilde{B}}_{21},\ -\tilde{B}_{22}A_{21N}).$$

Step 5. Evaluating the coefficients in the intermediate problem (2.13) by the formulas

$$q = c_{2B}R, \quad q^0 = c_{2B}x_{2B}^0, \quad Q = \bar{A}_{12B}R,$$

$$g = \bar{b}_1 - \bar{A}_{12}x_2^0 = \bar{b}_1 - \bar{A}_{12B}x_{2B}^0.$$

Step 6. Evaluating the components of the row vector t in constraint (2.15).

Step 7. Constructing the standard form of the aggregated problem (2.16).

To give a numerical example of the algorithm operation, consider the following linear programming problem with 4 equations and 12 variables as the base problem:

$$\begin{aligned}
&x_1 + 3x_2 - x_3 + 2x_4 - 4x_6 + 2x_7 + 2x_9 - x_{10} + 3x_{12} \to \min, \\
&-x_1 + 4x_2 + x_3 - x_4 + 3x_6 - 2x_7 + x_9 + x_{10} + 2x_{12} = 1, \\
&x_3 - x_4 + x_5 + 2x_6 + 3x_8 + x_9 - x_{10} + 5x_{11} + x_{12} = 8, \\
&2x_1 - 4x_2 - x_3 - x_4 - x_6 + 4x_7 - 2x_9 + 3x_{11} + x_{12} = 3, \\
&x_1 - 2x_2 - 3x_4 - x_5 + x_7 - 4x_8 + 3x_{10} + x_{12} = 2, \\
&x_i \geq 0, \quad i = \overline{1, 12}.
\end{aligned} \tag{2.20}$$

Suppose that the matrix A_{11} of variable coefficients occupies the upper left corner of the matrix A of the conditions of problem (2.20) and has the form $A_{11} = (-1, 4)$, i.e., $m_1 = 1$ and $n_1 = 2$. Then A decomposes into the blocks

$$A_1 = \begin{pmatrix} -1 & 4 \\ 0 & 0 \\ 2 & -4 \\ 1 & -2 \end{pmatrix},$$

$$A_2 = \begin{pmatrix} 1 & -1 & 0 & 3 & -2 & 0 & 1 & 1 & 0 & 2 \\ 1 & -1 & 1 & 2 & 0 & 3 & 1 & -1 & 5 & 1 \\ -1 & -1 & 0 & -1 & 4 & 0 & -2 & 0 & 3 & 1 \\ 0 & -3 & -1 & 0 & 1 & -4 & 0 & 3 & 0 & 1 \end{pmatrix},$$

$$A_{21} = \begin{pmatrix} 0 & 0 \\ 2 & -4 \\ 1 & -2 \end{pmatrix},$$

$$A_{12} = \begin{pmatrix} 1, & -1, & 0, & 3, & -2, & 0, & 1, & 1, & 0, & 2 \end{pmatrix}.$$

It can be shown that problem (2.20) has the optimal solution

$$x^0 = (6,\ 0,\ 8,\ 1,\ 1,\ 0,\ 0,\ 0,\ 0,\ 0,\ 0,\ 0)^T.$$

The inverse basic matrix corresponding to the optimal basis is

$$\tilde{B} = \begin{pmatrix} -5 & 2 & -3 & 2 \\ -7 & 3 & -5 & 3 \\ -3 & 1 & -2 & 1 \\ 4 & -1 & 3 & -2 \end{pmatrix}.$$

Let us partition the coefficients according to Fig. 2.1:

$$c_1 = (1,\ 3), \quad c_{2B} = (-1,\ 2,\ 0), \quad x^0_{2B} = (8,\ 1,\ 1)^T,$$

$$A_{1B} = \begin{pmatrix} -1 \\ 0 \\ 2 \\ 1 \end{pmatrix}, \quad A_{2B} = \begin{pmatrix} 1 & -1 & 0 \\ 1 & -1 & 1 \\ -1 & -1 & 0 \\ 0 & -3 & -1 \end{pmatrix}, \quad A_{21N} = \begin{pmatrix} 0 \\ -4 \\ -2 \end{pmatrix},$$

$$\tilde{B}_{21} = \begin{pmatrix} -7 \\ -3 \\ 4 \end{pmatrix}, \quad \tilde{B}_{22} = \begin{pmatrix} 3 & -5 & 3 \\ 1 & -2 & 1 \\ -1 & 3 & -2 \end{pmatrix}.$$

Below, we perform step-by-step the evaluations involved in the algorithm described above.

Step 1. We have

$$(A_{1B},\ A_{2B}\tilde{B}_{21}) = \begin{pmatrix} -1 & -4 \\ 0 & 0 \\ 2 & 10 \\ 1 & 5 \end{pmatrix}.$$

As G_1, we take the matrix formed by the intersections of the first and second columns with the first and third rows; thus,

$$G_1 = \begin{pmatrix} -1 & -4 \\ 2 & 10 \end{pmatrix}.$$

The succeeding steps are performed for this matrix G_1.

Step 2. The matrix $\tilde{B}_{21}$ is of full rank; therefore, its skeleton decomposition is

$$\tilde{B}_{21} = \tilde{\tilde{B}}_{21} = \begin{pmatrix} -7 \\ -3 \\ 4 \end{pmatrix}.$$

Step 3. We have

$$\bar{A}_{11} = \begin{pmatrix} -1 & 4 \\ 2 & -4 \end{pmatrix}, \quad \bar{A}_{12B} = \begin{pmatrix} 1 & -1 & 0 \\ -1 & -1 & 0 \end{pmatrix}, \quad \bar{b} = \begin{pmatrix} 1 & 3 \end{pmatrix}.$$

Step 4. We form the matrices

$$\tilde{B}_{22}A_{21N} = \begin{pmatrix} 14 \\ 6 \\ -8 \end{pmatrix}, \quad R = \begin{pmatrix} -7 & 7 & -14 \\ -3 & 3 & -6 \\ 4 & -4 & 8 \end{pmatrix}.$$

Step 5. The coefficients of problem (2.13) are

$$q = (1,\ -1,\ -2), \quad q^0 = -6,$$

$$Q = \begin{pmatrix} -4 & 4 & -8 \\ 10 & -10 & 20 \end{pmatrix}, \quad g = \begin{pmatrix} -6 \\ 12 \end{pmatrix}.$$

Step 6. The components of the row vector $t = (t_1, t_2, t_3)$ are

$$t_1 = \max\left\{\frac{7}{8},\ \frac{3}{1}\right\} = 3, \quad t_2 = 4,$$

$$t_3 = \max\left\{\frac{14}{8},\ \frac{6}{1}\right\} = 6.$$

Step 7. Introducing the slack aggregation variable $z \geq 0$, we state the aggregated base problem (2.16) for (2.20) as

$$x_1 + 3x_2 + p_1 - p_2 + 2p_3 - 6 \to \min,$$
$$-x_1 + 4x_2 - 4p_1 + 4p_2 - 8p_3 = -6,$$
$$2x_1 - 4x_2 + 10p_1 - 10p_2 + 20p_3 = 12,$$
$$3p_1 + 4p_2 + 6p_3 + z = 1,$$
$$x_1,\ x_2,\ p_1,\ p_2,\ p_3,\ z \geq 0.$$

The column of the coefficients of the variable p_3 is a positive linear combination of the other columns; therefore, this variable can be eliminated, and the problem takes the final form

$$x_1 + 3x_2 + p_1 - p_2 - 6 \to \min,$$
$$-x_1 + 4x_2 - 4p_1 + 4p_2 = -6,$$
$$2x_1 - 4x_2 + 10p_1 - 10p_2 = 12, \tag{2.21}$$
$$3p_1 + 4p_2 + z = 1,$$
$$x_1,\ x_2,\ p_1,\ p_2,\ z \geq 0.$$

Problem (2.21) has only three equations and five variables, which reduces the complexity of evaluating its optimal value under a variation of the coefficients in the matrix A_{11} in comparison with (2.20).

2.5. The Degenerate Case

If more than n hyperplanes meet at an optimal vertex x^0 of the base problem (2.3), this vertex is called degenerate. Some components of the vectors x^0_{1B} and x^0_{2B} then vanish.

The method for constructing the affine variety Π_y described in Section 2.2 is based on a representation with the use of the inverse matrix $\tilde{B}$ of the edges of the feasible solution polyhedron of the extended problem (2.4) incident to the vertex $u_1 = x^0_1$, $v = 0$, $u_2 = x^0_2$. In the degenerate case, there may exist several optimal bases at the vertex x^0 of problem (2.3), which correspond to different inverse basic matrices $\tilde{B}^{(1)}, \ldots, \tilde{B}^{(w)}$, and the edges of the feasible solution polyhedron of problem (2.4) incident to the vertex

$u_1 = x_1^0$, $v = 0$, $u_2 = x_2^0$ can then be represented with the use of various inverse basic matrices.

Depending on ΔA_{11}, the following three situations can arise:

(a) among the columns of the matrix $A_1 + \Delta A_{11}$ in the extended problem (2.4) is a column such that including it in the original optimal basis of problem (2.3) reduces the objective value in one step;

(b) there is no such column in problem (2.4) for any basis of the degenerate vertex $u_1 = x_1^0$, $v = 0$, $u_2 = x_2^0$;

(c) there is such a column in the extended problem (2.4) for a different basis of the degenerate vertex $u_1 = x_1^0$, $v = 0$, $u_2 = x_2^0$.

In situation (a), all the arguments used in the consideration of the nondegenerate case remain valid, except that formula (2.14) only applies to those i's for which $(x_{2B}^0)^i \neq 0$. In the definition of the row vector t, we replace the set I_j with

$$\tilde{I}_j = \{i \mid r^{ij} < 0,\ (x_{2B}^0)^i > 0\};$$

then the partially aggregated base problem (2.16) takes the form

$$\begin{aligned}
&c_1 x_1 + q^0 + qp \to \min,\\
&\bar{A}_{11} x_1 + Qp = g,\\
&tp + z^0 = 1, \qquad\qquad\qquad\qquad\qquad\qquad\qquad\qquad (2.22)\\
&-\sum_j r^{ij} p^j + z^i = 0, \quad i \in \{i \mid (x_{2B}^0)^i = 0\},\\
&x_1 \geq 0,\ p \geq 0,\ z^0 \geq 0,\ z^i \geq 0,\ i \in \{i \mid (x_{2B}^0)^i = 0\}.
\end{aligned}$$

This problem differs from (2.16) by the presence of the nonaggregated inequalities describing the nonnegativity constraints on the basic variables that vanish under the original choice of the optimal basis in the base problem (2.3). Without loss of generality, we can assume that the number of such basic variables in real-life systems is small in comparison with the dimension of the base problem and, therefore, the number of variables and equations in the partially aggregated problem is still much smaller than $\min\{m, n\}$.

The occurrence of situation (b) means that the point $u_1 = x_1^0$, $v = 0$, $u_2 = x_2^0$ is optimal in problem (2.4), and therefore, applying the variation ΔA_{11}, we do not obtain a problem from family (2.1) with a smaller optimal objective value.

Situation (c) is most difficult to analyze both with and without the use of partially aggregated problems, because constructing the sought direction requires searching through the inverse basic matrices (we denote them by $\tilde{B}^{(1)}, \dots, \tilde{B}^{(w)}$).

This situation can be handled in two ways. First, instead of the matrices $\tilde{B}_{21}$ and $\tilde{B}_{22}$, we can use the matrices $\tilde{S}_{21}$ and $\tilde{S}_{22}$ formed by linearly independent columns of the corresponding matrices of all bases at the optimal vertex x^0 of problem (2.3). This increases drastically the dimension of the partially aggregated problem (2.16); in unfavorable circumstances, this dimension may become commensurate with n, which makes the use of partially aggregated problems ineffective.

The second way is to use successively partially aggregated problems of form (2.22) constructed for different optimal bases of problem (2.3). It can be shown that, if the variation ΔA_{11} yields a problem of family (2.1) with a smaller optimal objective value, then there exists a variation $\Delta A'_{11} \in \mathfrak{U}$ such that, on the one hand, it does not change the solutions to the base and partially aggregated base problems, and on the other hand, the variation $\Delta A_{11} - \Delta A'_{11}$ changes the basis and gives rise to situation (a). Thus, pursuing the second way, we should try to find a variation ΔA_{11} that gives a more preferred solution to problem (2.22); if we fail to do that, then we should examine partially aggregated problems constructed for other bases.

The transition back to a variation improving a solution to the base problem from a variation improving an optimal solution to the partially aggregated problem is performed similarly to the nondegenerate case, because the disaggregation procedure transforms the points from the feasible solution set of problem (2.22), as well as the points from the feasible solution set of (2.22), into points from the feasible solution set of problem (2.3).

Remark 2.1. If the assumption

$$\max\{m_1, n_1\} \ll \min\{m, n\}$$

does not hold, it might be more convenient to aggregate only the variables and the nonnegativity constraints on the basic variables (if the number of equations is comparatively small), or only the equations (if the number of variables in the base problem is comparatively small).

2.6. Various Forms of the Base Problem

This section covers several topics related to transferring the partial aggregation technique to some representations of constraints in the base linear programming problem different from the standard representation. We consider the special features of the application of this aggregation method depending on the particular statement of the problem. Such an examination is necessary because, as a rule, it is more convenient to analyze the base problem in its natural statement determined by its meaning rather than to reduce it to the standard form. The reduction to the standard form increases the dimension of the problem and, thereby, expenditures for aggregation. At the same time, the application of algorithms oriented toward the natural statement of the problem may save a substantial amount of computation.

We start the exposition of the special features of the application of the partial aggregation method to particular representations of the constraints with the problem known as the linear programming problem with two-sided constraints, which is frequently encountered in planning and control and therefore deserves a separate consideration. This problem differs from the classical one by the presence of additional constraints on the variables, namely, of the upper bounds on their feasible values.

Such a problem can be stated in the form

$$\begin{aligned} & c_1x_1 + c_2x_2 \to \min, \\ & A_{11}x_1 + A_{12}x_2 = b_1, \\ & A_{21}x_1 + A_{22}x_2 = b_2, \\ & 0 \le x_1 \le d^1, \quad 0 \le x_2 \le d^2. \end{aligned} \tag{2.23}$$

The feasible set $X(A_{11})$ differs from that in (2.3) only by the presence of the constraints $x \le d$, where $d = \begin{pmatrix} d^1 \\ d^2 \end{pmatrix}$. In principle, by introducing slack variables, this problem can easily be reduced to the standard form considered in the preceding sections. However, the standard form of the problem with two-sided constraints involves n variables and $m + n$ equations, which is larger than in (2.23). For this reason, such a reduction is computationally inexpedient, especially at large m and n values. There is a special version of the modified simplex method [20, 32] for solving linear programming problems in form (2.23), where the constraints $x \le d$ are not taken into account explicitly and the matrix to be inverted at every step has size $m \times m$.

These considerations allow us to use the special structure of problem (2.23) in aggregation.

Let x^0 be an optimal solution to problem (2.23). As an optimal basis [20], we take a system of m linearly independent columns of the constraint matrix A corresponding to components $0 < x_j^0 < d_j$. If x^0 corresponds to a nondegenerate vertex of the polyhedron $X(A_{11})$, then such a basis exists, and the components specified above form a subvector $x_B^0 > 0$ of basic variables. The remaining (nonbasic) components of the optimal solution are equal to one of their boundary values and form a subvector $x_N^0 > 0$.

Suppose that x^0 corresponds to a nondegenerate vertex. Making the same assumptions as in Sections 2.1–2.4 and using the notation and decompositions introduced therein, consider the aggregating variety

$$\Pi_y = \left\{x \mid x_{2B} = x_{2B}^0 + Rp,\ x_{2N} = x_{2N}^0\right\},$$

whose dimension does not exceed $2(m_1 + n_1) \ll \min(m, n)$. Since (2.23) is equivalent to a standard linear programming problem, Lemma 2.1 and all the related results remain valid for (2.23).

It is only necessary to justify Lemma 2.2 (it says that $x^* \in \Pi_y$), whose proof uses the assumption that all nonbasic components x_{2N}^0 are zero. In the case under consideration, some of the nonbasic components may be positive (equal to their upper bounds). Nevertheless, Lemma 2.2 remains valid.

To show this, it suffices to construct a vector p^* such that the point x^* defined in the proof of Lemma 2.1 has the representation

$$x_{2B}^* = x_{2B}^0 + Rp^*, \quad x_{2N}^* = x_{2N}^0.$$

Let us do this. It follows from the constructive definition of x^* that

$$(A_1 + \Delta A_1^* \alpha^*)x_1^* + A_2 x_2^* = b, \quad x_{2N}^* = x_{2N}^0. \tag{2.24}$$

Let us apply the relations

$$\begin{aligned}
\tilde{B}_2 A_1 x_1^* &= \tilde{B}_2 a^s (x_{1s}^* - x_{1s}^0) + \tilde{B}_2 A_{1N} x_{1N}^0, \\
\tilde{B}_2 b &= x_{2B}^0 + \tilde{B}_2 A_N x_N^0, \\
\tilde{B}_2 A_2 x_2^* &= x_{2B}^* + \tilde{B}_2 A_{2N} x_{2N}^0,
\end{aligned} \tag{2.25}$$

where x_{1s}^* and x_{1s}^0 are the s-th components of the vectors x_1^* and x_1^0, and construct the vectors

$$\xi^* = -\begin{pmatrix} Ta_1^s \\ e_s \end{pmatrix}(x_{1s}^* - x_{1s}^0) - T\begin{pmatrix} \alpha_s^* \Delta a_1^s \\ 0 \end{pmatrix} x_{1s}^*,$$

$$p^* = \begin{pmatrix} \xi_+^* \\ -\xi_-^* \end{pmatrix}.$$

Here, ξ_+^* and ξ_-^* are the projections of ξ^* on the positive and negative orthants, respectively. Using relations (2.24) and (2.25), we obtain

$$\begin{aligned} x_{2B}^0 + Rp^* = x_{2B}^0 - \tilde{B}_2 a^s (x_{1s}^* - x_{1s}^0) - \alpha_s^* \tilde{B}_2 \Delta a^s x_{1s}^* &= \\ = x_{2B}^0 - \tilde{B}_2 A_1 x_1^* - \alpha_s^* \tilde{B}_2 \Delta a^s x_{1s}^* + \tilde{B}_2 A_{1N} x_{1N}^0 &= \\ = x_{2B}^0 - \tilde{B}_2 (A_1 + \Delta A_1^* \alpha^*) x_1^* + \tilde{B}_2 A_{1N} x_{1N}^0 &= \\ = x_{2B}^0 - \tilde{B}_2 (b - A_2 x_2^*) + \tilde{B}_2 A_{1N} x_{1N}^0 &= \\ = - \tilde{B}_2 A_N x_N^0 + x_{2B}^* + \tilde{B}_2 A_{2N} x_{2N}^0 + \tilde{B}_2 A_{1N} x_{1N}^0 &= x_{2B}^*, \end{aligned}$$

as required.

Thus, the theoretical conditions required to apply the partial aggregation method are fulfilled. The intersection of the feasible set $X(A_{11})$ of problem (2.23) with the manifold Π_y is the convex set

$$\begin{aligned} &A_{11} x_1 + A_{12B} Rp = b_1 - A_{12} x_2^0, \\ &A_{21} x_1 + A_{22B} Rp = b_2 - A_{22} x_2^0, \\ &0 \le x_1 \le d^1, \quad 0 \le x_{2B}^0 + Rp \le d_{2B}, \end{aligned}$$

in the space of variables $\binom{x_1}{p}$; the subvector d_{2B} is formed by the components of d corresponding to the variables x_{2B}.

Let us employ the idea of constraint reduction described in Sections 2.3 and 2.4. It consists in, first, eliminating linearly dependent equations and, then, approximating the domain

$$\{p \mid -Rp \le x_{2B}^0,\ Rp \le d_{2B} - x_{2B}^0\}$$

by an inscribed simplex of the form

$$\{p \mid tp \le 1,\ p \ge 0\}.$$

Put

$$D = \begin{pmatrix} -R \\ R \end{pmatrix}, \quad \xi = \begin{pmatrix} x_0 \\ d_{2B} - x_{2B}^0 \end{pmatrix}.$$

If x^0 is a nondegenerate vertex of the polyhedron $X(A_{11})$, then the vector ξ is strictly positive, and, as in Section 2.4, we can aggregate the system $Dp \le \xi$

with the use of the row vector t with coefficients

$$t_j = \begin{cases} \max\limits_{i \in I_j}(D_{ij}/\xi_i) & \text{if} \quad I_j = \{i \mid D_{ij} > 0\} \neq \varnothing, \\ 0 & \text{if} \quad I_j = \varnothing. \end{cases}$$

As a result, we obtain the following aggregated base problem with two-sided constraints:

$$\begin{aligned} & c_1 x_1 + qp + c_2 x_2^0 \to \min, \\ & \bar{A}_{11} x_1 + Qp = \bar{g}, \\ & 0 \le x_1 \le d^1, \quad tp + z = 1, \quad p \ge 0, \quad z \ge 0. \end{aligned} \tag{2.26}$$

By construction, it contains no more than $2(m_1+n_1)+1$ variables and no more than $m_1 + n_1 + 1$ equations. Thus, the presence of additional constraints of the form $x_1 \le d^1$, $x_2 \le d^2$ does not affect the bounds of the dimension of the partially aggregated problem (2.26), which remain the same as those for the standard form; this is due to the simple structure of these constraints and to the special features of the method for aggregating them. The aggregated base problem (2.26) can be used to construct an aggregated parametric problem similar to (2.17). The problems corresponding to various $\bar{A}_{11} + \Delta\bar{A}_{11} \in \mho$ in this parametric problem can be solved by applying versions of the simplex method designed for handling problems with two-sided constraints.

Let us turn to base problems with mixed and interval-type constraints. We start with the base linear programming problem with mixed constraints

$$\begin{aligned} & c_1^1 x_1 + c_2^1 x_2 \to \min, \\ & A'_{11} x_1 + A'_{12} x_2 \le b'_1, \\ & A''_{11} x_1 + A''_{12} x_2 = b''_1, \\ & A'_{21} x_1 + A'_{22} x_2 \le b'_2, \\ & A''_{21} x_1 + A''_{22} x_2 = b''_2, \\ & x_1 \ge 0, \quad x_2 \ge 0, \end{aligned} \tag{2.27}$$

which contains m constraints (not counting $x_1 \ge 0$ and $x_2 \ge 0$) and n variables. Subject to variation are the coefficients of the $m_1 \times n_1$ matrix $\begin{pmatrix} A'_{11} \\ A''_{11} \end{pmatrix}$.

Let us introduce a nonnegative $\bar{m}_1$-vector y_1 of slack variables for the inequalities with variable coefficients ($\bar{m}_1 \leq m_1$) and a vector $y_2 \geq 0$ for the inequalities with invariable coefficients. Put

$$A_{11} = \begin{pmatrix} A'_{11}, & E \\ A''_{11}, & 0 \end{pmatrix}, \quad A_{12} = \begin{pmatrix} A'_{12}, & 0 \\ A''_{12}, & 0 \end{pmatrix}, \quad A_{21} = \begin{pmatrix} A'_{21}, & 0 \\ A''_{21}, & 0 \end{pmatrix},$$

$$A_{22} = \begin{pmatrix} A'_{22}, & E \\ A''_{22}, & 0 \end{pmatrix}, \quad b_1 = \begin{pmatrix} b'_1 \\ b''_1 \end{pmatrix}, \quad c_1(c'_1, 0), \quad c_2 = (c'_2, 0),$$

where all submatrices and subvectors have consistent dimensions.

The change of variables

$$z_1 = \begin{pmatrix} x_1 \\ y_1 \end{pmatrix}, \quad z_2 = \begin{pmatrix} x_2 \\ y_2 \end{pmatrix}.$$

reduces problem (2.27) to the standard form

$$\begin{aligned} & c_1 z_1 + c_2 z_2 \to \min, \\ & A_{11} z_1 + A_{12} z_2 = b_1, \\ & A_{21} z_1 + A_{22} z_2 = b_2, \quad z_1,\ z_2 \geq 0. \end{aligned} \tag{2.28}$$

In reality, not all coefficients of the $m_1 \times (n_1 + \bar{m}_1)$ matrix A_{11} are subjected to variation, but we can require that A_{11} remain completely unaggregated in the course of constructing a partially aggregated problem. Then all the components of the vector $y_1 \geq 0$ retain their original form.

Let us apply the partial aggregation procedure to problem (2.28) under the assumption made above.

In the notation used below, the subscript B refers to the basic variables, which correspond to the optimal basis of problem (2.28), and N refers to the nonbasic variables. For simplicity, we assume that the variables and inequalities in (2.27) are enumerated in such a way that

$$x_1^{\mathrm{T}} = (x_{1B}^{\mathrm{T}},\ x_{1N}^{\mathrm{T}}), \quad y_1^{\mathrm{T}} = (y_{1B}^{\mathrm{T}},\ y_{1N}^{\mathrm{T}}), \quad z_{2B}^{\mathrm{T}} = (x_{2B}^{\mathrm{T}},\ y_{2B}^{\mathrm{T}}),$$

$$z_{2N}^{\mathrm{T}} = (x_{2N}^{\mathrm{T}},\ y_{2N}^{\mathrm{T}}), \quad z_2^{\mathrm{T}} = (z_{2B}^{\mathrm{T}},\ z_{2N}^{\mathrm{T}}).$$

Let $\bar{m}_{1B}$, n_{1B}, and n_{1N} denote the dimensions of the vectors y_{1B}, x_{1B}, and x_{1N}; then, for the decompositions specified above, the basic matrix B of the optimal basis of (2.28) and its inverse $\tilde{B} = B^{-1}$ have the forms

$$B = \begin{bmatrix} \overset{n_{1B}}{L_1} & \vdots & \overset{\bar{m}_{1B}}{E} & \vdots & L_3 \\ \cdots & \cdots & \cdots & \cdots & \cdots \\ L_2 & \vdots & 0 & \vdots & L_4 \end{bmatrix}, \quad \tilde{B} = \begin{bmatrix} \overset{\bar{m}_{1B}}{0} & \vdots & \overset{m-\bar{m}_{1B}}{F_1} \\ \cdots & \cdots & \cdots \\ E & \vdots & F_2 \\ \cdots & \cdots & \cdots \\ 0 & \vdots & F_3 \end{bmatrix}. \tag{2.29}$$

Let $z_1^0 = \begin{pmatrix} x_1^0 \\ y_1^0 \end{pmatrix}$ and $z_2^0 = \begin{pmatrix} x_2^0 \\ y_2^0 \end{pmatrix}$ be an optimal solution to (2.28). According to the results obtained in Sections 2.3 and 2.4, the aggregation of variables in (2.28) reduces to constructing the affine manifold

$$\Pi_y = \left\{ \begin{pmatrix} z_1 \\ z_2 \end{pmatrix} \middle| z_{2B} = z_{2B}^0 + Rp, \ z_{2N} = z_{2N}^0 \right\}, \tag{2.30}$$

where $R = (\tilde{\tilde{B}}_{21}, \ -\tilde{\tilde{B}}_{21}, \ -\tilde{B}_{22}A_{21N})$.

The intersection of the feasible set of problem (2.28) by this affine manifold is the polyhedron specified by the constraints

$$\begin{aligned} &A_{11}z_1 + A_{12B}Rp = b_1 - A_{12}z_2^0, \\ &A_{21}z_1 + A_{22B}Rp = b_2 - A_{22}z_2^0, \\ &z_1 \geq 0, \quad z_{2B}^0 + Rp \geq 0. \end{aligned}$$

Aggregating these constraints followed by eliminating the variables y_1 reduces the partially aggregated base problem to the problem with mixed constraints

$$\begin{aligned} &c_1'x_1 + c_{2B}'R_xp + c_2'x_2^0 \to \min, \\ &A_{11}'x_1 + A_{12B}'R_xp \leq b_1' - A_{12}'x_2^0, \\ &A_{11}''x_1 + A_{12B}''R_xp = b_1'' - A_{12}''x_2^0, \\ &\bar{Q}_1x_1 + \bar{Q}_2p = \bar{g}, \quad tp \leq 1, \\ &x_1 \geq 0, \quad p \geq 0. \end{aligned} \tag{2.31}$$

Here, R_x consists of the rows in R corresponding to the variables x_{2B}.

To estimate the dimension of problem (2.31), note that, due to the special features of the decomposition

$$\tilde{B} = \begin{pmatrix} \tilde{B}_{11}, & \tilde{B}_{12} \\ \tilde{B}_{21}, & \tilde{B}_{22} \end{pmatrix},$$

of the inverse basic matrix into blocks and to the structure of the matrix $\tilde{B}$ itself (see (2.29)), the submatrix $\tilde{B}_{21}$ has size $(m - n_{1B} - \bar{m}_{1B}) \times m_1$ and contains at least $\bar{m}_{1B} \leq m_1$ columns with zero components. Therefore, $\operatorname{rank}(\tilde{B}_{21}) \leq m_1 - \bar{m}_{1B}$, and the number of constraints in (2.31) (not counting $x_1 \geq 0$ and $p \geq 0$) equals

$$\operatorname{rank}(A_{1B}, A_{2B}\tilde{B}_{21}) + 1 \leq (n_{1B} + \bar{m}_{1B}) + (m_1 - \bar{m}_{1B}) + 1 \leq m_1 + n_1 + 1.$$

At the same time, the dimension of the manifold Π_y does not exceed

$$n_1 + m_1 + 2m_1 + n_1 = 2n_1 + 3m_1,$$

subtracting the dimension of y_1 from this value, we obtain the bound $2n_1 + 3m_1 - m_1 = 2(n_1 + m_1)$ for the number of variables in (2.31). Since the partially aggregated problem is written in the same form as the base problem, its dimensions obey the same estimates as those obtained for the problems of standard form.

To conclude, consider the problem with interval-type constraints

$$\begin{aligned} & c_1 x_1 + c_2 x_2 \to \min, \\ & \underline{b}_1 \leq A_{11}x_1 + A_{12}x_2 \leq \bar{b}_1, \\ & \underline{b}_2 \leq A_{21}x_1 + A_{22}x_2 \leq \bar{b}_2, \\ & x_1 \geq 0, \quad x_2 \geq 0, \end{aligned} \tag{2.32}$$

where $\underline{b}_1 < \bar{b}_1$, $\underline{b}_2 < \bar{b}_2$, the constraint matrix A has size $m \times n$, and the assumptions of Section 2.1 hold.

The modifications of the partial aggregation method considered above do not applicable to problem (2.32) as it is stated. The reduction of this problem to the standard form would increase the number of variables to $2m + m$ and the number of constraints to $2m$, and it would change the structure of the coefficient variations: instead of ΔA, we would have to consider less convenient variations of form $\begin{pmatrix} \Delta A \\ -\Delta A \end{pmatrix}$. At the same time, it is desirable that the constraints with variable coefficients in the partially aggregated problem have the interval form, as in (2.32).

Let us exploit the special features of the form of the original problem (2.32). Introducing a nonnegative vector $y = (y_1^{\mathrm{T}}, y_2^{\mathrm{T}})^{\mathrm{T}}$ of slack variables, where y_1 is of size m_1, and applying the change of variables $z_1^{\mathrm{T}} = (x_1^{\mathrm{T}}, y_1^{\mathrm{T}})$, $z_2^{\mathrm{T}} = (x_2^{\mathrm{T}}, y_2^{\mathrm{T}})$, we reduce (2.32) to an equivalent problem with two-sided constraints of the form

$$\begin{aligned} &(c_1, 0)z_1 + (c_2, 0)z_2 \to \min, \\ &(A_{11},\, E_{m_1})z_1 + (A_{12},\, 0)z_2 = \bar{b}_1, \\ &(A_{21},\, 0)z_1 + (A_{22},\, E_{m-m_1})z_2 = \bar{b}_2, \\ &x_1 \geq 0, \quad x_2 \geq 0, \quad 0 \leq y_1 \leq (\bar{b}_1 - \underline{b}_1), \quad 0 \leq y \leq (\bar{b}_2 - \underline{b}_2). \end{aligned} \tag{2.33}$$

We require that not only x_1, but also the variables y_1 and the corresponding constraints $0 \leq y_1 \leq (\bar{b}_1 - \underline{b}_1)$ in the partially aggregated problem retain their original forms. To perform such an aggregation, we use the results obtained in aggregating problems (2.23) and (2.27).

Let $z_1^0 = \begin{pmatrix} x_1^0 \\ y_1^0 \end{pmatrix}$ and $z_2^0 = \begin{pmatrix} x_2^0 \\ y_2^0 \end{pmatrix}$ be a solution to (2.33) corresponding to a nondegenerate vertex, and let the basis be formed by the coefficients of the system of equations where the variables and equalities are enumerated in such a way that the decompositions specified for problem (2.28) hold. Using the information about the inverse basic matrix $\tilde{B}$, we construct manifold (2.30).

The intersection of the feasible set of (2.33) with the manifold (2.30) gives a system of constraints of the form

$$\begin{aligned} &A_{11}x_1 + y_1 + Q^1 p = g^1, \\ &A_{21}x_1 + Q^2 p = g^2, \\ &x_1 \geq 0, \quad 0 \leq y_1 \leq (\bar{b}_1 - \underline{b}_1), \\ &x_{2B}^0 + R_x p \geq 0, \quad 0 \leq y_{2B}^0 + R_y p \leq (\bar{b}_{2B} - \underline{b}_{2B}), \end{aligned} \tag{2.34}$$

where R_x and R_y denote the submatrices formed by the rows of R that correspond to the vectors x_{2B} and y_{2B}, respectively. The subvectors $\bar{b}_{2B}$ and $\underline{b}_{2B}$ of the vectors $\bar{b}_2$ and $\underline{b}_2$ in their turn correspond to the interval constraints that are not active at the optimal point of the base problem (2.32).

As usual, we aggregate the system of equations in (2.34) by eliminating redundant equations; to aggregate the system of inequalities containing the variables p, we write it in the form $Dp \leq \xi$, where

$$D = \begin{pmatrix} -R_x \\ R_y \\ -R_y \end{pmatrix}, \quad \xi = \begin{pmatrix} x_{2B}^0 \\ \bar{b}_{2B} - \underline{b}_{2B} - y_{2B}^0 \\ y_{2B}^0 \end{pmatrix}.$$

By the definition of basis [20] (or support plan [32]), in the problem with two-sided constraints, the inequalities $0 < y_{2B}^0 < (\bar{b}_{2B} - \underline{b}_{2B})$ must hold; therefore, the vector ξ is positive, and the coefficients of the row vector t can be evaluated with the use of the formulas derived above for the problem with two-sided constraints.

As a result, after eliminating the variables $0 \leq y_1 \leq (\bar{b}_1 - \underline{b}_1)$, we obtain the partially aggregated base problem

$$c_1 x_1 + qp + c_2 x_2^0 \to \min,$$

$$\underline{b}_1 - A_{12} x_2^0 \leq A_{11} x_1 + Q^1 p \leq \bar{b}_1 - A_{12} x_2^0,$$

$$\tilde{A}_{21} x_1 + \tilde{Q}^2 p = \tilde{g}^2,$$

$$tp \leq 1, \quad x_1 \geq 0, \quad p \geq 0.$$

The number of variables in this problem is bounded by $2(m_1 + n_1)$, and the number of constraints does not exceed $m_1 + n_1 + 1$. Moreover, the constraints with variable coefficients have the same interval form as in (2.32).

Thus, applying an appropriate modification of the aggregation method, we can manage to perform aggregation in such a way that the aggregated problem have the same form as the base problem and the bounds on its dimension be invariant with respect to the form of the base problem. The desire to preserve the form of the constraints is caused not only by aesthetic considerations. If the constraints have similar forms, then the original and partially aggregated problems can be solved and analyzed by the same available computational algorithms designed for the particular statement of the problem, such as the known versions of the simplex method.

2.7. An Example of Varying One Element in the Constraint Matrix

Below, we exemplify the considerations of this chapter by the variation of only one element a_{11} in matrix A.

Let x^0 be the optimal solution to problem (2.3) corresponding to a nondegenerate vertex of the feasible polyhedron. The dimension of the partially aggregated problem (2.11) is then very small, which makes it possible to analytically study the behavior of the optimal objective value of this problem depending on Δa_{11}.

Let us discard the first condition of (1.3) from the definition of the parametric problem aggregation consistent with respect to parameter variations given in Section 1. This condition ensured that the improving variations of the aggregated problem were also improving for the original problem. Removing it allows us not to include the constraints corresponding to the nonnegativity constraints on the basic variables of the original problem in the partially aggregated problem. Thus, problem (2.11) takes the form

$$\begin{aligned} &c_1x_1 + c_{2B}(x_{2B}^0 + \tilde{\tilde{B}}_{21}\vartheta - \tilde{B}_{22}A_{21N}\nu) \to \min, \\ &(\bar{A}_{11} + \Delta A_{11})x_1 + \bar{A}_{12B}(\tilde{\tilde{B}}_{21}\vartheta - \tilde{\tilde{B}}_{22}A_{21N}\nu) = \bar{b}_1 - \bar{A}_{12}x_2^0, \\ &x_1 \geq 0, \quad \nu \geq 0. \end{aligned} \tag{2.35}$$

This problem contains no more than $m_1 + n_1 = 2$ equations and no more than $m_1 + 2n_1 = 3$ variables. Taking this into account, we rewrite problem (2.35) as

$$\begin{gathered} {}_1x_1 + q^0 + q_1\vartheta + q_2\nu \to \min, \\ (a_{11} + \Delta a_{11})x_1 + q_{11}\vartheta + q_{12}\nu = g_1, \\ \bar{a}_{21}x_1 + q_{21}\vartheta + q_{22}\nu = g_2, \\ x_1 \geq 0, \quad \nu \geq 0. \end{gathered} \tag{2.36}$$

(the coefficients are written in scalar notation).

First, consider the case where the column containing the element a_{11} is basic. We then have $A_{21N} = 0$ and, therefore, $q_2 = q_{12} = q_{22} = 0$.

Let us represent the determinant of the system of equations in problem (2.36) as

$$D = (a_{11} + \Delta a_{11})q_{21} - \bar{a}_{21}q_{11} = q_{21}\Delta a_{11} - d_0,$$

where d_0 is the determinant of the system of equations for $\Delta a_{11} = 0$. The method for aggregating the equations implies that $d_0 \neq 0$.

If $D \neq 0$, then the optimal value f of the objective function of (2.36) depends on the variation Δa_{11} as

$$f = \frac{d_1 \Delta a_{11} - d_2}{q_{21} \Delta a_{11} - d_0},$$

where

$$d_0 = \bar{a}_{21} q_{11} - a_{11} q_{21}, \quad d_1 = q_1 q_2 + q^0 q_{21},$$

$$d_2 = d_0 q^0 + c_1 q_{11} g_2 + q_1 \bar{a}_{21} g_1 - c_1 q_{21} g_1 - q_1 a_{11} g_2.$$

The partial derivative

$$\frac{\partial f}{\partial(\Delta a_{11})} = \frac{q_{21} d_2 - d_1 d_0}{D^2}$$

of this function vanishes under the condition

$$\delta = q_{21} d_2 - d_1 d_0 = 0, \tag{2.37}$$

which does not involve Δa_{11}. Hence, if condition (2.37) holds, then f does not depend on Δa_{11}, and therefore, the optimal objective value of the base problem (2.3) cannot be improved by varying the parameter a_{11}.

If condition (2.37) is violated, then the equation

$$\frac{d_1 \Delta a_{11} - d_2}{q_{21} \Delta a_{11} - d_0} = \frac{d_2}{d_0}$$

in Δa_{11}, which sets the condition that the optimal objective value of problem (2.36) coincides with that of the base problem (2.3) at $\Delta a_{11} = 0$, has the unique solution $\Delta a_{11} = 0$. Therefore, if $\delta > 0$, then decreasing the coefficient a_{11} does reduce the optimal objective value of (2.3) and increasing it does not. If $\delta < 0$, then, vice versa, increasing the coefficient a_{11} does reduce the optimal objective value of (2.3) and decreasing it does not. Certainly, these assertions are only valid under the condition that problem (2.3) remains solvable under the variation Δa_{11}.

Now, consider the case where the column containing a_{11} is not basic. Then problem (2.36) has the optimal solution $x_1 = 0$, $\vartheta = 0$, $\nu = 0$.

If the optimal vertex does not change under the variation Δa_{11}, then neither does the optimal solution.

The introduction of the column $\begin{pmatrix} a_{11} + \Delta a_{11} \\ \bar{a}_{21} \end{pmatrix}$ in the basis of problem (2.36) makes the optimal objective value smaller if the characteristic difference Δ_s of this column is positive and the variation of the basic variable ν is nonnegative, i.e., if

$$\Delta_B = (q_1, q_2) \begin{pmatrix} q_{11}^*, & q_{12}^* \\ q_{21}^*, & q_{22}^* \end{pmatrix} \begin{pmatrix} a_{11} + \Delta a_{11} \\ \bar{a}_{21} \end{pmatrix} - c_1 = h_1 \Delta a_{11} - h_2 > 0, \quad (2.38)$$

$$(q_{21}^*, q_{22}^*) \begin{pmatrix} a_{11} + \Delta a_{11} \\ \bar{a}_{21} \end{pmatrix} = q_{21}^* \Delta a_{11} + h_3 \geq 0,$$

where

$$\begin{pmatrix} q_{11}^*, & q_{21}^* \\ q_{21}^*, & q_{22}^* \end{pmatrix} = \begin{pmatrix} q_{11}, & q_{21} \\ q_{21}, & q_{22} \end{pmatrix}^{-1}; \quad h_1 = (q_1,\, q_2) \begin{pmatrix} q_{11}^* \\ q_{21}^* \end{pmatrix};$$

$$h_2 = c_1 - (q_1,\, q_2) \begin{pmatrix} q_{11}^*, & q_{12}^* \\ q_{21}^*, & q_{22}^* \end{pmatrix} \begin{pmatrix} a_{11} \\ \bar{a}_{21} \end{pmatrix}; \quad h_3 = (q_{21}^*,\, q_{22}^*) \begin{pmatrix} a_{11} \\ \bar{a}_{21} \end{pmatrix}.$$

Thus, a variation Δa_{11} that leaves problem (2.3) solvable leads to a decrease in the optimal objective value if and only if Δa_{11} satisfies conditions (2.38). The situations that can occur depending on h_1 and q_{21}^* are described in Table 2.1.

h_1	q_{21}^*	Conditions on Δa_{11}
$= 0$	any	there is no improving Δa_{11}
> 0	$= 0$	$h_2/h_1 < \Delta a_{11}$ if $h_3 \geq 0$
> 0	> 0	$\max\{h_2/h_1,\, -h_3/q_{21}^*\} < \Delta a_{11}$
> 0	< 0	$h_2/h_1 < \Delta a_{11} \leq -h_3/q_{21}^*$
< 0	$= 0$	$\Delta a_{11} < h_2/h_1$
< 0	> 0	$-h_3/q_{21}^* < \Delta a_{11} < h_2/h_1$
< 0	< 0	$\Delta a_{11} < \min\{h_2/h_1,\, -h_3/q_{21}^*\}$

Table 2.1.

So, in the case where only one element of the matrix A is varied, the use of consistent partially aggregated problems makes it possible to explicitly describe the set of all variations Δa_{11} decreasing the optimal value of the objective function.

§3. Parametric Convex Programming

The main goal of this section is to develop and generalize the method for aggregating parametric linear programming problems to certain classes of parametric convex programming problems. We consider problems described by functions convex with respect to their variables and depending on parameters in a special way.

3.1. Varying Parameters in the Linear Part of Constraints

The consistent aggregation approach described in the preceding section was in essence based only on the linearity of the constraints and on the concentration of the varying parameters (coefficients) in blocks of comparatively small sizes. The linearity of the other constraints and of the objective function was not used in aggregating the variables. This opens up the possibility of extending the obtained results to problems where both objective functions and unvariable constraints are convex.

We start with consideration of problems including unvariable convex constraints. Then, we show how the aggregation approach develop can be applied to parametrically analyze problems with convex objective functions.

Suppose that the original parametric problem has the form

$$\begin{aligned} &f(x) = cx \to \min, \\ &x \in X(A_{11} + \Delta A_{11}) = Z(A_{11} + \Delta A_{11}) \cap F, \end{aligned} \tag{3.1}$$

where

$$Z(A_{11} + \Delta A_{11}) = \{x \mid (A_{11} + \Delta A_{11})x_1 + A_{12}x_2 = b_1\},$$

$$F = \left\{ x \left| \begin{array}{l} A_{21}x_1 + A_{22}x_2 = b_2, \ x \geq 0, \\ q(x_1, x_2) \leq 0 \end{array} \right. \right\}.$$

Here, $x = \begin{pmatrix} x_1 \\ x_2 \end{pmatrix} \in \mathbf{R}^{n_1} \times \mathbf{R}^{n_2} = \mathbf{R}^n$ denotes the vector of variables, $q\colon \mathbf{R}^n \to \mathbf{R}^r$ is a differentiable convex vector function, and the parameters a and y are $m_1 \times n_1$ matrices $A_{11} + \Delta A_{11}$ and A_{11} from a certain feasible set $\mathfrak{U}$. We assume that $\max\{m_1, n_1\} \ll l, n$, where l is the number of linear equations in the description of the feasible set.

Setting $\Delta A_{11} = 0$ in (3.1), we obtain the base optimization problem

$$cx \to \min, \quad Ax = b, \quad x \geq 0, \quad q(x) \leq 0, \tag{3.2}$$

where $A = \begin{pmatrix} A_{11}, & A_{12} \\ A_{21}, & A_{22} \end{pmatrix}$, $b = \begin{pmatrix} b_1 \\ \\ b_2 \end{pmatrix}$, and $c = (c_1, c_2)$; this problem together with (3.1) is an object of our further considerations.

We aggregate problem (3.1) in such a way that gathering the variables preserve the original forms of the variables $x_1 \in \mathbf{R}^{n_1}$ and of the corresponding columns in the matrix condition from the linear part of the constraints, and that aggregating the constraints leave the inequalities $x_1 \geq 0$ and $q(x) \leq 0$ and the equations from the system $(A_{11} + \Delta A_{11})x_1 + A_{12}x_2 = b_1$ intact. The block $A_{11} + \Delta A_{11}$ retains its original form in the aggregated problem. As previously, we understand the consistency of the obtained aggregated problem and the original problem (3.1) in the sense of Definition 1.1.

To apply the results obtained in the preceding section for linear programming problems, we use a linearization method locally reducing the nonlinear problem (3.2) to a linear problem. Suppose that we know primal and dual optimal solutions x^0 and (π^0, ζ^0, ξ^0) to (3.2), where the components of the dual solution refer to the groups of constraints $Ax = b$, $x \geq 0$, and $q(x) \leq 0$, respectively. Let us introduce the sets of indices

$$I^A = \{i \mid q_i(x^0) = 0,\ \xi^0 > 0,\ i = \overline{1, r}\},$$

$$I^K = \{i \mid q_i(x^0) = 0,\ \xi^0 > 0,\ i = \overline{1, r}\},$$

$$J = \{1, 2, \dots, n\}, \quad J^A = \{j \mid x_j^0 = 0,\ \zeta_j^0 > 0,\ j \in J\},$$

$$J^K = \{j \mid x_j^0 = 0,\ \zeta_j^0 = 0,\ j \in J\}.$$

Definition 3.1. We say that problem (3.1) is nondegenerate at $\Delta A_{11} = 0$ if its feasible set $X(A_{11})$ satisfies the Slater regularity conditions and the dual solution (π^0, ζ^0, ξ^0) is unique.

Let H^A and H^K denote the matrices formed by the transposed gradients of the constraints with indices from the sets I^A and I^K, respectively. Note that, if $X(A_{11})$ is regular, then none of the rows in H^A and H^K can be zero. Leaving only those nonlinear constraints in (3.2) that correspond to I^A and linearizing them at the optimal point x^0, we obtain the linear programming problem

$$cx \to \min, \quad Ax = b, \quad x \geq 0, \quad H^A x \leq d, \tag{3.3}$$

where the right-hand side vector d has components $\nabla^T q_i(x^0)x^0$ for $i \in I^A$.

As is known, if $X(A_{11})$ is regular, then the point x^0 is an optimal solution to problem (3.3). Moreover, it is shown in [18] that, if (π^0, ζ^0, ξ^0) is unique, then the gradients of the system

$$Ax = b, \quad q_i(x) \leq 0, \quad i \in I^A, \quad x_j \geq 0, \quad j \in J^A,$$

evaluated at the point x^0 are linearly independent, and there exists a vector $\eta \in \mathbf{R}^n$ satisfying the relations

$$\begin{aligned} &A\eta = 0, \quad H^A\eta = 0, \quad H^K\eta < 0, \quad c\eta = 0, \\ &\eta_j = 0, \quad j \in J^A, \quad \eta_j > 0, \quad j \in J^K. \end{aligned} \tag{3.4}$$

This, in particular, implies that the columns of $\begin{pmatrix} A \\ H^A \end{pmatrix}$ with numbers from the set $J \setminus J^A$ form at least one nondegenerate matrix with the same number of rows. Among all these matrices, we choose one whose columns correspond to the largest number of positive components of x^0 (if there are several such matrices, we take any of them). We call the chosen matrix basic[1] and denote it by B; for the inverse matrix, we use the notation $\tilde{B} = B^{-1}$. By J_B, we denote the set of indices of the basic variables, which correspond to the columns of B. Hereafter, the superscripts A and K refer to the blocks of matrices or vectors corresponding to the sets I^A and I^K, respectively. Recall that the subscript B labels the blocks corresponding to the basic components (with indices from the set J_B) and N marks those corresponding to the nonbasic components.

In this section, we study some important properties of problem (3.1), using the linearized problem (3.3) as the main tool, and apply them to construct procedures for aggregating both variables and constraints.

As to the set $\mathfrak{V}$, we assume it to have property (2.2).

Under these assumptions, the following lemma is valid.

Lemma 3.1. *Suppose that problem (3.1) is nondegenerate at $\Delta A_{11} = 0$, the set $\mathfrak{V}$ has property (2.2), and $A_{11} + \Delta\tilde{A}_{11} \in \Gamma(A_{11})$. Then there exist a vector x^* and a scalar $\beta^* > 0$ such that, for any positive $\beta \leq \beta^*$, there is a $\Delta A_{11}(\beta)$ for which $A_{11} + \Delta A_{11}(\beta) \in \mathfrak{V}$,*

$$x(\beta) = (1 - \beta)x^0 + \beta x^* \in X(A_{11} + \Delta A_{11}(\beta)),$$

and $cx(\beta) < cx^0$.

---

[1] Traditionally, the basic matrix is chosen to have columns corresponding to the largest components of x^0.

Proof. Put

$$\tilde{x} = \operatorname{argmin}\{cx \mid x \in X(A_{11} + \Delta\tilde{A}_{11})\}.$$

By assumption, $c\tilde{x} < cx^0$.

Problem (3.3) is a linear programming problem; therefore, to analyze it at various preferred parameter values $A_{11} + \Delta\tilde{A}_{11} \in \Gamma(A_{11})$, we can apply a technique similar to that developed in Section 2. Namely, consider the extended problem

$$\begin{aligned} &\psi(w) = c(u+v) \to \min, \\ &Au + (A + \Delta\tilde{A})v = b, \\ &H^A(u+v) \le d, \quad u \ge 0, \quad v \ge 0, \end{aligned} \tag{3.5}$$

where $\omega = (u, v)$ is the vector of variables and

$$A + \Delta\tilde{A} = \begin{pmatrix} A_{11} + \Delta\tilde{A}_{11}, A_{12} \\ A_{21}, \quad\quad A_{22} \end{pmatrix}.$$

At the feasible point $\omega^0 = (x^0, 0)$ of problem (3.5), the objective function takes the value $\psi(\omega^0) = cx^0$, and at the point $\tilde{\omega} = (0, \tilde{x})$, it takes the value $\psi(\tilde{\omega}) = c\tilde{x} < cx^0$. The point $\tilde{\omega}$ is also feasible in (3.5), because $\tilde{x} \in X(A_{11} + + \tilde{A}_{11})$ by definition. Since the vector function $q(x)$ is convex, we have

$$H^A\tilde{x} - d = \nabla q^A(x^0)(\tilde{x} - x^0) \le q^A(\tilde{x}) - q^A(x^0) = q^A(\tilde{x}) \le 0.$$

Therefore, the point ω^0 is not optimal in (3.5), and, taking the matrix B defined above as the basic matrix corresponding to this point, we can find at least one column with negative characteristic difference among the first n_1 columns of the matrix $\begin{pmatrix} A + \Delta\tilde{A} \\ H^A \end{pmatrix}$ in problem (3.5) (the characteristic differences of the last $2n - n_1$ columns are nonnegative, because x^0 is optimal for (3.3)).

Suppose that this column has number s; it can be written as $\begin{pmatrix} a_s + \Delta\tilde{a}_s \\ h_s^A \end{pmatrix}$, where h_s^A and $a_s + \Delta\tilde{a}_s$ are the s-th columns of the matrices H^A and $A + \Delta\tilde{A}$, respectively. By definition, its characteristic difference (or dual estimate) is

$$\Delta_s^0 = c_s - c_B\tilde{B}\begin{pmatrix} a_s + \Delta\tilde{a}_s \\ h_s^A \end{pmatrix},$$

where c_s is the s-th component of c. Since $\Delta_s^0 < 0$, we might apply the geometric implementation of the simplex method to construct an edge $\Delta\tilde{\omega}$ going from ω^0 along which the value $c(u+v)$ decreases; this edge is uniquely determined by the basic matrix B and the vector $\begin{pmatrix} a_s + \Delta\tilde{a}_s \\ h_s^A \end{pmatrix}$. If all basic components $\omega_B^0 = x_B^0$ are positive, then this edge determines a descent direction, i.e., moving along this edge improves the objective value and does not lead outside the feasible set of (3.5). In the case under consideration (where $J^K \cap J_B \neq \varnothing$), some basic components vanish, which may become an obstruction to moving from ω^0 along the "improving" edge. For this reason, searching a suitable edge (which determines a descent direction) generally requires changing the basis B and examining all other bases at the given point ω^0.

To avoid the necessity for a search through the bases, it is necessary to "correct" the above-mentioned direction $\Delta\tilde{\omega}$ in such a way that it becomes a descent direction for (3.5) and fall inside the set

$$\{\omega \mid H^A(u+v) \leq 0,\ H^K(u+v) \leq 0\}.$$

The latter condition is related to the nonlinear constraints $q(u+v) \leq 0$.

Consider the following "perturbed" direction in $\mathbf{R}^{2n}$ specified by the set of points with coordinates

$$\begin{aligned} u_B(\beta) &= x_B^0 + \left[\delta\eta_B - B\begin{pmatrix} u_s + \Delta\tilde{a}_s \\ h_s^A \end{pmatrix} v_s - \varepsilon\tilde{B}\begin{pmatrix} 0 \\ \mu \end{pmatrix}\right]\beta, \\ u_N(\beta) &= x_N^0 + \delta\eta_N\beta, \quad v(\beta) = (0,\ldots,v_s\beta,\ldots,0)^T, \end{aligned} \tag{3.6}$$

where $\mu = (1,\ldots,1)^T \in \mathbf{R}^{|I_A|}$; the scalar value v_s is nonnegative; the values $\delta, \varepsilon \geq 0$, treated as perturbations, are external parameters of the ray (3.6); and $\beta \geq 0$ characterizes the motion along this ray from the point $\omega^0 = (x^0, 0)$.

Since $\Delta_s^0 < 0$ and, by assumption, relations (3.4) hold, we can always select sufficiently small numbers $v_s^0 > 0$, $\varepsilon^0 > 0$, and $\beta_1^* > 0$ and a sufficiently large number $\delta^0 > 0$ such that, first,

$$\begin{aligned} u_B(\beta_1^*) &= x_B^0 + \left[\delta^0\eta_B - \tilde{B}\begin{pmatrix} a_s + \Delta\tilde{a}_s \\ h_s^A \end{pmatrix} v_s^0 - \varepsilon^0\tilde{B}\begin{pmatrix} 0 \\ \mu \end{pmatrix}\right]\beta_1^* > 0, \\ H^K(u(\beta_1^*) &+ v(\beta_1^*)) - H^K x^0 < 0, \end{aligned} \tag{3.7}$$

and, secondly,

$$\Delta\psi = \Delta_s^0 v_s^0 - c_B\varepsilon^0 \begin{pmatrix} 0 \\ \mu \end{pmatrix} < 0.$$

Next, substituting (3.6) in (3.5) and taking into account (3.4), we easily conclude that all points from the interval $\omega(\beta)$ with $0 \le \beta \le \beta_1^*$ are feasible in problem (3.5), and

$$H^A(u(\beta_1^*) + v(\beta_1^*)) - d = -\varepsilon^0 \mu \beta_1^* < 0. \tag{3.8}$$

In addition, for $\beta > 0$, we have

$$\begin{aligned}\psi(\omega(\beta)) - \psi(\omega^0) = c(u(\beta) + v(\beta) - cx^0) &= \\ = c_B x_B^0 + c_B \left[\delta^0 \eta_\beta - \tilde{B} \begin{pmatrix} a_s + \Delta a_s \\ h_s^A \end{pmatrix} v_s^0 - \varepsilon^0 \tilde{B} \begin{pmatrix} 0 \\ \mu \end{pmatrix}\right] \beta &+ \\ + c_N x_N^0 + c_N \delta^0 \eta_N \beta + c_s v_s^0 \beta &= \\ = c\eta + \left[\Delta_s^0 v_s^0 - c_B \varepsilon^0 \tilde{B} \begin{pmatrix} 0 \\ \mu \end{pmatrix}\right] \beta = \Delta\psi\beta &< 0. \end{aligned} \tag{3.9}$$

Thus, the direction (3.6) satisfies all requirements set above. It remains to "project" it onto the original problem (3.1). Let us fix the chosen values v^0, ε^0, and δ^0 and consider the points $x(\beta) = u(\beta) + v(\beta)$ with $0 \le \beta \le \beta_1^*$ in the space of variables $x \in \mathbf{R}^n$ of problem (3.1); by construction, these points form a straight line segment in the nonnegative orthant. We also introduce the families

$$0 \le \alpha_s(\beta) = \frac{v_s(\beta)}{u_s(\beta) + v_s(\beta)} \le 1, \quad \text{where} \quad 0 \le \beta \le \beta_1^*,$$

of scalars and

$$(\Delta A(\beta)) = (0, \ldots, \Delta\tilde{a}_s \alpha_s(\beta), \ldots, 0)$$

of the corresponding matrices.

The above constructions imply that

$$\begin{aligned}(A + \Delta A(\beta))x(\beta) = Au(\beta) + (A + \Delta\tilde{A})v(\beta) &= \\ = Ax^0 - (a_s + \Delta\tilde{a}_s)v_s^0\beta + (a_s + \Delta\tilde{a}_s)v_s^0\beta &= b.\end{aligned}$$

for all nonnegative $\beta \le \beta_1^*$. In addition, according to (3.7) and (3.8), the directional derivatives at the point x^0 satisfy the inequalities

$$\nabla q^A(x^0)(x(\beta_1^*) - x^0) = H^A(u(\beta_1^*) + v(\beta_1^*)) - d < 0,$$

$$\nabla q^K(x^0)(x(\beta_1^*) - x^0) = H^K(u(\beta_1^*) + v(\beta_1^*)) - H^K x^0 < 0.$$

Since $q_i(x^0) = 0$ for $i \in I^A \cup I^K$ and $q_i(x^0) < 0$ for the other i's, these inequalities imply the existence of a sufficiently small number $\beta_2^* > 0$ such that $q(x(\beta)) < 0$ for all nonnegative $\beta \leq \beta_2^*$. So, the points $x(\beta)$ with $0 \leq \beta \leq \beta^* = \min\{\beta_1^*, \beta_2^*\}$ are feasible in problem (3.1) with the corresponding coefficients $A + \Delta A(\beta)$.

As to the values of the objective function at these points, (3.9) implies

$$cx(\beta) - cx^0 = c(u(\beta) + v(\beta) - x^0) < 0, \quad \beta > 0.$$

To conclude the proof of the lemma, we form the diagonal matrix $\alpha(\beta) =$ $= \operatorname{diag}\{0, \dots, \alpha_s(\beta), \dots, 0\}$ of order n_1 and put $\Delta A_{11}(\beta) = \Delta\tilde{A}_{11}\alpha(\beta)$ and $x^* = x(\beta^*)$. Since the set $\mathfrak{U}$ of feasible parameter values has property (3.2), the values $A_{11} + \Delta A_{11}(\beta)$ belong to $\mathfrak{U}$ for $0 \leq \beta \leq \beta^*$. At the same time, each point $x(\beta)$ can be represented as $x(\beta) = (1-\beta)x^0 + \beta x^*$; therefore,

$$x(\beta) \in X(A_{11} + \Delta A_{11}(\beta)), \quad cx(\beta) < cx^0, \quad 0 < \beta \leq \beta^*.$$

This completes the proof of the lemma.

Remark 3.1. It follows from the proof of Lemma 3.1 that, in the definition of ray (3.6), we can take $\delta^0 = 0$ if $x_B^0 > 0$ (i.e., if all basic components are positive) and $\varepsilon^0 = 0$ if the function q is affine.

The next property is concerned with the spatial arrangement of the points $x(\beta)$ mentioned in Lemma 3.1. It turns out that the points $x(\beta)$ constructed for all improving $A_{11} + \Delta A_{11} \in \Gamma(A_{11})$ lie in an affine manifold $\Pi_y \subset \mathbf{R}^n$ of a relatively small dimension (in comparison with that of $\mathbf{R}^n$). This allows us to deal only with points from Π_y, rather than from the entire space $\mathbf{R}^n$, in determining improving parameter values $A_{11} + \Delta A_{11} \in \Gamma(A_{11})$ for problem (3.1).

Let us rearrange the variables in problem (3.1) in such a way that the matrices A, $\tilde{B} = B^{-1}$, and H^A and the vector x admit the following decompositions similar to those introduced in Section 2:

$$A = \begin{pmatrix} A_{11B}, A_{11N}, A_{12B}, A_{12N} \\ A_{21B}, A_{21N}, A_{22B}, A_{22N} \end{pmatrix},$$

$$\tilde{B} = \begin{pmatrix} \tilde{B}_{11}, \tilde{B}_{12}, \tilde{B}_{13} \\ \tilde{B}_{21}, \tilde{B}_{22}, \tilde{B}_{23} \end{pmatrix} = \begin{pmatrix} \tilde{B}_1 \\ \tilde{B}_2 \end{pmatrix}, \quad x = \begin{pmatrix} x_{1B} \\ x_{1N} \\ x_{2B} \\ x_{2N} \end{pmatrix}, \tag{3.10}$$

$$H^A = \left(H_{1B}^A, H_{1N}^A, H_{2B}^A, H_{2N}^A\right).$$

It is assumed that all dimensions of blocks in the inverse matrix $\tilde{B}$ are consistent with the dimensions of the corresponding blocks of the matrix B, which has the structure

$$B = \begin{pmatrix} A_{11B}, A_{12B} \\ A_{21B}, A_{22B} \\ H_{1B}^A, \ H_{2B}^A \end{pmatrix}.$$

In particular, in the case under consideration, the submatrices A_{21N} and H_{1N}^A have sizes $(l-m_1)\times n_{1N}$ and $|I^A|\times n_{1N}$, respectively, and the submatrices $\tilde{B}_{11}$ and $\tilde{B}_{23}$ have sizes $n_{1B}\times m_1$ and $(l+|I^A|-n_{1B})\times|I^A|$, where $n_{1B}+n_{1N} = n_1$.

Let us form the matrices

$$R = \left(\tilde{B}_{21}, \ -\tilde{B}_{21}, \ -(\tilde{B}_{22}, \tilde{B}_{23})\begin{pmatrix} A_{21N} \\ H_{1N}^A \end{pmatrix}, \ -\tilde{B}_{23}\mu\right),$$

$$P_y = \begin{pmatrix} R, \eta_{2B} \\ 0, \ \eta_{2N} \end{pmatrix},$$

whose elements are uniquely determined by the coefficients of problem (3.3), and consider the manifold

$$\Pi_y = \{x \mid x_1 \in \mathbf{R}^{n_1}, \ x_2 = x_2^0 + P_y p, \ p \in \mathbf{R}^{n_A}\}. \tag{3.11}$$

The same arguments as in the preceding section can be used to show that the dimension n_A of Π_y does not exceed $2(n_1+m_1+1)$ by virtue of the choice of the matrix P_y. In addition, the structure of P_y is selected in such a way that the following lemma is valid.

Lemma 3.2. *Suppose that the assumptions of Lemma 3.1 hold. Then the points $x(\beta)$ constructed in the proof of Lemma 3.1 belong to Π_y.*

Proof. Prior to proceeding to the proof, let us explain what the assertion to be proved means as related to the extended problem (3.5). While Lemma 3.1 says that each improving $A_{11} + \Delta\tilde{A}_{11} \in \Gamma(A_{11})$ in (3.5) determines at least one descent direction of form (3.6), Lemma 3.2 actually states that the descent directions of form (3.6) determined by all $A_{11} + \Delta\tilde{A}_{11} \in \Gamma(A_{11})$ lie in an affine manifold of a small dimension.

To prove the lemma, it suffices to show that $x_2(\beta) = x_2^0 + P_y p(\beta)$ for some $p(\beta) \in \mathbf{R}^{n_A}$.

Let $\Delta\tilde{a}_s'$ be the subvector formed by the first m_1 components of the vector $\Delta\tilde{a}_s$ defined in the proof of Lemma 3.1. Put

$$\xi^* = -(A_{11N}e_s + \Delta\tilde{a}_s')v_s^0,$$

$$p_1^* = \begin{pmatrix} \xi_+^* \\ -\xi_-^* \end{pmatrix}, \quad p_2^* = \begin{pmatrix} e_s v_s^0 \\ \varepsilon^0 \\ \delta \end{pmatrix}, \quad p^* = \begin{pmatrix} p_1^* \\ p_2^* \end{pmatrix},$$

where the n_{1N}-vector e_s corresponding to the columns of A_{11N} is zero if $s \in J_B$ (i.e., if its number s is basic) and unit if s is nonbasic (1 then occupies the s-th position). The vectors ξ_+^* and ξ_-^* are the projections of ξ^* on the positive and negative orthants, respectively.

Put $p(\beta) = p^*\beta$. By definition, $x(\beta) = u(\beta) + v(\beta)$. Using (3.6), the notation introduced above, and the equalities $v_{2B}(\beta) = 0$ and $v_{2N}(\beta) = 0$ (which are implied by $s \le n_1$), we obtain

$$x_{2N}(\beta) = u_{2N}(\beta) + v_{2N}(\beta) = u_{2N}(\beta) = x_{2N}^0 + \eta_{2N}\delta^0\beta = x_{2N}^0 + (0, \eta_{2N})p^*\beta.$$

At the same time,

$$x_{2B}(\beta) = u_{2B}(\beta) + v_{2B}(\beta) = u_{2B}(\beta) = \\ = x_{2B}^0 - \tilde{B}_2 \begin{pmatrix} a_s + \Delta\tilde{a}_s \\ h_s^A \end{pmatrix} v_s^0\beta - \varepsilon^0\tilde{B}\begin{pmatrix} 0 \\ \mu \end{pmatrix}\beta + \delta^0\eta_{2B}\beta.$$

The representation

$$\tilde{B}_2 \begin{pmatrix} a_s + \Delta a_s \\ h_s^A \end{pmatrix} = \tilde{B}_{21}(A_{11N}e_s + \Delta\tilde{a}_s') + (\tilde{B}_{22}, \tilde{B}_{23})\begin{pmatrix} A_{21N} \\ H_{1N}^A \end{pmatrix} e_s,$$

gives

$$x_{2B}(\beta) = x_{2B}^0 + (\tilde{B}_{21}, -\tilde{B}_{21})p_1^*\beta - \\ - \left((\tilde{B}_{22}, \tilde{B}_{23})\begin{pmatrix} A_{21N} \\ H_{1N}^A \end{pmatrix}, \tilde{B}_2\begin{pmatrix} 0 \\ \mu \end{pmatrix}, -\eta_{2B} \right) p_2^*\beta = \\ = x_{2B}^0 + (R, \eta_{2B})p^*\beta = x_{2B}^0 + (R, \eta_{2B})p(\beta).$$

This completes the proof of the lemma.

Remark 3.2. The vectors $p(\beta)$ are nonnegative for all $\beta \geq 0$.

Remark 3.3. According to Remark 3.1, if $x_B^0 > 0$ and q is affine, we can take

$$R = \left(\tilde{B}_{21},\ -\tilde{B}_{21},\ -(\tilde{B}_{22},\ \tilde{B}_{23})\begin{pmatrix} A_{21N} \\ H_{1N}^A \end{pmatrix}\right) \quad \text{and} \quad P_y = \begin{pmatrix} R \\ 0 \end{pmatrix}.$$

Now, we proceed to aggregate the variables in (3.1) following the general aggregation scheme. The goal of aggregation is to construct the set

$$Z_y(A_{11} + \Delta A_{11}) \cap F_y = \left\{(x_1, p) \mid (x_1, x_2^0 + P_y p) \in Z(A_{11} + \Delta A_{11}) \cap F\right\}.$$

For this purpose, we apply a suitable change of variables to cut the feasible solutions to problem (3.10) by manifold (3.11). As a result, we obtain the intermediate problem

$$f_y(x_1, p) = c_1 x_1 + c_2 R p + c_2 x_2^0 \to \min,$$
$$(x_1, p) \in X_y(A_{11} + \Delta A_{11}) = Z_y(A_{11} + \Delta A_{11}) \cap F_y,$$

where

$$Z_y(A_{11} + \Delta A_{11}) = \{(x_1, p) \mid (A_{11} + \Delta A_{11})x_1 + A_{12} P_y p = g_1\},$$

$$F_y = \left\{ (x_1, p) \middle| \begin{array}{l} A_{21}x_1 + A_{22}P_y p = g_2,\ x_1 \geq 0, \\ x_2^0 + P_y p \geq 0,\ q(x_1, x_2^0 + P_y p) \leq 0 \end{array} \right\},$$

$$g_1 = b_1 - A_{12}x_2^0, \quad g_2 = b_2 - A_{22}x_2^0.$$

Aggregating the obtained constraints reduces to constructing a set $G_y \subset F_y$ of structure simpler than that of F_y (i.e., G_y involves a smaller number of constraints). Following the method described in Section 2, we can replace the system of equations

$$A_{21}x_1 + A_{22}P_y p = g_2$$

by the subsystem

$$\tilde{A}_{21}x_1 + \tilde{A}_{22}P_y p = \tilde{g}_2 \tag{3.12}$$

comprising a significantly smaller number of equations. This is achieved by eliminating linearly dependent equations. According to the results of Section 2, the number of rows in (3.12) does not exceed n_1 in the general case.

Let us aggregate the system $x_2^0 + P_y p \geq 0$. It is convenient to introduce the additional decomposition $x_{2B} = (x_{21B},\ x_{22B})$, where $x_{21B}^0 > 0$ and $x_{22B}^0 = 0$. It corresponds to the following decompositions of the matrix R and vector η_{2B}:

$$R = \begin{pmatrix} R_1 \\ R_2 \end{pmatrix}, \quad \eta_{2B} = \begin{pmatrix} \eta_{21B} \\ \eta_{22B} \end{pmatrix}.$$

Using these decompositions, we represent the set $\{p \mid x_2^0 + P_y p \geq 0\}$ as the intersection

$$\{p \mid x_{21B}^0 + (R_1,\ \eta_{21B})p \geq 0\} \cap \{p \mid (R_2,\ \eta_{22B})p \geq 0\} \cap \\ \cap \{p \mid x_{2N}^0 + (0,\ \eta_{2N})p \geq 0\}. \tag{3.13}$$

Since $x_{21B}^0 > 0$, Remark 3.1 and considerations of Section 2 allow us to replace the first set in this intersection by the embedded simplex

$$\{p \mid tp \leq 1,\ p \geq 0\} \subset \{p \mid x_{21B}^0 + (R_1,\ \eta_{21B})p \geq 0\},$$

where the coefficients in the nonnegative row vector t are evaluated by the formulas given in Section 2. The second set, which is a cone, is left unchanged. As to the third set, the relations $x_{2N}^0 \geq 0$ and $\eta_{2N} \geq 0$ imply the inclusion

$$\{p \mid p \geq 0\} \subset \{p \mid x_{2N}^0 + (0,\ \eta_{2N})p \geq 0\}.$$

Summarizing, we can leave the system $x_1 \geq 0$, $q(x_1, x_2^0 + P_y p) \leq 0$ in its original form and apply the above considerations to form the nonempty set

$$G_y = \left\{ (x_1, p) \left| \begin{array}{l} \tilde{A}_{21}x_1 + \tilde{A}_{22}P_y p = \tilde{g}_2,\ x_1 \geq 0, \\ tp \leq 1,\ p \geq 0,\ (R_2,\ \eta_{22B})p \geq 0, \\ q(x_1, x_2^0 + P_y p) \leq 0 \end{array} \right. \right\},$$

where by construction, $G_y \subset F_y$.

Thus, the aggregated parametric problem obtained at the final step of the aggregation procedure takes the form

$$\begin{aligned} & f_y(x_1,\ p) = c_1 x_1 + c_2 P_y p + c_2 x_2^0 \to \min, \\ & (x_1,\ p) \in X_y(A_{11} + \Delta A_{11}) = Z_y(A_{11} + \Delta A_{11}) \cap G_y. \end{aligned} \tag{3.14}$$

Let us compare the dimensions of problem (3.14) with those of (3.1). The original problem (3.1) contains n variables and $m = l + r$ constraints[2], including l linear equations. Let n_{22B} denote the dimension of the vector x_{22B} (i.e., the number of nonzero basic variables in x_2^0). The total number of zero basic components in an optimal solution, which characterizes the degree of degeneracy of this solution, is usually small, and we can therefore assume that $n_{22B} \ll n$. Using the estimates obtained in the preceding constructions and in Section 2, we easily derive the bounds $n_A \leq 2(n_1 + m_1 + 1)$ and $m_A \leq n_1 + m_1 + r + 1$, where n_A is the number of variables and m_A is the number of constraints in (3.14). By assumption, $\max\{m_1, n_1\} \ll l, n$, whence $n_A \ll n$. At the same time, if (3.1) is "almost linear," i.e., the number r of its nonlinear constraints is much smaller than m (as is often the case with problem having large dimensions, for example, when a nonlinear production function is used in linear input–output models), then $m_A \ll m$. Therefore, the dimensions of the aggregated parametric problem (3.14) are substantially smaller than those of the original problem (3.1).

It remains to prove the main result.

Theorem 3.1. *Under the assumptions of Lemma 3.1, the aggregation of the original parametric problem (3.1) into problem (3.14) is consistent at the point* A_{11}.

Proof. According to Remark 1.1, it suffices to establish the fulfillment of the second condition in the Definition 1.1 of consistency.

Suppose that $\Gamma(A_{11}) \neq \varnothing$ and $A_{11} + \Delta\tilde{A}_{11} \in \Gamma(A_{11})$. Let us apply Lemmas 3.1 and 3.2 and put

$$\beta = \bar{\beta} = \min\{\beta^*, [tp^*]^{-1}\} \leq \beta^*,$$

where $p^* \in \mathbf{R}^{n_A}$ is the vector defined in the proof of Lemma 3.2. For this $\bar{\beta}$, we set $\bar{x} = x(\bar{\beta})$ and $\Delta\bar{A}_{11} = \Delta A_{11}\alpha(\bar{\beta})$. By Lemma 3.1, we have $\bar{x} \in X(A_{11} + \Delta\bar{A}_{11})$ and

$$f(\bar{x}) < f(x^0) = \varphi(A_{11}). \tag{3.15}$$

At the same time, Lemma 3.2 and the definition of $\bar{\beta}$ imply $\bar{x}_2 = x_2^0 + P_y p^* \bar{\beta}$ and

$$(\bar{x}_1, p^*\bar{\beta}) \in X_y(A_{11} + \Delta\bar{A}_{11}).$$

[2]We do not count constraints of the type $x \geq 0$, which are simple from the point of view of simplex-like solution methods.

Note that

$$f_y(\bar{x}_1, p^*\bar{\beta}) = f(\bar{x}_1, x_2^0 + P_y p^*(\bar{\beta})) = f(\bar{x});$$

combining this with the relations (3.15) and $\varphi(A_{11}) = \varphi_y(A_{11})$, we obtain

$$\varphi_y(A_{11} + \Delta\bar{A}_{11}) \leq f_y(\bar{x}_1, p^*\bar{\beta}) < \varphi(A_{11}) = \varphi_y(A_{11}).$$

Thus, $A_{11} + \Delta\bar{A}_{11} \in \Gamma_A(A_{11})$, which proves the theorem.

Let us show how the results obtained above can be applied to analyze parametric problems with convex objective functions. Suppose that the original problem has the form

$$\begin{aligned} &f(x) = \max_{i=\overline{1,K}} f_i(x) \to \min, \\ &(A + \Delta A)x = b, \quad x \geq 0, \quad q(x) \leq 0. \end{aligned} \tag{3.16}$$

Here, $A+\Delta A = \begin{pmatrix} A_{11} + \Delta A_{11}, A_{12} \\ A_{21}, \qquad A_{22} \end{pmatrix}$ is the matrix of constraints and $f_i\colon \mathbf{R}^n \to \mathbf{R}$, where $i = \overline{1,K}$, are convex differentiable functions. As is known, the function $f(x) = \max_{i=\overline{1,K}} f_i$ is then convex but not necessarily differentiable.

Let us transform (3.16) into the equivalent parametric convex programming problem

$$\begin{aligned} &z \to \min, \\ &(A + \Delta A)x = b, \quad x \geq 0, \\ &f_i(x) - z \leq 0, \quad \forall i, \quad q(x) \leq 0, \end{aligned} \tag{3.17}$$

which belongs to the class of problems (3.1) under consideration.

Applying the aggregation technique described above to (3.17) and reducing the resulting problem of type (3.14) to the original minimax form, we can construct the following aggregated problem for (3.16):

$$\begin{aligned} &fy(x_1, p) = \max_{i=\overline{1,K}} f_i(x_1, x_2^0 + P_y p) \to \min, \\ &(A_{11} + \Delta A_{11})x_1 + A_{12}P_y p = g_1, \quad x_1 \geq 0, \quad p \geq 0, \\ &\bar{A}_{21}x_1 + \bar{A}_{22}P_y p = \bar{g}_2, \quad tp \leq 1, \\ &(R_2, \eta_{22B})p \geq 0, \quad q(x_1, x_2^0 + P_y p) \leq 0. \end{aligned} \tag{3.18}$$

If (3.17) is nondegenerate at $\Delta A_{11} = 0$, then, according to Theorem 3.1, problem (3.18) is consistent with (3.16) at the point A_{11}.

In the special case of $K = 1$, when the objective function $f(x)$ is convex and differentiable, there is no need in the passage to the equivalent statement (3.17); for the problem (3.16) as it is, the following theorem is then valid.

Theorem 3.2. *Suppose that $K = 1$, problem (3.16) is nondegenerate at $\Delta A_{11} = 0$, and $\mho$ has property (2.2). Then the aggregation of (3.16) into (3.18) is consistent at the point A_{11}.*

It follows from the aforesaid that the possibility of consistent aggregation is not related to the differentiability of the objective function of the problem. Rather, it is determined by the nondegeneracy condition (ensuring the uniqueness of the dual solution); in the case under consideration, this condition is sufficient.

To more convincingly demonstrate the necessity of the nondegeneracy requirement, consider the example

$$\begin{aligned} &f(x) = \min\{f_1(x), f_2(x)\} \to \max, \\ &(2 + \Delta A_{11})x_1 + \tfrac{3}{2}x_2 + x_3 = 1, \\ &x_1 \geq 0, \quad x_2 \geq 0, \quad x_3 \geq 0, \end{aligned} \tag{3.19}$$

where $A_{11} = 2$,

$$f_1(x) = x_1 + 2x_2 + x_3 - 1, \quad f_2(x) = 2x_1 + x_2 + x_3 - 1.$$

The equivalent statement of form (3.17) for problem (3.19) is

$$\begin{aligned} &x_4 \to \min, \\ &(2 + \Delta A_{11})x_1 + \tfrac{3}{2}x_2 + x_3 = 1, \\ &x_1 + 2x_2 + x_3 + x_4 \geq 1, \\ &2x_1 + x_2 + x_3 + x_4 \geq 1, \\ &x_1 \geq 0, \quad x_2 \geq 0, \quad x_3 \geq 0, \end{aligned} \tag{3.20}$$

and its dual is

$$\begin{aligned} &\lambda_1 + \lambda_2 + \lambda_3 \to \max, \\ &(2 + \Delta A_{11})\lambda_1 + \lambda_2 + 2\lambda_3 \leq 0, \\ &\tfrac{3}{2}\lambda_1 + 2\lambda_2 + \lambda_3 \leq 0, \\ &\lambda_1 + \lambda_2 + \lambda_3 \leq 0, \\ &\lambda_2 + \lambda_3 = 1, \quad \lambda_2 \geq 0, \quad \lambda_3 \geq 0. \end{aligned}$$

Put $\Delta A_{11} = 0$. A direct verification shows that the vector $x^0 = (0,0,1,0)^T$ is then a solution to (3.20), the optimal solutions to the corresponding dual problem form the set

$$\{\lambda \mid \lambda_1 = -1,\ \lambda_2 + \lambda_3 = 1,\ 0 \leq \lambda_2 \leq 1/2\},$$

and the optimal value is $\varphi(A_{11}) = x_4^0 = 0$. Note that $I^K = 0$ and $I^A = \{1,2\}$.

Thus, (3.20) is degenerate in the sense of definition (3.1). For this reason, Lemma 3.1 does not apply, and it is not always possible to construct a descent direction of form (3.6) only with the use of the columns subjected to variation for a fixed basic matrix B, because constructing such a direction may require changing the basis.

To show this, consider $J_B = \{1,3,4\}$ as the set of basic variables. The corresponding basic matrix B at the point x^0 and its inverse are

$$B = \begin{pmatrix} 2,1,0 \\ 1,1,1 \\ 2,1,1 \end{pmatrix}, \quad \tilde{B} = B^{-1} = \begin{pmatrix} 0, & -1, & 1 \\ 1, & 2, & -2 \\ -1, & 0, & 1 \end{pmatrix}.$$

The dual vector corresponding to this basic is $\lambda^0 = c_B\tilde{B} = (-1,0,1)$.

Suppose that the parameters in (3.20) are subjected to the variation $\Delta A_{11} = \Delta\tilde{A}_{11} = -1$. The changed system has the new optimal solution $\tilde{x} = (\frac{2}{5}, \frac{2}{5}, 0, -\frac{1}{5})^T$, the dual solution $\tilde{\lambda} = (-\frac{6}{5}, \frac{4}{5}, \frac{1}{5})$, and the optimal value $\varphi(A_{11} + \Delta\tilde{A}_{11}) = \varphi(1) = \tilde{x}_4 = -\frac{1}{5}$. Thus, $\varphi(1) < \varphi(A_{11})$, i.e., $2 + \Delta\tilde{A}_{11} \in \Gamma(A_{11})$. According to the proof of Lemma 3.1, the "perturbed" column

$$\begin{pmatrix} a_s + \Delta\tilde{a}_s \\ h_s^A \end{pmatrix} = \begin{pmatrix} 1 \\ 1 \\ 2 \end{pmatrix}$$

has the negative characteristic difference

$$\Delta_s^0 = c_s - c_B\tilde{B}\begin{pmatrix} a_s + \Delta\tilde{a}_s \\ h_s^A \end{pmatrix} = -1 < 0.$$

Consider direction (3.6). Taking into account Remark 3.3 and the non-degeneracy of problem (3.20) at $\Delta A_{11} = 0$, we can set $\varepsilon = 0$, $\delta = 0$, and $v_s = 1$. Then the motion along the ray (3.6) is impossible because of the first component of the vector

$$x_B^0 - \tilde{B}\begin{pmatrix} a_s + \Delta\tilde{a}_s \\ h_s^A \end{pmatrix}\beta = \begin{pmatrix} 0 \\ 1 \\ 0 \end{pmatrix} - \begin{pmatrix} 1 \\ -1 \\ 1 \end{pmatrix}\beta = \begin{pmatrix} -\beta \\ 1+\beta \\ -\beta \end{pmatrix},$$

which is not nonnegative at $\beta \geq 0$.

To construct a feasible descent direction of type (3.6), we must choose another basis at the optimal point $x^0 = (0,0,1,0)^T$. For example, if the basic set is $J_B = \{2,3,4\}$, then

$$B = \begin{pmatrix} 3/2, 1, 0 \\ 2,\ 1, 1 \\ 1,\ 1, 1 \end{pmatrix}, \quad \tilde{B} = \begin{pmatrix} 0, & 1, & -1 \\ 1, & -3/2, & 3/2 \\ -1, & 1/2, & 1/2 \end{pmatrix},$$

$$x_B^0 = (0,\ 1,\ 0)^T, \quad \Delta_s^0 = c_s - c_B\tilde{B}\begin{pmatrix} a_s + \Delta\tilde{a}_s \\ h_s^A \end{pmatrix} = -\frac{1}{2} < 0.$$

In this case, we have

$$x_B^0 - \tilde{B}\begin{pmatrix} a_s + \Delta\tilde{a}_s \\ h_s^A \end{pmatrix}\beta = \begin{pmatrix} 0 \\ 1 \\ 0 \end{pmatrix} - \begin{pmatrix} -1 \\ 5/2 \\ 1/2 \end{pmatrix}\beta = \begin{pmatrix} \beta \\ 1 - 5/2\,\beta \\ -1/2\,\beta \end{pmatrix}$$

and it is easy to show that direction (3.6) constructed with the use of the chosen matrix is feasible, and the objective function decreases along this direction.

Thus, in the degenerate case, constructing a descent direction may require searching through optimal bases at the point x^0. It is necessary to consider all directions of form (3.6) for various x_B^0, c_B, and $\tilde{B}$. But, as opposed to the nondegenerate case, such directions may be "too large in number" to be localized in an affine manifold Π_y of small dimension.

3.2. Parameters in Convex Constraints

Let us apply the results obtained in the preceding section to generalize the consistent aggregation method to convex problems of a more general form, which does not assume the linearity of variable constraints. As opposed to the cases considered above, the proofs of the main results are based on differential decompositions, an analysis of the Lagrange function of the parametric problem under examination, and the construction of feasible descent directions, rather than on simplex method ideas related to linear programming problems.

We aggregate not only linear, but also those nonlinear constraints that are not active at the aggregation point. In the case where the active constraints of the original problem constitute an insignificant part of the entire set of constraints (which is typical for nonlinear problems of large dimension), this approach makes it possible to substantially reduce the number of aggregated constraints in comparison with the bounds given in Section 3.1.

As the original problem, we consider the parametric problem

$$
\begin{aligned}
& f(x) \to \min, \\
& z \in Z(a) \cap F = Z(a) \cap Q \cap S \subset \mathbf{R}^n, \\
& Z(a) = \{x \mid g(r(a),\, x) \le 0\}, \\
& Q = \{x \mid q(x) \le 0\}, \quad S = \{x \mid h(x) \le 0\},
\end{aligned}
\tag{3.21}
$$

where the functions $g(r,\,\cdot)\colon \mathbf{R}^n \to \mathbf{R}^{m_1}$, $q\colon \mathbf{R}^n \to \mathbf{R}^{m_2}$, and $h\colon \mathbf{R}^n \to \mathbf{R}^{m_3}$ are convex on $\mathbf{R}^n$ for any $r \in \mathbf{R}^k$, $m_1 + m_2 + m_3 = m$, and the affect of the parameters on the constraints in (3.21) is determined by a continuous operator r from $\mathbf{R}^q$ to $\mathbf{R}^k$.

The specified structure of the function $g(r, x)$ is selected because, in practice, it is not always possible to directly change the coefficients r involved in the description of the original model. These coefficients may be "controlled" indirectly, by means of available parameters related to them by some functional dependence (which is sometimes very complicated or even implicit). For example, the change in the rate of expenditures, which reflects the change in production technology, depends on the general level of the organization of production at the enterprise, the amount of money allotted for the introduction of new techniques, reconstruction, re-equipment, etc., and it must be balanced according to these factors. So, even when the problem is considered in a nonlinear setting, it still seems plausible that the functional dependences mentioned above can be directly included in the constraints, for instance, in the form $r(a)$, as in (3.21).

The sets $Z(a)$, Q, and S are analogs of the sets formed by the constraints $(A+\Delta A)x = b$, $q(x) \le 0$, and $x \ge 0$, respectively (these sets are considered in the preceding section); to a certain degree, this determines the specification of the structure of the "invariable" domain F in the form of the intersection $Q \cap S$. Such a specification does not cause loss of generality, because Q and S can coincide with $\mathbf{R}^n$ and be thereby missing.

Consider a point $y \in \mho$. Suppose that, for this point, a primal optimal solution x_y to problem (3.21) and the corresponding dual optimal solution $\pi_y = (u_y,\, w_y,\, v_y) \ge 0$ are known. By analogy with Definition 3.1, we call problem (3.21) nondegenerate at the point y if $X(y)$ is Slater regular and π_y exists and is unique.

Hereafter, we assume that

1. The functions f, g, h, and $g(r(y),\,\cdot)$ are differentiable at x_y;

2. The functions $g(\cdot, x)$ are affine on $\mathbf{R}^k$ and monotone with respect to each component r_j with $j = \overline{1,k}$ at every $x \in S$;

3. The functions $\frac{\partial g(r(y), \cdot)}{\partial r_j}$ $(j = \overline{1,k})$ are affine on S.

The assumption that the constraints are monotone with respect to the coefficients r_j is not too strong. For instance, it holds for the linear programming problems in standard form, which are widely used in practice. In more general situations, this assumption means that the dependence of the constraints on the coefficients in the model does not change its sign within the set of the physically admissible values of variables.

Below, we successively specify and give meaning to the basic constructions of the aggregation procedure as applied to problem (3.21); then, we prove the consistency of the constructed aggregation. For this purpose, we introduce sets of indices

$$I_i = \{1, \ldots, m_i\}, \qquad i = \overline{1,3},$$

$$I_1^A = \{i \mid g_i(r(y),\, x_y) = 0,\ u_{yi} > 0\},$$

$$I_1^K = \{i \mid g_i(r(y),\, x_y) = 0,\ u_{yi} = 0\},$$

$$I_2^A = \{i \mid g_i(x_y) = 0,\ w_{yi} > 0\}, \quad I_2^K = \{i \mid g_i(x_y) = 0,\ w_{yi} = 0\},$$

$$I_3^A = \{i \mid h_i(x_y) = 0,\ v_{yi} > 0\}, \quad I_3^K = \{i \mid h_i(x_y) = 0,\ v_{yi} = 0\},$$

$$I_i^N = I_i \backslash I_i^K \backslash I_i^A, \quad i = \overline{1,3}, \quad I^A = \bigcup_{i=1}^{3} I_i^A, \quad I^K = \bigcup_{i=1}^{3} I_i^K, \quad I^N = \bigcup_{i=1}^{3} I_i^N;$$

in what follows, we use the superscripts A, K, and N to label the subvectors formed by the components whose numbers belong to I^A, I^K, and I^N, respectively.

We start with aggregating variables. Let H denote the $|I^A| \times n$ matrix

$$H = \begin{pmatrix} \nabla_x g^A(r(y),\, x_y) \\ \nabla q^A(x_y) \\ \nabla h^A(x_y) \end{pmatrix}.$$

Suppose that problem (3.21) is nondegenerate at the point y. Then [18] all the gradients in H are linearly independent, and therefore, H has a right inverse H^*, for which $HH^* = E$, where E is the identity matrix of the corresponding size.

Let $H^* = (H_1^*, H_2^*, H_3^*)$ be the decomposition of H^* into the blocks of sizes $n \times m_1^A$, $n \times m_2^A$, and $n \times m_3^A$ corresponding to the decomposition of the matrix H; as (1.4), we take the manifold

$$\Pi_y = \{x \mid x = x_y + P_y p,\ p \in \mathbf{R}^{n_y}\},$$
$$P_y = (H_1^*,\ -H_1^*,\ -H_2^*\mu^2,\ -H_3^*,\ \eta), \tag{3.22}$$

where $\mu^2 = (1, \dots, 1)^T$ and $\eta \in \mathbf{R}^n$ is a vector for which

$$H\eta = 0, \quad \nabla_x g^K(r(y), x_y)\eta < 0, \quad \nabla g^K(x_y)\eta < 0, \quad \nabla h^K(x_y)\eta < 0$$

(its existence follows from the nondegeneracy assumption [18]).

The aggregation of variables corresponding to the intersection of $\mathbf{R}^n$ with manifold (3.22) yields the sets

$$Z_y(a) = \{p \mid g(r(a), x_y + P_y p) \le 0\},$$

$$F_y(a) = \{p \mid q(x_y + P_y p) \le 0,\ h(x_y + P_y p) \le 0\}$$

in the space $L_y = \mathbf{R}^{n_y}$. The dimension of L_y is determined by the structure of the operator P_y, which, in its turn, is chosen from formal considerations so as to ensure the aggregation consistency (a rigorous justification is given later on).

If $I^K = \varnothing$ and the constraints are not supposed to be aggregated, we can take

$$P_y = -(H_1^*,\ H_2^*\mu^2,\ H_3^*).$$

If, in addition, some groups of constraints are missing in (3.21), i.e., if $Q = \mathbf{R}^n$ or $S = \mathbf{R}^n$, then the terms of P_y related to the matrices H_2^* or H_3^*, respectively, also disappear.

Let us aggregate the constraints. We denote the basic vectors in the space L_y by s^i $(i = \overline{1, n_y})$ and put

$$l_i = \max \left\{ l \ge 0 \,\middle|\, \begin{array}{l} q^N(x_y + P_y s^i l) \le 0, \\ h^N(x_y + P_y s^i l) \le 0 \end{array} \right\} \tag{3.23}$$

for each $i = \overline{1, n_y}$. By meaning, $x_y \in \operatorname{int}\{x \mid q^N(x) \le 0,\ h^N(x) \le 0\}$; hence, $l_i > 0$ for $i = \overline{1, n_y}$. Let us show that the simplex $\{p \mid \bar{l}p \le 1, p \ge 0\}$, where $\bar{l} = (l_1^{-1}, \dots, l_{n_y}^{-1})$, is an inner approximation for the set formed by the constraints nonactive at the point x_y.

Lemma 3.3. *The following inclusion holds:*

$$\{p \mid \bar{l}p = 1,\ p \geq 0\} \subset \{p \mid q^N(x_y + P_y p) \leq 0,\ h^N(x_y + P_y p) \leq 0\}.$$

Proof. Take an arbitrary point $p^* \in \{p \mid \bar{l}p \leq 1,\ p \geq 0\}$. It can be represented as a convex combination of the vertices p^i $(i = \overline{0, n_y})$ of the simplex under consideration; i.e., there exist $\alpha_i^* \geq 0$ such that

$$\sum_{i=0}^{n_y} \alpha_i^* = 1, \quad p^* = \sum_{i=0}^{n_y} \alpha_i^* p^i.$$

By construction, the vertices of the simplex are the origin $p^0 = 0$ and the points $p^i = s^i l_i$ with $i = \overline{1, n_y}$; therefore, $p^* = \sum_{i=1}^{n_y} \alpha_i^* s^i l_i$.

Next, since $q^N(x_y) < 0$ and $q^N(x_y + P_y s^i l_i) \leq 0$ for $i = \overline{0, n_y}$, the convexity of the function q implies the inequality

$$q^N(x_y + P_y p^*) = q^N\left(x_y \sum_{i=0}^{n_y} \alpha_i^* + P_y \sum_{i=0}^{n_y} \alpha_i^* p^i\right) =$$

$$= q^N\left(\sum_{i=0}^{n_y} \alpha_i^*(x_y + P_y p^i)\right) \leq \sum_{i=0}^{n_y} \alpha_i^* q^N(x_y + P_y p^i) =$$

$$= \alpha_0^* q^N(x_y) + \sum_{i=1}^{n_y} \alpha_i^* q^N(x_y + P_y s^i l_i) \leq 0.$$

Similarly, $h^N(x_y + P_y p^*) \leq 0$. Thus, $p^* \in \{p \mid q^N(x_y + P_y p) \leq 0\}$, which proves the lemma.

Remark 3.4. The proof of Lemma 3.3 uses the simplest inner approximation of the set

$$F^N = \{p \mid q^N(x_y + P_y p) \leq 0,\ h^N(x_y + P_y p) \leq 0\}$$

by a set of a simpler structure (by a simplex in the case under consideration). This is a special implementation of a more flexible approach, where the constraints in F^N are first divided into an arbitrary number of groups, and then each domain formed by a separate group of constraints is approximated by a simplex $\{p \mid \bar{l}^i p \leq 1,\ p \geq 0\}$ constructed by analogy with the simplex used above. The entire set F^N is approximated by the intersection of these simplices, i.e., by the set $\{p \mid \bar{l}^i p \leq 1,\ i = \overline{1,K},\ p \geq 0\}$, where K is the number of constraint groups. The division into groups can be based on formal considerations or determined by the meaning of the problem. It should only be taken into account that an increase in the number of groups improves the approximation but complicates the structure of the approximating set. Note also that, instead of simplices, sets of a different structure can be used.

Now, as G_y, we take the set

$$G_y = \left\{ p \;\middle|\; \begin{array}{l} q^A(x_y + P_y p) \leq 0,\ h^A(x_y + P_y p) \leq 0, \\ q^K(x_y + P_y p) \leq 0,\ h^K(x_y + P_y p) \leq 0, \\ \bar{l}p \leq 1,\ p \geq 0 \end{array} \right\}. \tag{3.24}$$

It is easy to verify that Lemma 3.3 directly implies the required inclusion $G_y \subseteq F_y$ for this set.

As a result, the aggregated parametric problem assumes the form

$$\begin{aligned} & f_y(p) = f(x + P_y p) \to \min, \\ & g(r(a),\, x_y + P_y p) \leq 0, \quad \bar{l}p \leq 1, \quad p \geq 0, \\ & q^A(x_y + P_y p) \leq 0, \quad h^A(x_y + P_y p) \leq 0, \\ & q^K(x_y + P_y p) \leq 0, \quad h^K(x_y + P_y p) \leq 0. \end{aligned} \tag{3.25}$$

Let us estimate its dimension. By construction, the dimension of manifold (3.22) is $n_y \leq 2m_1 + m_3 + 2$, and the number of constraints in (3.25) (not counting the "simple" constraints $p \geq 0$) is $m_y = m_1 + m_2^A + m_2^K + m_3^A + {}$ $+ m_3^K + 1$. As a rule, the number m_1 of the constraints subjected to variation is much smaller than m, and the number of active (at the optimal point) constraints is also comparatively small in nonlinear problems of large dimensions. Therefore, we can assume that $m_y \ll \max\{m, n\}$. If, in addition, $m_3 \ll m$ in (3.21) (as in the optimization problems considered above, including the linear programming problems), then we have $\max\{m_y, n_y\} \ll \max\{m, n\}$; i.e., the dimension of the aggregated problem (3.25) is then much lower than that of the original problem (3.21).

Let us prove the consistency of the aggregation performed above. We assume that the set $\mathfrak{U}$ has the following property relative to the point y

under examination: together with each $a \in \mho$, the parameter values $a(\alpha)$ whose images $r(a(\alpha))$ can be represented as

$$r(a(\alpha)) = (E - \alpha)r(y) + \alpha r(a), \quad \alpha = \operatorname{diag}\{\alpha_i\}, \quad 0 \leq \alpha_1 \leq 1, \quad i = \overline{1,k}, \tag{3.26}$$

i.e., as componentwise "mixtures" of the old vector $r(y)$ and the new vectors $r(a)$ corresponding to this a, are feasible.

Theorem 3.3. *If problem (3.21) is nondegenerate at a point y and $\mho$ has property (3.26), then the aggregation of (3.21) into (3.25) is consistent at the point y.*

Proof. According to Remark 1.1, we only have to verify the fulfillment of the second condition in (1.3). For this purpose, we need the following auxiliary notation and assertions.

To avoid cumbersome formulas, we shall assume without loss of generality that the vector functions $g(\cdot, x)$ are not merely monotone but increasing with respect to r_j for all $j \in \overline{1,k}$ (see assumption 2 about the constraints). In the case under consideration, this means that $V_j(x) = \frac{\partial g(r(y), x)}{\partial r_j} \geq 0$ for $j \in \overline{1,k}$.

Consider $x_a \in X(a)$, $r_a = r(a)$, $r_y = r(y)$, and $\Delta r(a) = r_a - r_y$ and let $V_{ij}(x)$ denote the i-th component of $V_j(x)$ for $i \in I_1$.

Let us introduce the sets of indices

$$J = \{1, \dots, q\},$$

$$J(a) = \{j \in J \mid \Delta r_j(a) < 0\},$$
$$J_+(a) = \{j \in J(a) \mid u_y V_j(x_y) > 0\},$$
$$J_0(a) = \{j \in J(a) \mid u_y V_j(x_y) = 0\};$$

we then have $J(a) = J_+(a) \cup J_0(a)$. Recall that the notation (u_y, w_y, v_y) stands for the components of the dual optimal solution to problem (3.21) at the parameter value y.

Lemma 3.4. *Suppose that the set $X(y)$ is Slater regular and $J_0(a) \neq \varnothing$. Then, for each $j \in J_0(a)$, there exists an $|I_1| \times |I_3^A|$ matrix $\Lambda^j \geq 0$ such that*

$$\begin{aligned} V_j(x)\Delta r_j(a) &\leq -\Lambda^j \nabla h^A(x_y)(x - x_y)\Delta r_j(a), \\ u_j V_j(x)\Delta r_j(a) &= -u_y \Lambda^j \nabla h^A(x_y)(x - x_y)\Delta r_j(a) \end{aligned} \tag{3.27}$$

for all $x \in S$.

Proof. Indeed, we have $V_j(x) \geq 0$ for $x \in S = \{x \mid h(x) \leq 0\}$; therefore, if the function V_{ij} vanishes at x_y, then it attains its maximum over S at this point. The regularity of $X(y)$, the affinity of V_{ij}, and the necessary optimality conditions imply the existence of a Kuhn–Tucker vector $\lambda_{ij} \geq 0$ such that

$$V_{ij}(x) = -\lambda_{ij} \nabla h(x_y)(x - x_y), \quad \lambda_{ij} h(x_y) = 0, \quad x \in \mathbf{R}^n.$$

Let us rewrite this relation in the form

$$V_{ij}(x) = -\lambda_{ij}^A \nabla h^A(x_y)(x - x_y) - \lambda_{ij}^K \nabla h^K(x_y)(x - x_y).$$

By assumption, $h(x) \leq 0$ and h is convex; therefore, under the conditions of the lemma, we have

$$\nabla h^K(x_y)(x - x_y) \leq h(x) - h(x_y) = h(x) \leq 0;$$

since λ_{ij}^K is nonnegative, we conclude that

$$V_{ij}(x) \geq -\lambda_{ij}^A \nabla h^A(x_y)(x - x_y).$$

Let $\Lambda^j \geq 0$ be the matrix whose i-th row coincides with λ_{ij}^A if $V_{ij}(x_y) = 0$ and is zero if $V_{ij}(x_y) > 0$ for $i \in I_1$. Taking into account that $\Delta r_j(a) < 0$ and $u_j V_j(x_j) = 0$ for $j \in J_0(a)$, we obtain relations (3.27). This completes the proof of the lemma.

Consider the row $|I_3^A|$-vector

$$\theta(a) = v_y^A + u_y \Lambda, \quad \text{where} \quad \Lambda = \sum_{j \in J_0(a)} \Lambda^j \Delta r_j(a)$$

(we use the notation introduced above).

Lemma 3.5. *Suppose that the set $X(y)$ is Slater regular and $a \in \Gamma(y)$ is a point such that $J_+(a) = \varnothing$. Then $J_0(a) \neq \varnothing$ and there exists an $s \in I_3^A$ for which $\theta_s(a) < 0$.*

Proof. Under the regularity conditions, the relations

$$\nabla f(x_y) + u_y \nabla_x g(r_y, x_y) + w_y \nabla q(x_y) + v_y \nabla h(x_y) = 0 \tag{3.28}$$

and

$$u_y g(r_y, x_y) = 0, \quad w_y q(x_y) = 0, \quad v_y h(x_y) = 0 \tag{3.29}$$

hold at the optimal point x_y.

We have $x_a \in X(a)$. Therefore, the convexity of the corresponding functions and (3.29) imply that, first,

$$u_y \nabla_r g(r_y, x_a)\Delta r(a) + u_y \nabla_x g(r_y, x_y)(x_a - x_y) \leq$$
$$\leq u_y[g(r_a, x_a) - g(r_y, x_a) + g(r_y, x_a) - g(r_y, x_y)] =$$
$$= u_y[g(r_a, x_a) - g(r_y, x_y)] = u_g g(r_a, x_a) \leq 0,$$

and, secondly,

$$w_y \nabla q(x_y)(x_a - x_y) \leq w_y[q(x_a) - q(x_y)] = w_y q(x_a) \leq 0.$$

This, the convexity of f, and (3.27) yield

$$\Delta\varphi = \varphi(a) - \varphi(y) = f(x_a) - f(x_y) \geq \nabla f(x_y)(x_a - x_y) =$$
$$= -(u_y \nabla_x g(r_y, x_y) + w_y \nabla q(x_y) + v_y \nabla h(x_y)(x_a - x_y) \geq$$
$$\geq u_y \nabla_r g(r_y, x_a)\Delta r(a) - v_y \nabla h(x_y)(x_a - x_y).$$

Using the second relation in (3.27), the definition of the vector $\theta(a)$, and the emptiness of the set $J_+(a)$, we obtain

$$\Delta\varphi \geq u_y \sum_{j \in J_0(a)} V_j(x_a)\Delta r_j(a) - v_j \nabla h(x_y)(x_a - x_y) =$$
$$= -(u_y \Lambda + v_y^A)\nabla h^A(x_y)(x_a - x_y) = -\theta(a)\nabla h^A(x_y)(x_a - x_y).$$

We have

$$\nabla h^A(x_y)(x_a - x_y) \leq h^A(x_a) - h^A(x_y) = h^A(x_a) \leq 0,$$

$\theta(a)h^A(x_y) = 0$, and $\varphi(a) - \varphi(y) < 0$; hence, there exists an $s \in I_3^A$ such that $\theta_s(a) < 0$, which proves the lemma.

As applied to problem (3.21), the components of the vector $\theta(a)$ can be regarded as analogs of the characteristic differences evaluated with respect to the nonbasic columns, which indicate the direction and rate of the change of the objective function along the corresponding edge of the constraint polyhedron.

The further constructions aim at constructing directions in $\mathbf{R}^k \times \mathbf{R}^n$ that are feasible for problem (3.21); they include directions such that the values of the partial derivatives of f along them are determined by the values of the corresponding components of $\theta(a)$. This actually ensures the existence of at least one feasible direction (along which the objective function decreases).

Let us proceed to study the most important properties of manifold (3.22). We put

$$T_0(a) = \Big\{ z \in \mathbf{R}^{|I^A|} \;\Big|\; z^1 = \sum_{j \in J_+(a)} V_j^A(x_y)\Delta r_j(a),\ z^2 = 0,\ z^3 = 0 \Big\},$$

if $J_+(a) \neq \varnothing$ and

$$T_s(a) = \Big\{ z \in \mathbf{R}^{|I^A|} \;\Big|\; z^1 = \Lambda^A e_s,\ z^2 = 0,\ z^3 = e_s \Big\}, \quad s \in I_3^A,$$

if $J_0(a) \neq \varnothing$; here, e_s is the s-th basis vector in $\mathbf{R}^{|I_3^A|}$ and $z = (z^1, z^2, z^3) \in \mathbf{R}^{|I_1^A|} \times \mathbf{R}^{|I_2^A|} \times \mathbf{R}^{|I_3^A|}$. Finally, we put $T(a) = \bigcup_{s=0}^{m_3^A} T_s(a)$.

In $\mathbf{R}^k \times \mathbf{R}^n$, consider the family of pairs $(\psi(\beta), x(\beta))$ of the form

$$W_a(z, \varepsilon, \delta, \varphi^0(\cdot)) = \begin{cases} \psi(\beta) = r_y + (r_a - r_y)^- - \varphi^0(\beta), \\ x(\beta) = x_y - H^*(z + \varepsilon\mu)\beta + \delta\eta\beta. \end{cases} \tag{3.30}$$

Here,

$$\mu = (\mu^1, \mu^2, \mu^3) = (1, \ldots, 1)^T \in \mathbf{R}^{|I_1^A|} \times \mathbf{R}^{|I_2^A|} \times \mathbf{R}^{|I_3^A|}.$$

The function $\varphi^0 \colon \mathbf{R} \to \mathbf{R}$, the scalars $\delta \geq 0$ and $0 \leq \varepsilon \leq 1$, and the vector $z \in T(a)$ are external parameters of this family and the scalar $\beta \geq 0$ is its internal parameter; $(r_a - r_y)^-$ is the projection of the vector $r_a - r_y$ on the negative orthant. If the function φ^0 is affine, then each fixed set of external parameters determines a direction through the point $(\psi(0), x_y)$ in the space $\mathbf{R}^k \times \mathbf{R}^n$.

Let $M = \{x \mid x = x_y + P_y p,\ p \in G_y\} \subset \mathbf{R}^n$ be the image of the set G_y defined by (3.24) under the mapping $p \mapsto x = x_y + P_y p$.

Lemma 3.6. Suppose that problem (3.21) is nondegenerate at the point y. Then there exist scalars $\delta = \delta^M > 0$ and $\beta^M(\varepsilon) > 0$ such that, if $0 < \varepsilon \leq 1$, then the points $x(\beta)$ defined in (3.30) belong to M for all nonnegative $\beta \leq \beta^M(\varepsilon)$ and any $z \in T(a)$.

Proof. For the sake of generality, we assume that $I^K \neq \varnothing$. Since $0 \leq \varepsilon \leq 1$ and the set $T(a)$ is bounded, there exist vectors $\tau^q(a)$ and $\tau^h(a)$ such that

$$\begin{aligned} -\nabla q^K H^*(z + \varepsilon\mu) &\leq \tau^q(a), \\ -\nabla h^K H^*(z + \varepsilon\mu) &\leq \tau^h(a) \end{aligned} \tag{3.31}$$

for any $z \in T(a)$.

Therefore,

$$q^A(x(\beta)) = \nabla q^A(x_y)(x(\beta) - x_y) + o(\beta) =$$
$$= \nabla q^A(x_y)(-H^*(z + \varepsilon\mu) + \delta\eta)\beta + o(\beta) =$$
$$= -(z^2 + \varepsilon\mu^2)\beta + o(\beta) = -\varepsilon\mu^2\beta + o(\beta)$$

and, taking into account (3.31), we obtain

$$q^K(x(\beta)) = \nabla q^K(x_y)(-H^*(z + \varepsilon\mu) + \delta\eta)\beta = o(\beta) \leq$$
$$\leq (\tau^q(a) + \delta\nabla q^K(x_y)\eta)\beta + o(\beta).$$

Similar relations hold for the function h. Therefore, by virtue of the inequalities

$$\nabla q^K(x_y)\eta < 0, \quad \nabla h^K(x_y)\eta < 0, \quad q^N(x_y) < 0, \quad h^N(x_y) < 0,$$

which follow from the nondegeneracy of problem (3.31) at y, there exist a $\delta = \delta^M > 0$ and a sufficiently small scalar $\beta_1^M(\varepsilon) > 0$ such that, whenever $0 < \varepsilon \leq 1$ and $0 \leq \beta \leq \beta_1^M(\varepsilon)$, we have

$$q(x(\beta)) \leq 0, \quad h(x(\beta)) \leq 0, \tag{3.32}$$

or, equivalently, $x(\beta) \in F = Q \cap S$.

Next, by virtue of the special choice of family (3.30), the points $x(\beta)$ belong to $\{x \mid x = x_y + P_y p,\ p \geq 0\}$. This is why the affine manifold Π_y was defined in form (3.22). Now, taking into account (3.30), we represent every vector $p(\beta) \in \mathbf{R}^{n_y}$ such that $x(\beta) - x_y = P_y p(\beta)$ in the form $p(\beta) = p(1)\beta$. Employing the finiteness of $T(a)$, we can find a $\beta_2^M(\varepsilon) > 0$ such that, whatever $z \in T(a)$, the points $p(\beta)$ belong to $\{p \mid \bar{l}p \leq 1,\ p \geq 0\}$ for all nonnegative $\beta \leq \beta_2^M(\varepsilon)$.

To complete the proof of the lemma, it remains to recall (3.32) and put $\beta^M(\varepsilon) = \min\{\beta_1^M(\varepsilon), \beta_2^M(\varepsilon)\}$.

Combining the results obtained above, we arrive at the following assertion.

Lemma 3.7. *If problem (3.21) is nondegenerate at y and $a \in \Gamma(y)$, then there exist $a(\beta) \in \mathbf{R}^q$ and parameters $\beta^* > 0$, z, ε, δ, and $\varphi^0(\cdot)$ of family (3.30) such that $a(\beta) \in \mho$, $x(\beta) \in Z(a(\beta)) \cap M$, and $f(x(\beta)) < f(x_y)$ for $0 < \beta \leq \beta^*$.*

Proof. According to Lemma 3.6, for $0 < \varepsilon \le 1$, there exist $\delta^M, \beta^M(\varepsilon) > 0$ such that the points $x(\beta)$ belong to M for any $0 \le \beta \le \beta^M(\varepsilon)$ and $z \in T(a)$. Therefore, $V_j(x(\beta)) \ge 0$ for $j \in J$ by assumption 3.

Consider two cases.

(1). The set $J_+(a)$ is nonempty. In this case, we set $z \in T_0(a)$ and $\varphi^0(\beta) = \beta$ in (3.30). Then

$$g^A(\psi(\beta), x(\beta)) = g^A(\psi(\beta), x(\beta)) - g^A(r_y, x_y) =$$
$$= \nabla_r g^A(r_y, x_y)(r_a - r_y)^- \beta + \nabla_x g^A(r_y, x_y)(x(\beta) - x_y) + o(\beta) =$$
$$= \sum_{j \in J_+(a)} V_j^A(x_y)\Delta r_j(a)\beta - (z^1 + \varepsilon\mu^1)\beta + o(\beta) =$$
$$= \Big(\sum_{j \in J_+(a)} V_j^A(x_y)\Delta r_j(a) - z^1\Big)\beta - \varepsilon\mu^1\beta + o(\beta) = -\varepsilon\mu^1\beta + o(\beta).$$

At the same time,

$$g^K(\psi(\beta), x(\beta)) = \nabla_r g^K(r_y, x_y)(r_a - r_y)^- \beta +$$
$$+ \nabla_x g^K(r_y, x_y)H^*(z + \varepsilon\mu)\beta + \nabla_x g^K(r_y, x_y)\eta\delta\beta + o(\beta).$$

We have $\nabla_x g^K(r_y, x_y)\eta < 0$ for $0 \le \varepsilon \le 1$; therefore, selecting a sufficiently large $\delta = \delta^g$, we can assert that

$$g^K(\psi(\beta), x(\beta)) \le \tau^g(a)\beta + o(\beta) \tag{3.33}$$

for some $\tau^g(a) < 0$.

In addition, according to the relations (3.28), (3.30), $z \in T_0(a)$, and $H\eta = 0$, we have

$$f(x(\beta)) - f(x_y) = \nabla f(x_y)(x(\beta) - x_y) + o(\beta) =$$
$$= [u_y^A \nabla_x g^A(r_y, x_y) + w_y^A \nabla q^A(x_y) + v_y^A \nabla h^A(x_y)](x(\beta) - x_y) + o(\beta) =$$
$$= [u_y^A(z^1 + \varepsilon\mu^1) + w_y^A(z^2 + \varepsilon\mu^2) + v_y^A(z^3 + \varepsilon\mu^3)]\beta + o(\beta) =$$
$$= \beta \sum_{j \in J_+(a)} u_y V_j(x_y)\Delta r_j(a) + \varepsilon C_y\beta + o(\beta),$$

where $C_y = u_y^A\mu^1 + w_y^A\mu^2 + v_y^A\mu^3$.

Combining the obtained decompositions and taking into account the definition of $J_+(a)$ and the condition $g^N(x_y) < 0$, we arrive at the conclusion that there exist a $\delta = \delta^g$ and sufficiently small $\varepsilon > 0$ and $0 < \beta^g(\varepsilon) \le 1$ such that

$$g(\psi(\beta), x(\beta)) \le 0, \quad f(x(\beta)) < f(x_y), \quad 0 < \beta \le \beta^g(\varepsilon). \tag{3.34}$$

(2). The set $J_+(a)$ is empty. In this case, according to Lemma 3.5, $J_0(a) \neq \varnothing$ and $\theta_s(a) < 0$ for some s. Put $\varphi^0(\beta) \equiv 1$ and take a $z \in T_s(a)$. We then have

$$g^A(\psi(\beta), x(\beta)) = g^A(\psi(\beta), x(\beta)) - g^A(r_y, x(\beta)) + g^A(r_y, x(\beta)) - \\ - g^A(r_y, x_y) = \sum_{j \in J_0(a)} V_j(x(\beta))\Delta r_j(a) + \nabla_x g^A(r_y, x_y)(x(\beta) - x_y) + o(\beta).$$

Substituting (3.27) and (3.30) in the last expression yields

$$g^A(\psi(\beta), x(\beta)) \leq - \sum_{j \in J_0(a)} \Lambda^j \Delta r_j(a) \nabla h^A(x_y)(x(\beta) - x_y) - (z^1 + \varepsilon\mu^1)\beta + o(\beta) =$$

$$= \sum_{j \in J_0(a)} \Lambda^j \Delta r_j(a)(z^3 + \varepsilon\mu^3)\beta - (z^1 + \varepsilon\mu^1)\beta + o(\beta) \leq$$

$$\leq \Big(\sum_{j \in J_0(a)} \Lambda^j \Delta r_j(a) z^3 - z^1 \Big)\beta - \varepsilon\mu^1\beta + o(\beta) =$$

$$= (\Lambda z^3 - z^1)\beta - \varepsilon\mu^1\beta + o(\beta) = -\varepsilon\mu^1\beta + o(\beta).$$

Next, as in the first case, it can be shown that a relation similar to (3.33) holds and

$$f(x(\beta)) - f(x_y) = \nabla f(x_y)(x(\beta) - x_y) + o(\beta) = \\ = (u_y^A z^1 + v_y^A z^3)\beta + \varepsilon C_y \beta + o(\beta) = \theta_s(a)\beta + \varepsilon C_y\beta + o(\beta)$$

Thus, (3.34) is valid in this case too.

It remains to define the parameters $a(\beta) \in \mho$. Recalling the definition of the vectors $\psi(\beta)$ in (3.30) and analyzing the proof, we see that the values $\psi(\beta)$ $(0 < \beta \leq \beta^g(\varepsilon) \leq 1)$ chosen above can be represented in the form

$$\psi(\beta) = (E - \alpha^\beta)r_y + \alpha^\beta r(a),$$

where $\alpha^\beta = \operatorname{diag}\{\alpha_i^\beta\}$ and $0 \leq \alpha_i^\beta \leq 1$ for $i \in \overline{1,k}$. Taking into account (3.26), we can define $a(\beta)$ to be the preimages of the vectors $\psi(\beta)$ under the mapping r.

Finally, taking into account Lemma 3.6, selecting z and ε in accordance with the above considerations, and setting $\delta = \max\{\delta^M, \delta^g\}$ and $\beta^* = \min\{\beta^M(\varepsilon), \beta^g(\varepsilon)\}$, we obtain the required result and arrive at the end of the proof of Lemma 3.7.

To complete the proof of consistency, suppose that $\Gamma(y) \neq \varnothing$ in (3.21), i.e., there exists an $\bar{a} \in \Gamma(y)$. Then, according to Lemma 3.7, we can find an $a(\bar{\beta})$ and an $x(\bar{\beta})$ such that for all $a(\bar{\beta}) \in \mho$, $x(\bar{\beta}) \in Z(a(\bar{\beta})) \cap M$, we have $f(x(\beta)) < f(x_y) = \varphi(y)$. This means that problem (3.25) has a feasible point $\bar{p} \in X_y(a(\bar{\beta}))$ for which $f_y(\bar{p}) = f(x(\bar{\beta})) < \varphi(y)$. Thus, $\varphi_y(a(\bar{\beta})) \leq f_y(\bar{p}) < \varphi(y)$. By the choice of Π_y, we have $\varphi(y) = \varphi_y(y)$; hence, $\varphi_y(a(\bar{\beta})) < \varphi_y(y)$ and, by definition, $\bar{a}(\bar{\beta}) \in \Gamma_A(y)$. The second condition in (1.3) is fulfilled. This proves the theorem.

If necessary, the condition for that the set $\mho$ can be stated in a different form, with the use of the parameters a rather than of their images $r(a)$. Let us consider this in more detail.

Suppose that, at the base point y, together with each $a \in \mho$, the parameters

$$a(\alpha) = (E - \alpha)y + \alpha a, \text{ where } \alpha = \operatorname{diag}\{\alpha_i\} \text{ and } 0 \leq \alpha_i \leq 1 \text{ for } i \in \overline{1,q}, \tag{3.35}$$

are feasible. In this case, the aggregation consistency theorem remains valid under certain additional assumptions about the mapping r.

Theorem 3.4. *Suppose that r is a continuous linear mapping from R^q to R^q of the form $r(a) = \big(r_1(a_1), r_i(a_2), \ldots, r_q(a_q))\big)$ and problem (3.21) is nondegenerate at y. Then, if $\mho$ has property (3.35), the aggregation of problem (3.21) into (3.25) is consistent at the point y.*

Proof. This theorem is proved similarly to Theorem 3.3; the only difference is in the construction of the feasible parameters $a(\beta) \in \mho$. It follows from the proof of Theorem 3.3 that

$$\psi_i(\beta) = r_i(y_i)(1 - \varphi^0(\beta)) + r_i(a_i)\varphi^0(\beta),$$

where either $\varphi^0(\beta) \in \{0, 1\}$ or $0 \leq \varphi^0(\beta) = \beta \leq 1$. In the former case, the preimage of the component $\psi_i(\beta)$ is y_i or a_i, and in the latter, if β is sufficiently small, $\psi_i(\beta)$ has a preimage $a_i(\beta)$ such that $y_i \leq a_i(\beta) \leq$ $\leq a_i$ (because r is continuous). In any case, the vector $a(\beta)$ formed by the preimage components mentioned above belongs to the set $\mho$ provided that it has property (3.35). This proves Theorem 3.4.

We conclude the consideration of problem (3.21) with a numerical example illustrating the special features of the suggested consistent aggregation technique. Suppose that the original parametric problem has the form

$$\begin{aligned} &(x_2 - 2)^2 - 2x_1 - x_3 + (x_4 - 1)^4 \to \min, \\ &r(a)x_1 + x_2^2 + x_3 + 4x_4 \leq 1, \\ &x_1 \geq 0, \quad x_1 \leq 2x_2, \end{aligned} \tag{3.36}$$

where $r(a) = \frac{1}{8}(a^2 - 12)a + 2$ for $a \in \mathbf{R}$. The number of variables in this problem is $n = 4$, and the number of constraints is $m = 3$.

The results of the parametric analysis of this problem, which consists in constructing the primal and dual optimal solutions and the optimal value as functions of $z = r(a)$, are given in Table 3.1. From this table, taking into account that $z = r(a) \geq 2$ at $a \geq 2\sqrt{3}$ and $-2\sqrt{3} \leq a \leq 0$, we obtain the optimum function

$$\varphi(a) = \begin{cases} 4 - \left(2 - \frac{1}{8}(a^2 - 12)a\right)^2/2 & \text{if} \quad a \leq -2\sqrt{3} \quad \text{or} \quad 0 \leq a \leq 2\sqrt{3}, \\ 2 & \text{if} \quad a \geq 2\sqrt{3} \quad \text{or} \quad -2\sqrt{3} \leq a \leq 0, \end{cases}$$

which is plotted in Fig. 3.1 by a solid line.

	$z \leq 2$	$z \geq 2$
x_1	$4 - z$	0
x_2	$(4 - z)/2$	1
x_3	$1 - z(4 - z) - -(4 - z)^2/4$	0
x_4	0	0
λ_1	1	1
λ_2	0	$z - 2$
λ_3	$2 - z$	0
φ	$4 - (4 - z)^2/2$	2

Table 3.1.

Consider the point $y = 2$, which is a point of strict local (but not global) minimum for the nonlinear function φ. Note that, in the case under consideration, the traditional approximations of φ, including the first- and second-order approximations constructed at y, do not allow us to find a preferred point where φ takes a value smaller than $\varphi(y) = -4$. The reason for this is that y is a point of global minimum for these approximations, because $\varphi'(y) = 0$ and $\varphi''(y) > 0$. At the same time, according to conditions (1.3), the preferred points can be found by approximating the function φ at y with the optimum function of the aggregated parametric problem (3.25) obtained from (3.21) by the consistent aggregation method.

Indeed, at $a = 2$, problem (3.36) has the optimal solution $x_y = (4, 2, -3, 0)^T$, and the Jacobi matrix is

$$H = \begin{pmatrix} 0, & 4, & 1, 4 \\ 1, & -2, & 0, 0 \end{pmatrix}.$$

Aggregating only the variables, we choose the right inverse

$$H^* = \begin{pmatrix} 0, 0, 1, 0 \\ 1, 0, 0, 0 \end{pmatrix}^T$$

and make the change

$$x = \begin{pmatrix} 4 \\ 2 \\ -3 \\ 0 \end{pmatrix} + \begin{pmatrix} 0\,1 \\ 0\,0 \\ 1\,0 \\ 0\,0 \end{pmatrix} \begin{pmatrix} p_1 \\ p_2 \end{pmatrix} = \begin{pmatrix} 4 + p_2 \\ 2 \\ -3 + p_1 \\ 0 \end{pmatrix} = \begin{pmatrix} \xi_1 \\ 2 \\ \xi_2 \\ 0 \end{pmatrix}$$

in (3.36). As a result, we obtain the aggregated parametric problem

$$\begin{aligned} &-2\xi_1 - \xi_2 + 1 \to \min, \\ &r(a)\xi_1 + \xi_2 + 3 \le 0, \quad \xi_1 \ge 0, \quad \xi_1 \le 4, \end{aligned} \tag{3.37}$$

with $n_y = 2$ variables and $m_y = 3$ constraints. In this problem, $r(a) = \frac{1}{8}(a^2 - 12)a + 2$, as in (3.37). From Table 3.2, which contains a solution to problem (3.37) represented as functions of $z = r(a)$, we obtain the optimum function

$$\varphi_y(a) = \begin{cases} \dfrac{1}{2}(a^2 - 12)a + 4 & \text{if } \ a \le -2\sqrt{3} \quad \text{or} \quad 0 \le a \le 2\sqrt{3}, \\ 4 & \text{if } \ a \ge 2\sqrt{3} \quad \text{or} \quad -2\sqrt{3} \le a \le 0 \end{cases}$$

of this problem; its graph is shown in Fig. 3.1 by the dashed line.

	$z \le 2$	$z \ge 2$
ξ_1	4	0
ξ_2	$-3 - 4z$	-3
λ_1	1	1
λ_2	0	$z - 2$
λ_3	$2 - z$	0
λ_4	$4z - 4$	4

Table 3.2.

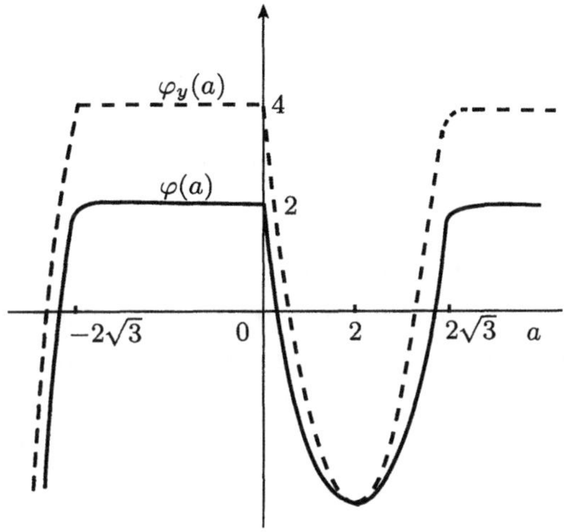

Fig. 3.1.

It is easy to see that $y = 2$ is a point of local (but not global) optimum for the constructed function φ_y, as well as for φ. For example, at $a = -6$, $\varphi_y(a) = -68 < \varphi_y(y) = -4$. The consistency condition requires that $\varphi(a) < \varphi(y)$; the fulfillment of this inequality is seen directly from the graphs in Fig. 3.1.

§4. Special Classes of Problems

In this section, we apply the general aggregation scheme to some special classes of parametric optimization problems. The specific features of particular problems play an essential role in these implementations. In some cases, for instance, in studying quadratic programming problems (Section 4.1), these features allow us to strengthen earlier results. In other cases, they make us to weaken the setting of the aggregation problem to obtain constructive results; this refers, in particular, to integer programming problems considered in Sections 4.4 and 4.6. The consistent aggregation of infinite-dimensional problems (Section 4.2) and of a number of optimal control problems (Section 4.3) uses some ideas from Section 3.

4.1. Stronger Consistency Conditions. Quadratic Programming

The consistency conditions given in Section 1 mean that, if the original parametric problem admits a parameter variation improving the optimal value, then the aggregated problem also admits an improving variation, and any improving variation in the aggregated problem leads to an improvement in the optimal value of the original parametric problem. Thus, the passage to the aggregated problem preserves the information on the possibility in principle of improving the optimal value in the original problem by varying its parameters.

It should however be mentioned that aggregation does involve information loss. If the original problem has no certain special features, then lowering its dimension costs either the loss of a part of useful information or excessive computational expenditures for the aggregation proper. For this reason, when we set an aggregation problem, it is very important to known what information can be ignored and what retained in passing to the aggregated problem.

A consistent aggregated problem correctly reflects the direction of variation in the optimal value of the original problem under a change in parameters, but the precise value of this variation is lost. In other words, it allows us to find parameter variations improving the optimal value, but it does not provide the possibility of comparing such improving variations with each other; this to a certain degree reduces the effectiveness and convenience of processing the aggregated problem.

In this context, it is interesting to identify classes of parametric optimization problems that admit aggregation preserving not only the direction but also the magnitude of changes in the optimal value under parameter variations. This section studies one class of such problems. We show that the aggregated problem is quite adequate to the original problem, as concerns the influence of the parameters on the optimal objective value.

First, we refine the statement of the aggregation problem, and then, perform the necessary mathematical constructions.

Definition 4.1. We say that an aggregation of the original parametric problem (1.1) into a problem of form (1.2) is exact with respect to the parameter variations at a point y if $\varphi_y(a) = \varphi(a)$ for all $a \in \mho$.

It is easy to see that the problem of form (1.2) obtained as a result of such an aggregation is consistent with (1.1) in the sense of Definition 1.1, i.e., the conditions of exact aggregation are stronger than (1.3).

Let us make several general remarks on the construction of an aggregated problem with the required properties. As the original problem, we consider

$$
\begin{aligned}
&f(a,x) \to \min, \\
&g(a,x) = 0, \quad q(a,x) \le 0.
\end{aligned} \tag{4.1}
$$

Here, the functions $f(a,\,\cdot\,)\colon \mathbf{R}^n \to \mathbf{R}$ and $q(a,\,\cdot\,)\colon \mathbf{R}^n \to \mathbf{R}^r$ are convex and $g(a,\,\cdot\,)\colon \mathbf{R}^n \to \mathbf{R}^m$ is affine in $\mathbf{R}^n$ for any $a \in \mho$.

To state the conditions under which problem (4.1) can be exactly aggregated, consider its Lagrange function

$$L(a,x,\pi) = f(a,x) + ug(a,x) + wq(a,x),$$

where $\pi = (u,w) \in \mathbf{R}^m \times \mathbf{R}^r_+$[3] is the vector of dual variables (or Lagrange multipliers). Let (x_a, π_a), where $\pi_a = (u_a, w_a)$, denote a saddle point of L in the set of values $(x,u,w) \in \mathbf{R}^n \times \mathbf{R}^m \times \mathbf{R}^r_+$ (we assume that such a point exists for any $a \in \mho$). By virtue of the well-known relation between the optimality conditions and the existence of a saddle point, the points x_a and π_a are optimal solutions to the primal problem (4.1) and its dual, respectively, at the parameter value a [12].

Now, suppose that problem (4.1) has the following property: for any $a \in \mho$, the saddle points (x_a, π_a) belong to some affine manifold

$$\Pi = \{x \mid x = x_y + P_y p\} \times \{\pi \mid u = u_y + vH_y,\ w \in \mathbf{R}^r\} \tag{4.2}$$

of dimension lower than that of the extended space of variables $\mathbf{R}^n \times \mathbf{R}^m \times \mathbf{R}^r$ of the original problem. Making the change of variables

$$x = x_y + P_y p, \quad u = u_y + vH_y,$$

in L, we define the "aggregated" Lagrange function

$$
\begin{aligned}
L_y(a,p,\omega) &\triangleq x_y + P_y p,\ u_y + vH_y,\ w) = \\
&= f(a,\, x + P_y p) + (u_y + vH_y)g(a,\, x_y + P_y p) + wq(a,\, x_y + P_y p),
\end{aligned}
$$

where $\omega = (v,w) \in \mathbf{R}^{m_y} \times \mathbf{R}^r_+$. This function has fewer arguments than L, and it can be written in the form

$$L_y(a,p,w) = f_y(a,p) + vg_y(a,p) + wq_y(a,p).$$

[3]As previously, we treat the dual variables as a row vector of the corresponding dimension.

Here,

$$f_y(a,p) = f(a,\ x_y + P_y p) + u_y g(a,\ x_y + P_y p),$$

$$g_y(a,p) = H_y g(a,\ x_y + P_y p), \quad q_y(a,p) = q(a,\ x_y + P_y p),$$

and, by construction, the functions $f_y(a,\ \cdot): \mathbf{R}^{n_y} \to \mathbf{R}$ and $q_y(a,\ \cdot): \mathbf{R}^{n_y} \to \mathbf{R}^r$ are convex and $g_y(a,\ \cdot): \mathbf{R}^{n_y} \to \mathbf{R}^{m_y}$ is affine in $\mathbf{R}^{n_y}$ for any $a \in \mho$, because the functions $f(a,\ \cdot)$, $q(a,\ \cdot)$ and $g(a,\ \cdot)$ in the original problem (4.1) are so.

The function L_y is obtained from L by an affine transformation and can therefore be treated as the Lagrange function of some aggregated parametric problem, which generally belongs to the same class of problems as the original problem (4.1). Based on the form of L_y, we can state this aggregated problem as

$$\begin{aligned} &f_y(a,p) \to \min, \\ &g_y(a,p) = 0, \quad q_y(a,p) \leq 0. \end{aligned} \tag{4.3}$$

Thus, if the mentioned localization of the saddle points of the original problem (4.1) indeed takes place, then this problem can be directly aggregated into problem (4.3) with the use of the procedure described above. It turns out that, for any $a \in \mho$, the set of optimal solutions to the original problem belongs to the set of optimal solutions to the aggregated problem disaggregated in a special way, and the optimal values of these problems coincide. Let us show this. By $X^0(a)$ and $X_y^0(a)$, we denote the sets of optimal solutions to problems (4.1) and (4.3), respectively.

Theorem 4.1. *If $a \in \mho$, then, for any $x_a \in X^0(a)$, there exists a $p_a \in X_y^0(a)$ such that $x_a = x_y + P_y p_a$.*

Proof. For any fixed $a \in \mho$, the optimality of x_a in problem (4.1) implies the existence of a vector $\pi_a = (u_a, w_a) \in \mathbf{R}^m \times \mathbf{R}_+^r$ such that the pair (x_a, π_a) is a saddle point for the Lagrange function L, i.e.,

$$L(a, x_a, \pi) \leq L(a, x_a, \pi_a) = f(a, x_a) \leq L(a, x, \pi_a). \tag{4.4}$$

for any $x \in \mathbf{R}^n$ and $\pi \in \mathbf{R}^m \times \mathbf{R}_+^r$. By assumption (see (4.2)), $(x_a, \pi_a) \in \Pi$; hence, there exist p_a and v_a for which

$$x_a = x_y + P_y p_a, \quad u_a = u_y + vH_y.$$

Put $\omega_a = (v_a, w_a)$. According to (4.4), for any $p \in \mathbf{R}^{n_y}$ and $\omega = (v, w) \in \mathbf{R}^{m_y} \times \mathbf{R}^r_+$, we have

$$L_y(a, p_a, \omega) \triangleq L(a, x_a, \pi) \leq L(a, x_a, \pi_a) \triangleq$$
$$\triangleq L_y(a, p_a, \omega_a) \leq L(a, x, \pi_a) \triangleq L_y(a, p, \omega_a),$$

which means that (p_a, ω_a) is a saddle point of the Lagrange function L_y. Therefore, p_a is optimal in the aggregated problem (4.3). This completes the proof of the theorem.

Corollary 4.1. *Under the assumptions made above, the aggregation of problem (4.1) into (4.3) is exact in the sense of Definition 4.1.*

This follows directly from the method for constructing problem (4.3), which ensures that

$$\varphi(a) = f(a, x_a) = f(a,\ x_y + P_y p_a) = f_y(a, p_a) = \varphi_y(a).$$

Note also that the optimality of some point p in problem (4.3) is not sufficient for the optimality of the disaggregated vector $x = x_y + P_y p$ in (4.1). For this reason, in the general case, a solution to the original problem cannot be directly reconstructed from a given solution to the aggregated problem. This can however be achieved under certain additional assumptions.

Theorem 4.2. *Suppose that $X^0(a) \neq \varnothing$ for some $a \in \mho$. If the matrix H_y in (4.2) is identity or problem (4.1) has a unique optimal solution at this a, then $X^0(a) = \{x \mid x = x_y + P_y p,\ p \in X^0_y(a)\}$.*

Proof. By virtue of Theorem 4.1, it suffices to show that $\{x \mid x = x_y + {} + P_y p,\ p \in X^0_y(a)\} \subset X^0(a)$.

By condition, $X^0(a)$ contains at least one point x_a. According to Theorem 4.1, for this x_a, inequalities (4.4) hold, and there exist $p_a \in X^0_y(a)$ and $\omega_a = (v_a, w_a)$ such that $x_a = x_y + P_y p_a$ and the pair (p_a, ω_a) is a saddle point for the function L_y.

Take an arbitrary point $\bar{p} \in X^0_y(a)$. To this point, there corresponds a vector $\bar{\omega} = (\bar{v}, \bar{w}) \in \mathbf{R}^{m_y} \times \mathbf{R}^r_+$ such that

$$L_y(a, \bar{p}, \omega) \leq L_y(a, \bar{p}, \bar{\omega}) \leq L_y(a, p, \bar{\omega}) \tag{4.5}$$

for all $p \in \mathbf{R}^{n_y}$ and $\omega \in \mathbf{R}^{m_y} \times \mathbf{R}^r_+$. Applying inequalities of type (4.5) to the pairs (p_a, ω_a) and $(\bar{p}, \bar{\omega})$, we can show that $(\bar{p}, \omega_a)$ is also a saddle point, i.e.,

$$L_y(a,\bar{p},\omega) \le L_y(a,\bar{p},\omega_a) \le L_y(a,p,\omega_a)$$

for all $p \in \mathbf{R}^{n_y}$ and $\omega \in \mathbf{R}^{m_y} \times \mathbf{R}^r_+$.

Set $\bar{x} = x_y + P_y\bar{p}$ and $\bar{\pi} = (u_y + \bar{v}H_y,\ \bar{w})$. Comparing the last inequality with (4.4) and taking into account that, by condition, either $\bar{x} = x_a$ or $\pi = (u_y + v,\ w)$, we obtain

$$L(a,\bar{x},\pi) \le L(a,\bar{x},\pi_a) \le L(a,x,\pi_a), \quad x \in \mathbf{R}^n, \quad \pi \in \mathbf{R}^m \times \mathbf{R}^r_+.$$

Thus, $\bar{x} \in X^0(a)$. This completes the proof of the theorem.

Note that the uniqueness of an optimal solution required in the statement of Theorem 4.2 is ensured by, e.g., the strict convexity of the objective function $f(a,\,\cdot)$. The assumption that the matrix H_y is identity means actually that aggregation does not affect the constraints of the original problem.

The aggregation scheme described above is based on assumption (4.2), which is often violated. For (4.2) to hold, problem (4.1) must have a special structure ensuring that the optimal solutions x_a corresponding to various parameter values $a \in \mho$ belong to a certain "kernel" in the space $\mathbf{R}^n$. This allows us to restrict optimization to this kernel in seeking a solution to (4.1) at every fixed $a \in \mho$; to obtain a solution, it is then sufficient to find an optimal solution $p_a \in X^0_y(a)$ to the aggregated problem (4.3) and construct a point of the form $x_a = x_y + P_y p_a$, which is optimal in the original problem (4.1) under certain conditions. Theorem 4.1 ensures the existence of at least one point $p_a \in X^0_y(a)$ with the required properties in the case of $X^0(a) \ne \varnothing$, while Theorem 4.2 states conditions under which a solution to the optimization problem (4.1) can be constructed with the use of an arbitrary $p \in X^0_y(a)$.

Clearly, the main difficulty is to constructively define manifold (4.2). In the general case, for the class of problems under consideration, the saddle points cannot always be localized in a low-dimensional affine manifold. The possibility of such a localization essentially depends not only on the particular forms of the functions f, g, and q, which describe the original problem, but also on the number of simultaneously varied parameters. Below, we examine a narrower class of parametric optimization problems to which the described aggregation technique applies.

As (4.1), consider the parametric quadratic programming problem

$$\begin{aligned} &f(x) = d^T x + \tfrac{1}{2}x^T C x \to \min, \\ &(A + \Delta A)x = b, \quad qx \le h, \end{aligned} \tag{4.6}$$

where C is a positive definite matrix (i.e., $x^T Cx > 0$ for $x \neq 0$); $A + \Delta A$ and q are $m \times n$ and $r \times n$ matrices, respectively; b and h are vectors of dimensions m and r, respectively; and $\operatorname{rank}(A + \Delta A) = m$ for any feasible variations ΔA.

In this section, we examine the influence of the variable parameters of problem (4.6) on the optimal objective value under the following assumptions:

1. The matrix $A + \Delta A$ of coefficients has the structure

$$A + \Delta A = \begin{pmatrix} A_{11} + \Delta A_{11}, A_{12} \\ A_{21}, \quad A_{22} \end{pmatrix},$$

where the block $A_{11} + \Delta A_{11}$ of variable coefficients has size $m_1 \times n_1$;

2. For any $A_{11} + \Delta A_{11} \in \mho$, the feasible domain $X(A_{11} + \Delta A_{11})$ of problem (4.1) is nonempty.

In what follows, we label some objects by the subscripts a or y to indicate that they refer to the parameter values $a = A_{11} + \Delta A_{11}$ or $y = A_{11}$, respectively.

Proceeding to examine problem (4.6), we first note that, if $X(A_{11} + + \Delta A_{11}) \neq \varnothing$, then the corresponding optimization problem from family (4.6) has both primal and dual optimal solutions x_a and π_a. Knowing a dual solution π_a, we can get rid of the inequalities in the system of constraints by replacing (4.6) with the equivalent problem of the form

$$\begin{aligned} &(d^T + w_a q)x + \tfrac{1}{2}x^T Cx - w_a h \to \min, \\ &(A + \Delta A)x = b. \end{aligned} \tag{4.7}$$

The special structure of problem (4.7) allows us to explicitly derive its solution from the original data based on the necessary and sufficient conditions for this class of problems [27]. Namely, under the assumptions made about the matrices C and $A + \Delta A$, the dual and primal solutions to (4.7) are given by

$$u_a^T = -\left((A + \Delta A)C^{-1}(A + \Delta A)^T\right)^{-1}\left[b + (A + \Delta A)C^{-1}(d + q^T w_a^T)\right], \tag{4.8}$$

$$x_a = -C^{-1}\left[(A + \Delta A)^T u_a^T + d + q^T w_a^T\right]. \tag{4.9}$$

Let us show that, in this case, the solutions can be localized, i.e., the vectors x_a and u_a lie in affine manifolds of comparatively low dimensions for any $a = A_{11} + \Delta A_{11} \in \mho$. To prove this, we decompose the matrices as

$$A = \begin{pmatrix} A_1 \\ A_2 \end{pmatrix}, \quad C^{-1} = (\tilde{C}_1, \tilde{C}_2) = \begin{pmatrix} \tilde{C}_{11}, \tilde{C}_{12} \\ \tilde{C}_{21}, \tilde{C}_{22} \end{pmatrix},$$

$$(AC^{-1}A^T)^{-1} = By = (B_y^1,\ B_y^2),$$

where the blocks B_y^1, $\tilde{C}_1$, and $\tilde{C}_{11}$ have sizes $m \times m_1$, $n \times n_1$, and $n_1 \times n_1$, respectively; $A_1 = (A_{11},\ A_{12})$; and $A_2 = (A_{21},\ A_{22})$. Bearing in mind these decompositions, consider the matrix

$$G_a = \begin{pmatrix} G_a^1, E_{m_1} \\ G_a^2,\ \ 0 \end{pmatrix},$$

where

$$G_a^1 = A_1\tilde{C}_1\Delta A_{11}^T + \frac{1}{2}\Delta A_{11}\tilde{C}_{11}\Delta A_{11}^T, \quad G_a^2 = A_2\tilde{C}_1\Delta A_{11}^T.$$

The so-defined matrix G_a has size $m \times 2m_1$, and its blocks G_a^1 and G_a^2 depend on the parameters $A_{11} + \Delta A_{11}$. The following lemma is valid.

Lemma 4.1. *For any $A_{11}+\Delta A_{11} \in \mho$, there exists a $2m_1 \times m$ matrix F_a such that*

$$\left((A + \Delta A)C^{-1}(A + \Delta A)^T\right)^{-1} = B_y - B_y G_a F_a. \tag{4.10}$$

Proof. It is easy to verify that the matrix to be inverted in the left-hand side can be represented as the sum

$$(A + \Delta A)C^{-1}(A + \Delta A)^T = AC^{-1}A^T + D_a + D_a^T,$$

where $D_a = \begin{pmatrix} G_a^1, 0 \\ G_a^2, 0 \end{pmatrix}$. Define $\sigma = \begin{pmatrix} 0, & E_{m_1} \\ E_{m_1}, & 0 \end{pmatrix}$. Then

$$AC^{-1}A^T + D_a + D_a^T = AC^{-1}A^T + G_a\sigma G_a^T,$$

and we can apply either the Krohn decomposition method [29] or the rank annihilation method [6] for inverting matrices; according to these methods, we have

$$\left(AC^{-1}A^T + G_a\sigma G_a^T\right)^{-1} = B_y - B_yG_a\left(\sigma^{-1} + G_a^TB_yG_a\right)^{-1}G_a^TB_y.$$

Setting $F_a = \left(\sigma^{-1} + G_a^TB_yG_a\right)^{-1}G_a^TB_y$, we immediately obtain the required assertion. This completes the proof of the lemma.

We continue the constructions. Let us introduce the matrix $(A_2\tilde{C}_1,\ A_2C^{-1}q^T)$ and its skeleton decomposition

$$ST = (A_2\tilde{C}_1,\ A_2C^{-1}q^T), \tag{4.11}$$

where S and T are, by definition, matrices of full rank. Consider also the matrices

$$H_y = (B_y^1,\ B_y^2S)^T, \quad P_y = (\tilde{C}_1,\ C^{-1}q^T,\ C^{-1}A^TH_y^T) \tag{4.12}$$

Their coefficients are uniquely determined by the original data of problem (4.6) with $\Delta A = 0$ and, therefore, do not depend on the variation ΔA. By construction, H_y contains no more than m_1+n_1+r rows, and the number of columns in P_y does not exceed $2(n_1+r)+m_1$.

Finally, let x_y and u_y denote the primal and dual optimal solutions to the base problem for (4.7). The following lemma is valid.

Lemma 4.2. *For any $A_{11} + \Delta A_{11} \in \mho$, there exist vectors p_a and v_a such that*

$$x_a = x_y + P_yp_a, \quad u_a = u_y + v_aH_y$$

in (4.8) and (4.9).

Proof. First, we prove the second relation. Formulas (4.8) and (4.9) imply that, for some matrix F_a, we have

$$u_a^T = -(B_y - B_yG_aF_a)\left[b + (A+\Delta A)C^{-1}(d + q^Tw_a^T)\right],$$

or

$$u_a^T = -(B_y - B_yG_aF_a)\alpha,$$

where $\alpha = b + (A+\Delta A)C^{-1}(d+q^Tw_a^T)$. Note that

$$u_y^T = -B_y\left[b + AC^{-1}(d+q^Tw_y^T)\right].$$

Therefore, the difference $u_a^T - u_y^T$ can be conveniently written as

$$u_a^T - u_y^T = B_y\left[G_aF_a\alpha - \Delta AC^{-1}(d+q^Tw_a^T) - AC^{-1}q^T(w_a^T - w_y^T)\right].$$

Let us introduce the additional notation

$$v_a^1 = (G_a^1,\ E_{m_1})F_a\alpha - (\Delta A_{11},\ 0)C^{-1}(d+q^Tw_a^T) - A_1C^{-1}q^T(w_a^T - w_y^T),$$

$$v_a^2 = T\begin{pmatrix} (\Delta A_{11}^T,\ 0)F_a(\alpha) \\ w_y^T - w_a^T \end{pmatrix}, \qquad v_a = \begin{pmatrix} v_a^1 \\ v_a^2 \end{pmatrix}^T.$$

Taking into account the preceding calculations, the structure of the matrices G_a, A, and ΔA, and the relations (4.11) and

$$B_y A C^{-1} q^T = B_y^1 A_1 C^{-1} q^T + B_y^2 A_2 C^{-1} q^T,$$

we obtain

$$u_a - u_y = (B_y^1 v_a^1 + B_y^2 S v_a^2)^T = v_a H_y,$$

which can be verified directly.

Let us turn to the relation $x_a = x_y + P_y p_a$. It is proved similarly, based on expression (4.9), according to which

$$x_a - x_y = -C^{-1}\Delta A^T u_a^T - C^{-1} q^T (w_a^T - w_y^T) - C^{-1} A^T (u_a - u_y)^T.$$

Setting

$$p_a^1 = -(\Delta A_{11}^T,\ 0)u_a^T, \quad p_a^2 = \begin{pmatrix} w_y^T - w_a^T \\ -v_a^T \end{pmatrix}, \qquad p_a = \begin{pmatrix} p_a^1 \\ p_a^2 \end{pmatrix}$$

and applying the equality $u_a = u_y + v_a H_y$ proved above, we obtain

$$x_a - x_y = \tilde{C}_1 p_a^1 + C^{-1}(q^T,\ A^T H_y^T) p_a^2 = P_y p_a.$$

This completes the proof of the lemma.

Remark 4.1. By construction, the rows of the matrix H_y are linearly independent, i.e., this matrix is of full rank. The matrix P_y in (4.12) can also be made to have linearly independent columns by replacing it with the left matrix in the skeleton decomposition.

Lemma 4.2 gives an affirmative answer to the question about the localization of the solutions to the original problem. We only have to determine the dimension n_Π of the affine manifold (4.2), (4.12) and compare it with the dimensions of problem (4.6). The bounds for the sizes of the matrices P_y and H_y given above show that

$$n_\Pi \le 2(n_1 + r) + m_1 + (m_1 + n_1 + r) + r = 3n_1 + 2m_1 + 4r.$$

Thus, if $\max(m_1, n_1, r) \ll \min(m, n)$ in (4.6), then $n_\Pi \ll m + n + r$, i.e., the dimension of the manifold is much less than that of the extended space of variables of the original problem. In this case, the aggregation is quite effective.

It remains to write the aggregated problem itself. The Lagrange function for (4.6) is

$$L(a, x, \pi) = d^T x + \frac{1}{2} x^T C x + u((A + \Delta A)x - b) + w(qx - h),$$

where $x \in \mathbf{R}^n$ and $\pi = (u, w) \in \mathbf{R}^m \times \mathbf{R}^r_+$. Cutting the spaces $\mathbf{R}^n$ and $\mathbf{R}^m \times \mathbf{R}^r$ by the manifold $\{x \mid x = x_y + P_y p\}$ and $\{\pi \mid u = u_y + vH_y\}$, we obtain the required aggregated problem from (4.6) in the form

$$\begin{aligned} &f_y(a, p) = f_y(p) + \Delta f_y(a, p) \to \min, \\ &(Q + \Delta Q)p = \Delta b_y, \quad q_y p \le h_y, \end{aligned} \tag{4.13}$$

where

$$f_y(p) = d_y^T p + \frac{1}{2} p^T C_y p + c_y, \quad \Delta f_y(a, p) = \Delta d^T p + \Delta c,$$

$$Q + \Delta Q = H_y(A + \Delta A)P_y, \quad \Delta b_y = -H_y \Delta A x_y, \quad q_y = qP_y, \quad h_y = h - qx_y.$$

The coefficients in the objective function are

$$d_y^T = d^T P_y + x_y^T C P_y + u_y A P_y,$$

$$C_y = P_y^T C P_y, \quad c_y = d^T x_y + \frac{1}{2} x_y^T C x_y,$$

$$\Delta d^T = u_y \Delta A P_y, \quad \Delta c = u_y \Delta A x_y.$$

By virtue of Corollary 4.1, the aggregation of problem (4.6) into (4.13) is exact. By construction, the number of variables in (4.13) does not exceed $2(n_1 + r) + m_1$ and the number of constraints, $n_1 + m_1 + 2r$.

Since the rows in H_y are linearly independent, the matrix $Q + \Delta Q$ in the system of equations is of full rank at any ΔA. Indeed, this matrix has the structure

$$Q + \Delta Q = (Q_1, \, Q_2, \, H_y(A + \Delta A)C^{-1} A^T H_y^T),$$

where $Q_1 = H_y(A + \Delta A)\tilde{C}_1$ and $Q_2 = H_y(A + \Delta A)C^{-1} q^T$. The submatrix $H_y(A + \Delta A)C^{-1} A^T H_y^T$ is nondegenerate, because A, $A + \Delta A$, and H_y are matrices of full rank and C is nondegenerate. This directly implies that $Q + \Delta Q$ has full rank. If P_y also has full rank (see Remark 4.1), then $C_y = P_y^T C P_y$ is positive definite. In this case, the parametric problem (4.13) belongs to precisely the same class of problems as the original problem (4.6).

Remark 4.2. In the case where the assumption $r \ll \min(m, n)$ about the number of inequalities in problem (4.6) does not hold (say, if the system $qx \leq h$ is the nonnegativity condition $x \geq 0$), then the application of the method described above is inexpedient, and problem (4.6) can be aggregated with the use of the procedure described in Section 3.1 (though, the property $\varphi_y(a) = \varphi(a)$ is then lost). Nevertheless, the combination of the methods considered in this section and in Section 3 provides a fairly flexible and efficient tool for parametrically analyzing quadratic programming problems, for it makes it possible to aggregate (4.6) over a wide range of values of r.

To conclude, we make several observations that allow us to generalize the method described above to problems with nonlinear inequality-type constraints. As is known, in higher-dimensional linear problems, the matrix of conditions is sparse, i.e., the fraction of its nonzero coefficients is comparatively small and may amount to only several percent. The higher-dimensional nonlinear problems also have specific features. As mentioned in [28], in the majority of applications, the number of the nonlinear constraints and variables determining the nonlinear parts of the constraints in the problem is also small in comparison with their total number. Taking this into account, consider the following problem with "almost linear" constraints instead of (4.6):

$$\begin{gathered} f(x) = d^T x + \tfrac{1}{2} x^T C x \to \min, \\ (A + \Delta A)x = b, \quad qx + \psi(P_N x) \leq h, \end{gathered} \tag{4.14}$$

where $\psi \colon \mathbf{R}^N \to \mathbf{R}^r$ is a differentiable convex function and P_N is the projector onto a subspace of dimension $N \ll n$. (Problems whose objective functions also have nonlinear terms reduce to form (4.14) too.)

Suppose that (4.14) admits a regular linearization [5] at an optimal point $x_a \in X^0(A_{11} + \Delta A_{11})$ and a dual optimal solution $\pi_a = (u_a, w_a)$ is known. Then (4.4) is equivalent to the quadratic programming problem

$$\begin{gathered} (d^T + w_a q + \psi_a)x + \tfrac{1}{2} x^T C x + \mathrm{const}_a \to \min, \\ (A + \Delta A)x = b, \end{gathered} \tag{4.15}$$

where $\psi_a = w_a \nabla\psi(P_N x_a) P_N$ is a row vector and $\mathrm{const}_a = w_a[\psi(P_N x_a) - \nabla\psi(P_N x_a) P_N x_a - h]$ is a scalar.

Problem (4.15) differs from (4.7) in that the vector of the coefficients in the linear part of its objective function includes an additional "variable" term ψ_a depending on the parameter $a = A_{11} + \Delta A_{11}$. As a result, in aggregation, instead of ST and P_y in (4.11) and (4.12), we have to consider constructions of the form

$$ST = (A_2\tilde{C}_1,\ A_2C^{-1}q^T,\ A_2C^{-1}P_N^T),$$

$$P_y = (\tilde{C}_1,\ C^{-1}q^T,\ C^{-1}P_N^T,\ C^{-1}A^TH_y^T).$$

The corresponding aggregated problem obtained from (4.14) looks like

$$\begin{aligned} &f_y(a,p) = f_y(p) + \Delta f_y(a,p) \to \min, \\ &(Q + \Delta Q)p = \Delta b_y, \quad q_y p + \psi_y(p) \le h_y. \end{aligned} \tag{4.16}$$

Here, $\psi_y(p) = \psi(P_N x_y + P_n P_y p)$; the number of constraints does not exceed $n_1 + m_1 + 2r + N$ and the number of variables, $2(n_1 + r + N) + m_1$. Thus, under the assumptions made, the dimensions of (4.6) are still much less than those of (4.14).

4.2. Linear-Convex Problems in Banach Spaces

In this section, we transfer the consistent aggregation method developed for finite-dimensional problems to linear-convex parametric optimization problems in Banach spaces. Intuitively, such an extension is possible because the results of Section 3.2 are based on the optimality conditions and on the notion of right inverse operator, which have natural generalizations over infinite-dimensional spaces. Note also that, historically, the desire to carry over the results on finite-dimensional problems to Banach spaces was caused by practical demands rather than was a goal by itself, for such a generalization makes it possible, in particular, to apply methods of mathematical programming to optimal control problems.

Suppose given Banach spaces L and Y, a functional $f\colon L \to \mathbf{R}$, and mappings $g\colon \mathbf{R}^q \times L \to \mathbf{R}^{m_1}$, $q\colon L \to Y$, and $h\colon L \to \mathbf{R}^{m_2}$, where $m_1 + m_2 = m$. Let us fix a point $y \in \mho \subset \mathbf{R}^q$ and consider the parametric problem

$$\begin{aligned} &f(x) \to \min, \\ &g(a,x) \le 0, \quad h(x) \le 0, \quad q(x) = 0. \end{aligned} \tag{4.17}$$

In the case under consideration, $a \in \mho$ and $X(a) = Z(a) = \cap S \cap Q$, where

$$Z(a) = \{x \mid g(a,x) \le 0\}, \quad S = \{x \mid h(x) \le 0\}, \quad Q = \{x \mid q(x) = 0\}.$$

Suppose that

1. the function f and the mappings $g(y,\,\cdot\,)$ and h are Gateaux differentiable at a point $x_y \in X^0(y)$ and convex in L, and the mapping q is affine and Gateaux differentiable at the point x_y;

2. the mapping $g(\,\cdot\,, x)$ is affine in $\mathbf{R}^q$ and monotone with respect to a_j for $j = \overline{1,q}$ at any $x \in S$; and

3. the vector functions $V_j(x) = \frac{\partial g(y,x)}{\partial a_j}$ with $j = \overline{1,q}$ are affine in $\mathbf{R}^n$.

Thus, (4.17) is an analogue of the finite-dimensional parametric problem (3.21), and when the parameter values $a = y$ are fixed, it is a convex programming problem in a Banach space. Its distinguishing feature is that the space of variables x is generally infinite-dimensional, and the constraint $q(x) = 0$ is operator.

Pursuing the analogy with the considerations of Section 3.2, let us describe the basic constructions of the consistent aggregation scheme as applied to problem (4.17). We only consider aggregation of variables, because the methods for aggregating constraints were given in the preceding sections in detail.

Prior to proceeding to construct the manifold (1.4), we introduce some notation. Consider the mapping

$$\Phi = \begin{pmatrix} g \\ h \\ q \end{pmatrix} : \mathbf{R}^q \times L \to \mathbf{R}^m \times Y$$

and remove those constraints $\Phi(a,x) \le 0$ that do not vanish at $a = y$, $x = x_y$. We denote the mapping corresponding to the remaining (active) constraints by

$$\Phi^A = \begin{pmatrix} g^A \\ h^A \\ q \end{pmatrix} : \mathbf{R}^q \times L \to \mathbf{R}^{m_1^A} \times \mathbf{R}^{m_2^A} \times Y = \mathbf{R}^{m^A} \times Y;$$

thus, $m^A = m_1^A + m_2^A$,

$$\Phi^A(a,x) = (g^A(a,x),\ h^A(x),\ g(x)),$$

and $\Phi^A(y, x_y) = 0$.

Let $\Lambda_\Gamma\colon L \to \mathbf{R}^{m^A} \times Y$ denote the Gateaux derivative of Φ^A with respect to x at the point $a = y$, $x = x_y$, i.e.,

$$\Lambda_\Gamma x = \begin{pmatrix} \dfrac{\partial g^A(y, x_y)}{\partial x} x \\ \dfrac{\partial h^A(x_y)}{\partial x} x \\ \dfrac{\partial q(x_y)}{\partial x} x \end{pmatrix}. \tag{4.18}$$

To describe the basic constructions determining the structure of Π_y, we assume that the Gateaux derivative

$$\Lambda_\Gamma = \frac{\partial \Phi^A(y, x_y)}{\partial x} \in \mathcal{L}(L,\ \mathbf{R}^{m^A} \times Y)$$

has a right inverse, that is, an operator $\Lambda_\Gamma^*\colon \mathbf{R}^{m^A} \times Y \to L$ such that

$$\Lambda_\Gamma \Lambda_\Gamma^* z = z \quad \text{for all} \quad z = (z^1, z^2, z^3) \in \mathbf{R}^{m_1^A} \times \mathbf{R}^{m_2^A} \times Y.$$

Using this operator, we define

$$\Pi_y = \Big\{ x \,\Big|\, x = x_y + \sum_{i=1}^{m^A} x^i(y) p_i \Big\}, \tag{4.19}$$

where $x^i(y) = -\Lambda_\Gamma^*(e_i, 0)$ for $i = 1, 2, \ldots, m^A$ ($e_i \in \mathbf{R}^{m^A}$ are the unit basis vectors). Thus, the manifold Π_y defined by (4.19) is finite-dimensional and isomorphic to the space $\mathbf{R}^{m^A}$. Its dimension equals m^A.

Now, we can aggregate the original problem following the approach described above. Take a $p = (p_1, \ldots, p_{n_y})^T$, where $n_y = m^A$. Cutting the space L by manifold (4.19) in problem (4.17), we obtain the aggregated parametric problem

$$\begin{aligned} f_y(p) &= f\Big(x_y + \sum_{i=1}^{n_y} x^i(y) p_i\Big) \to \min, \\ g_y(a, p) &= g\Big(a,\ x_y + \sum_{i=1}^{n_y} x^i(y) p_i\Big) \le 0, \\ h_y(p) &= h\Big(x_y + \sum_{i=1}^{n_y} x^i(y) p_i\Big) \le 0, \\ q_y(p) &= q\Big(x_y + \sum_{i=1}^{n_y} x^i(y) p_i\Big) = 0 \end{aligned} \tag{4.20}$$

with parameter vector $a \in \mho$.

Fixing any $a \in \mho$ in (4.20), we again obtain a convex programming problem, but it is stated in the finite-dimensional space $\mathbf{R}^{n_y}$, as opposed to the infinite-dimensional problem (4.17). This is the essence of aggregating variables. We only have to prove that the aggregation is consistent in the sense of (1.3). For the class of problems under consideration, we can assume that the feasible parameter set $\mho$ has property (3.35).

We say that a set $X(y)$ is regular if it contains a point x such that $g(y,x) < 0$ and $h(x) < 0$.

Definition 4.2. We say that problem (4.17) is nondegenerate at a point y if $X(y)$ is regular and Λ_Γ is an epimorphism (i.e., a surjective mapping).

Theorem 4.3. *If problem (4.17) is nondegenerate at a point $y \in \mho$ and $\mho$ has property (3.35), then the aggregation of the parametric problem (4.17) into problem (4.20) is consistent at the point y.*

Proof. The proof is basically the same as for Theorem 3.3, and we only give its outline. We use the notation of Section 3.2 without specifying it.

The Lagrange function of problem (4.17) is

$$L(a,x,\pi) = f(x) + ug(a,x) + vh(x) + wq(x),$$

where $\pi = (u,v,w) \in \mathbf{R}_+^{m_1} \times \mathbf{R}_+^{m_2} \times Y^*$ denotes the row vector of Lagrange multipliers and Y^* stands for the space dual to Y. We have

$$x_y \in X^0(y) = \operatorname{Argmin}\{f(x) \mid x \in X(y)\};$$

hence, under the conditions of the theorem, the optimality conditions [11] imply that, at $a = y$, there exists a Kuhn–Tucker vector $\pi_y = (u_y, v_y, w_y)$ such that

$$\frac{\partial L(y,x_y,\pi_y)}{\partial x} = 0, \quad u_y g(y,x_y) = 0, \quad v_y h(x_y) = 0, \tag{4.21}$$

where the derivative with respect to x is Gateaux.

Put $\Delta a_j = (a-y)_j$ for $j \in J$ and $J(a) = \{j \in J \mid \Delta a_j < 0\}$. As in Section 3.2, we assume without loss of generality that $V_j(x) = \frac{\partial g(y,x)}{\partial a_j} \geq 0$ for $j \in \overline{1,q}$. Following the same line of reasoning as in the proof of Theorem 3.3 and taking into account the convexity of (4.17) at $a = y$, we can show that, if $J_0(a) \neq \varnothing$, then there exist matrices $\Lambda^j \geq 0$ with $j \in J_0(a)$ such that

$$\begin{aligned} V_j(z)\Delta a_j &\leq -\Lambda^j \frac{\partial h^A(x_y)}{\partial x} x \Delta a_j, \quad x \in S, \\ U_y V_j(z)\Delta a_j &= -u_y \Lambda^j \frac{\partial h^A(x_y)}{\partial x} x \Delta a_j. \end{aligned} \tag{4.22}$$

For the row m_2^A-vector

$$\theta(a) = (\theta_1(a), \dots, \theta_{m_2^A}(a)) = v_y^A + \sum_{j \in J_0(a)} u_y \Lambda^j \Delta a_j$$

the following assertion (an analogue of Lemma 3.5) is valid.

Lemma 4.3. *If the set $X(y)$ is regular and $a \in \Gamma(y)$, then at least one of the following relations holds:*

$$J_+(a) \neq \varnothing, \quad u_y \sum_{j \in J_+(a)} V_j(x_j) \Delta a_j < 0;$$

$$J_0(a) \neq \varnothing, \quad \theta_s(a) < 0 \quad \text{for some} \quad s \in \{1, \dots, m_2^A\}.$$

The proof of the theorem is continued by considering the family $W_a(z, \varepsilon, \varphi^0(\cdot))$ of the pairs $(a(\beta), x(\beta)) \in \mathbf{R}^q \times L$ having the form

$$\begin{aligned} a(\beta) &= y + (a - y)^- \varphi^0(\beta), \\ x(\beta) &= x_y - \Lambda_\Gamma^*(z + \varepsilon\mu)\beta, \end{aligned} \tag{4.23}$$

where

$$\mu = (\mu^1, \mu^2, 0) \in \mathbf{R}^{m_1^A} \times \mathbf{R}^{m_2^A} \times Y, \quad \mu^1 = (1, \dots, 1)^T,$$
$$\mu^2 = (1, \dots, 1)^T, \quad z = (z^1, z^2, 0) \in \mathbf{R}^{m_1^A} \times \mathbf{R}^{m_2^A} \times Y,$$

$\varphi^0(\beta)$ is a scalar function, and $(a - y)^-$ is the projection of $(a - y)$ on the negative orthant.

It is easy to see that, for any z and ε, the points $x(\beta)$ defined by (4.23) belong to manifold (4.19). At the same time, if $a \in \Gamma(y)$, then the parameters z, ε, and $\varphi^2(\cdot)$ can be selected so that the corresponding directions defined by (4.23) satisfy the condition

$$(a(\beta),\ x(\beta)) \in \mathbf{R}^q \times X(a(\beta)),$$

for all sufficiently small $\beta \geq 0$, and the objective function f decreases along these directions. Namely, the following lemma is valid.

Lemma 4.4. *If problem (4.17) is nondegenerate at a point y and $a \in \Gamma(y)$, then there exist z, ε, $\varphi^0(\cdot)$, and $\beta^* > 0$ such that $a(\beta) \in \mho$, $x(\beta) \in X(a(\beta)) \cap \Pi_y$, and $f(x(\beta)) < f(x_y)$ for all $0 < \beta \leq \beta^*$.*

Proof. First, note that, under the conditions of the lemma, there exists a right inverse operator Λ_Γ^*.

Lemma 4.4 is proved constructively by analogy with Lemmas 3.6 and 3.7. The proof is based on relations (4.21) and (4.22). It reduces to the formation of z, ε, and $\varphi^0(\cdot)$ such that, first,

$$q(x(\beta)) = 0, \quad h^A(x(\beta)) = -\varepsilon\mu^2\beta + o(\beta),$$

$$g^A(a(\beta), x(\beta)) = -\varepsilon\mu^1\beta + o(\beta),$$

and, secondly,

$$f(x(\beta)) - f(x_y) = \Delta\varphi\beta + C_y\varepsilon\beta + o(\beta),$$

where $C_y = u_y^A\mu^1 + v_y^A\mu^2$ and

$$\Delta\varphi = \begin{cases} u_y \sum\limits_{j\in J_+(a)} V_j(x_y)\Delta a_j, & J_+(a) \neq \varnothing, \\ \theta_j(a), & J_0(a) \neq \varnothing. \end{cases}$$

These decompositions hold if the vector z is defined by

$$z = \begin{cases} z^1 = \sum\limits_{j\in J_+(a)} V_j(x_y)\Delta a_j, & z^2 = 0 \quad \text{if} \quad J_+(a) \neq \varnothing \\ z^1 = \sum\limits_{j\in J_0(a)} \Lambda^j e_s\Delta a_j, & z^2 = e_s \quad \text{if} \quad J_0(a) \neq \varnothing, \end{cases}$$

and the function φ^0, by

$$\varphi^0(\beta) = \begin{cases} \beta, & J_+(a) \neq \varnothing, \\ 1, & J_0(a) \neq \varnothing. \end{cases}$$

Now, the validity of the assertion of the lemma for sufficiently small $\varepsilon > 0$ and $0 < \beta^* \leq 1$ follows from Lemma 4.3 and the special features of the function φ^0.

In essence, Lemma 4.4 means that it is sufficient to construct an aggregated problem only at the points of manifold (4.19); such a restriction does not cause loss of information on the improving variations in problem (4.17).

The proof of consistency can be completed by repeating the arguments used at the end of the proof of Theorem 3.3.

The technique for aggregating parametric optimization problems described above is based on the assumption that the Gateaux derivative Λ_Γ has a right inverse Λ_Γ^*. In view of this, let us consider methods for constructing such a right inverse operator.

Suppose that $L = \mathbf{R}^n$, $Y = \mathbf{R}^k$, $r = m^A + k$, and all functions are differentiable in the usual sense. Then the operator $\Lambda_\Gamma \colon \mathbf{R}^n \to \mathbf{R}^r$ can be identified with the $r \times n$ matrix whose columns are the transposed gradients of the active constraints. Let $\bar{r}$ denote the rank of the matrix operator A. Then the assumption that Λ_Γ is an epimorphism[4] is equivalent to the relation $\bar{r} = r \leq n$, i.e., to the linear independence of the gradients of the constraints of (4.17) active at the point x_y for $a = y$.

In this case, the operator Λ_Γ^* can be constructed in different ways. For instance, we can set $\Lambda_\Gamma^* = A^+$, where $A^* = A^T(AA^T)^{-1}$ is the pseudoinverse of A [8]. A somewhat different approach is as follows. Since the matrix A is of full rank under the assumptions made, it has a (generally nonunique) nonsingular submatrix B of rank r. Without loss of generality, we can assume that $A = (B, D)$. Then, as Λ_Γ^*, we can take $\Lambda_\Gamma^* = \begin{pmatrix} B^{-1} \\ 0 \end{pmatrix}$. Depending on the choice of B, different Λ_Γ^* are obtained. Actually, this form of Λ_Γ^* is used in Sections 2 and 3 to aggregate linear programming problems; there, B is the basic matrix corresponding to an optimal vertex of the constraint polyhedron.

Now, suppose that $Y = \mathbf{R}^k$, as above, but L is an arbitrary Banach space. It is known [5] that, if $\bar{r}$ is the rank of the operator $\Lambda_\Gamma \in \mathcal{L}(L, \mathbf{R}^r)$, then the dimension of the quotient space $L/\operatorname{Ker}\Lambda_\Gamma$ of the space L modulo the kernel $\operatorname{Ker}\Lambda_\Gamma = \{x \mid \Lambda_\Gamma x = 0,\ x \in L\}$ equals $\bar{r}$, and L contains an $\bar{r}$-dimensional subspace $L^{\bar{r}}$ such that each element $x \in L$ has a unique representation

$$x = x^* + y, \quad x^* \in L^{\bar{r}}, \quad y \in \operatorname{Ker}\Lambda_\Gamma.$$

Since $\bar{r} = r$ (because Λ_Γ is an epimorphism), we can establish a one-to-one correspondence between $\mathbf{R}^r$ and L^r, and there exists an operator $\Lambda^{-1} \in \mathcal{L}(\mathbf{R}^r, L^r)$ for which

$$\Lambda_\Gamma \Lambda^{-1} z = z, \quad z \in \mathbf{R}^r, \quad \Lambda^{-1}\Lambda_\Gamma x = x^*, \quad x \in L.$$

The operator $\Lambda_\Gamma^* = \Lambda^{-1}$ is, obviously, a right inverse. Note that the particular form of Λ^{-1} (and, therefore, of Λ_Γ^*) depends both on L^r and on the choice of the basis in L^r.

[4]In reality, as is shown in Section 3 for finite-dimensional problems, this requirement can be replaced by the weaker condition that the dual solution is unique.

As an example, consider the integral operator

$$\Lambda_\Gamma u(\cdot) = \begin{pmatrix} \int_0^1 tu(t)\,dt \\ \int_0^1 t^2 u(t)\,dt \end{pmatrix}, \tag{4.24}$$

in the Banach space $L_2[0,1]$ of functions $u(t)$ square integrable in $\mathbf{R}^2$. According to the above considerations, $L_2[0,1]$ has a two-dimensional subspace L^2 such that $L_2[0,1] = L^2 + \operatorname{Ker}\Lambda_\Gamma$. Here, $\operatorname{Ker}\Lambda_\Gamma$ consists of the elements $\psi(t) \in L_2[0,1]$ orthogonal to the linearly independent functions t and t^2, and the elements of L_2 can be represented as linear combinations $\alpha_1 t + \alpha_2 t^2$.

It is easy to see that, in the case under consideration, we can seek the right inverse operator $\Lambda_\Gamma^*: \mathbf{R}^2 \to L_2[0,1]$ in the form

$$\Lambda_\Gamma^* x = (t, t^2)Gx,$$

where $x = \begin{pmatrix} x_1 \\ x_2 \end{pmatrix} \in \mathbf{R}^2$ and the normalizing 2×2 matrix G is selected so that

$$\Lambda_\Gamma \Lambda_\Gamma^* = \begin{pmatrix} 1, 0 \\ 0, 1 \end{pmatrix}. \tag{4.25}$$

By virtue of (4.24), condition (4.25) is equivalent to the relation

$$\begin{pmatrix} \int_0^1 t^2 dt, & \int_0^1 t^3 dt \\ \int_0^1 t^3 dt, & \int_0^1 t^4 dt \end{pmatrix} G = \begin{pmatrix} 1/2, 1/3 \\ 1/3, 1/4 \end{pmatrix} G = \begin{pmatrix} 1, 0 \\ 0, 1 \end{pmatrix}.$$

whence

$$G = \begin{pmatrix} 1/2, 1/3 \\ 1/3, 1/4 \end{pmatrix}^{-1} = \begin{pmatrix} 18, -24 \\ -24, \ \ 36 \end{pmatrix},$$

$$\Lambda_\Gamma^* x = (18t - 24t^2)x_1 + (36t^2 - 24t)x_2.$$

Finally, suppose that the L and Y in (4.17) are arbitrary Banach spaces. Then, if Λ_Γ is an epimorphism, it has a right inverse under fairly general assumptions [1], but more constructive results can be obtained only in special cases. Below, we describe one of them.

Since q is affine, the solution set of the operator equation $q(x) = 0$ in (4.17) can be represented as the sum $x_y + \operatorname{Ker} q$. If we managed to give a constructive parametric description of $x_y + \operatorname{Ker} q$, then, substituting this representation in the constraints $g(a, x) \leq 0$ and $h(x) \leq 0$, we could "eliminate" $q(x) = 0$ and thereby obtain a problem with a finite number of functional constraints, which has already been considered.

The main problem is to obtain a constructive parametric representation of the solutions to the equation $q(x) = 0$. In the general case, this is impossible. It is however known that, in optimal control problems, operator equations are used to write systems of differential constraints. If such a system is linear with respect to the state variables x and has the form

$$\dot{x}(t) = A(t)x(t) + b(u(t), t), \quad x(t_0) = x^0, \tag{4.26}$$

then it can be "solved" with respect to x with the use of the Cauchy formula; i.e., we can express the variables x in terms of the controls u and thereby exclude the system of constraints (4.26) from the further consideration. This argument is used to reduce certain optimal control problems linear in the state variables to Lyapunov-type problems [1].

To conclude, we give a numerical example of using the consistent aggregation method for $L = \mathbf{R}^n$. As (4.17), consider the problem

$$\begin{aligned} f(x) &= 5(x_1 - 2)^2 + 2x_2 + (x_3 - 4)^2 - 6x_4 \to \min, \\ g(a, x) &= a(2x_1 - x_2) + (x_1 + x_3 + x_4)^2 - 9 \leq 0, \\ h(x) &= -2x_1 + x_2 \leq 0 \end{aligned} \tag{4.27}$$

with parameter $a \in \mho \subset \mathbf{R}$.

Let us fix the parameter value $y = 3$. At $a = y = 3$, problem (4.27) has the optimal solution $x_y = (1, 2, 1, 1)^T$; the corresponding Kuhn–Tucker vector is $\pi_y = (u_y, v_y) = (1, 1)$, and the optimal value is $\varphi(y) = f(x_y) = 12$.

Let us aggregate this problem at the point $a = y$. For this purpose, note that, at x_y, both constraints are active and the operator Λ_Γ defined by (4.18) is given by the matrix

$$\Lambda_\Gamma = \begin{pmatrix} 12, & -3, 6, 6 \\ -2, & 1, 0, 0 \end{pmatrix}.$$

The right inverse operator Λ_Γ^* is determined not uniquely. Take, for example,

$$\Lambda_\Gamma^* = \begin{pmatrix} 0, 0, 1/6, 0 \\ 0, 1, 1/2, 0 \end{pmatrix}^T .$$

It is easy to see that $\Lambda_\Gamma \Lambda_\Gamma^* = \begin{pmatrix} 1, 0 \\ 0, 1 \end{pmatrix}$.

This, in the definition of the affine manifold,

$$x^1(y) = -(0,\ 0,\ 1/6,\ 0)^T, \quad x^2(y) = -(0,\ 1,\ 1/2,\ 0)^T,$$

and the manifold itself has the form

$$\Pi_y = \left\{ x \;\middle|\; x = \begin{pmatrix} 1 \\ 2 \\ 1 \\ 1 \end{pmatrix} - \begin{pmatrix} 0 \\ 0 \\ 1/6 \\ 0 \end{pmatrix} p_1 - \begin{pmatrix} 0 \\ 1 \\ 1/2 \\ 0 \end{pmatrix} p_2 \right\}.$$

The intersection of the feasible set of (4.27) with the constructed manifold determines the aggregated parametric problem in two variables

$$\begin{aligned} f_y(p) &= (p_1/6 + p_2/2 + 3)^2 - 2p_2 + 3 \to \min, \\ g_y(a, p) &= ap_2 + (p_1/6 + p_2/2 - 3)^2 - 9 \le 0, \\ h_y(p) &= -p_2 \le 0, \end{aligned} \tag{4.28}$$

where $a \in \mho$, as in (4.27). It can be shown that (4.28) with $a = y$ has the optimal solution $p_y = (0, 0)^T$ and the optimal value $\varphi_y(y) = f_y(p_y) = 12$.

The results of the parametric analysis of problems (4.27) and (4.28) aimed at examining the behavior of their optimum functions are given in Tables 4.1 and 4.2, where $\lambda_i(a)$ with $i = 1, 2$ denote the components of the dual solution.

Analyzing these data, we see that $\varphi(a) \le \varphi_y(a)$ at $a > 0$, as expected (see Remark 1.1). Let us consider various sets $\mho$ of feasible parameter values in order to show how a consistent aggregated problem can be used to examine the behavior of the optimum function of the original problem.

	$0 < a \leq 2$	$a \geq 2$
$x_1(a)$	1	1
$x_2(a)$	$2 + 9a/4 - 9/a$	2
$x_3(a)$	1	1
$x_4(a)$	$3a/2 - 2$	1
$\lambda_1(a)$	$2/a$	1
$\lambda_2(a)$	0	$a - 2$
$\varphi(a)$	$30 - 9a/2 - 18/a$	12

Table 4.1.

	$0 < a \leq 2$	$a \geq 2$
$p_1(a)$	$-3p_2(a) + 18(2-a)/(2+a)$	0
$p_2(a)$	$\left(9 - (6a/(2+a)^2)\right)/a$	0
$\lambda_1(a)$	$2/a$	1
$\lambda_2(a)$	0	$a - 2$
$\varphi_y(a)$	$3 + (12/(2+a))^2 - 2(9 - (6a/(2+a))^2)/a$	12

Table 4.2.

If $\mho = [2, \infty)$, then the point $y = 3$ is, obviously, a point of global optimum for the function φ_y on $\mho$, i.e., $\Gamma_A(y) = \varnothing$. We can apply the second condition in (1.3), according to which Γ_y must be empty. This is confirmed by the data given in Table 4.1; they show that the optimal value in the original problem (4.17) with $a \geq 2$ is attained at the same point x_y.

Consider $\mho = [1, \infty)$. For $1 \leq a < 2$, we have $\varphi_y(a) < \varphi_y(y) = 12$, i.e., these points a are preferred to y for the function φ_y and, therefore, $\Gamma_A \neq \varnothing$. In this case, the first condition from (1.3) applies, according to which $\varphi(a) < \varphi(y)$ at $1 \leq a < 2$.

Thus, the functions φ and φ_y have the same sets of preferred parameter values, i.e., $\Gamma(y) = \Gamma_A(y)$. In a more general case, only the one-sided inclusion

$\Gamma_A(y) \subset \Gamma(y)$ holds (which corresponds to the first consistency condition in (1.3)).

Note that, even in such a comparatively simple example, the analytical construction of the optimum function φ_y for the aggregated problem (4.28) is much simpler than for the original problem (4.27), because (4.28) has only two variables; nevertheless, φ_y allows us to qualitatively estimate the behavior of φ over the entire set of parameters a because of the aggregation consistency. This aggregation effect must be much more pronounced in higher-dimensional, especially infinite-dimensional, problems.

4.3. Optimal Control of Linear Systems under Terminal Constraints

Dynamic models and problems were one of the first objects examined from the point of view of constructing aggregated analogues. An extensive bibliography on setting the aggregation problem and developing aggregation methods for these classes of problems is contained in [30]. However, the parametric optimal control problems have long remained unstudied in this respect, although they play an important role in applications, such as appraisal of scientific and technical innovations.

In this section, we apply the consistent aggregation technique to one class of parametric optimal control problems for linear differential equations with terminal constraints. As a result, we obtain a method that makes it possible to aggregate the controls of the problem leaving the state variables intact. The theoretical base of the method is the Pontryagin maximum principle and the apparatus for consistently aggregating finite-dimensional optimization problems developed in the preceding sections.

Let $\mathcal{L}^n_\infty[0,T]$ be the space of n-dimensional functions with measurable components bounded almost everywhere on $[0,T]$, and let $W^n_{1,\infty}[0,T]$ be the Banach space of n-dimensional absolutely continuous functions on $[0,T]$ whose first derivatives belong to $\mathcal{L}^n_\infty[0,T]$.

Consider the parametric optimal control problem

$$J^0(u(\cdot)) = \gamma^0(x(T)) \to \min,$$

$$\dot{x}(t) = D(t)x(t) + b(u(t),\, t), \quad x(0) = x^0, \tag{4.29}$$

$$u(t) \in U(a(t),\, t) \in \mathbf{R}^m, \quad t \in [0,T],$$

$$J^i(u(\cdot)) = \gamma^i(x(T)) \le 0, \quad i = \overline{1,K}, \tag{4.30}$$

where $x(t)$, $u(t)$, and $a(t) \in \mho$ are the vector functions of phase coordinates, controls, and parameters of dimensions n, m, and q, respectively; $\mho \subset \mathcal{L}^q_\infty[0,T]$ is the set of feasible parameter values $y(\cdot)$; $\gamma^i\colon \mathbf{R}^n \to \mathbf{R}$ with $i = \overline{0,K}$ are convex functions differentiable in $\mathbf{R}^n$; $D(t)$ is an $n \times n$ matrix function with measurable components bounded almost everywhere on $[0,T]$; $b\colon \mathcal{L}^m_\infty[0,T] \times [0,T] \to \mathbf{R}^n$ is a vector function continuous with respect to u and measurable with respect to t on $[0,T]$; and $U\colon \mho \times [0,T] \to 2^{\mathbf{R}^m}$ is a point-to-set mapping superpositionally measurable in t, i.e., such that the mapping $t \mapsto U(a(t),t)$ is R-measurable on $[0,T]$ for any measurable function $a(\cdot)$. The last condition ensures the existence of a measurable selection [23] for the mapping $t \mapsto U(a(t),t)$.

Fixing parameter values $a(\cdot)$ in (4.29), (4.30), we obtain a standard optimal control problem. Let us give a more precise definition of optimality as applied to such a problem.

A pair $x(t)$, $u(t)$ is said to be a feasible process if $(x(t),u(t)) \in W^n_{1,\infty}[0,T] \times \mathcal{L}^m_\infty[0,T]$ and all constraints in (4.29), (4.30) are satisfied. The control $u(t)$ is then called feasible. A feasible process $x(t)$, $u(t)$ is considered optimal if there exists an $\varepsilon > 0$ such that, for any other feasible process $z(t)$, $v(t)$ with $\|z(t) - x(t)\|_W < \varepsilon$, we have $J^0(v(\cdot)) \geq J^0(u(\cdot))$. Here, $\|\cdot\|_W$ stands for the norm in $W^n_{1,\infty}[0,T]$.

To state an optimality criterion for processes, we denote the attainability set of system (4.29) in the class of feasible controls at time T by

$$S(a(\cdot)) = \left\{ x(T) \;\middle|\; \begin{array}{l} \dot{x}(t) = D(t)x(t) + b(u(t),\,t), \quad x(0) = x^0, \\ u(t) \in U(a(t),\,t), \quad t \in [0,T] \end{array} \right\}$$

and write problem (4.29), (4.30) in the equivalent form

$$\begin{aligned} &\gamma^0(x) \to \min, \\ &\gamma^i(x) \leq 0, \quad i = \overline{1,K}, \quad x \in S(a(\cdot)). \end{aligned} \tag{4.31}$$

Thanks to the special features of system (4.29) related to the linearity in state variables, the set $S(a(\cdot))$ is convex [23]. Recall that the functions γ^i ($i = \overline{0,K}$) are also convex; therefore, the finite-dimensional problem (4.31) is a convex programming problem.

We assume additionally that (4.31) satisfies the (Slater) regularity condition

$$\{x \in S(a(\cdot)) \mid \gamma^i(x) < 0,\ i = \overline{1,K}\} \neq \varnothing. \tag{4.32}$$

and introduce the set of indices

$$I(x) = \{0\} \cup \{i \mid \gamma^i(x) = 0,\ i = \overline{1,K}\}.$$

It is convenient to state the optimality criterion at a point x as follows.

Lemma 4.5. *Suppose that the regularity condition (4.32) holds. Then a feasible point x in problem (4.31) is optimal if and only if*

$$\min\left\{\max_{i\in I(x)} \nabla^T\gamma^i(x)(z-x) \mid z \in S(a(\cdot))\right\} = 0. \tag{4.33}$$

Proof. We apply the argument used in [36], according to which the optimality of x in (4.31) is equivalent to its optimality in the problem

$$\begin{aligned} &\xi(z) = \max\{\gamma^0(z) - \gamma^0(x),\ \gamma^i(z),\ i = \overline{1,K}\} \to \min, \\ &z \in S(a(\cdot)). \end{aligned} \tag{4.34}$$

Indeed, if x is not optimal in (4.34), then there exists a $z^1 \in S(a(\cdot))$ such that $\xi(z^1) < \xi(x)$, i.e., $\gamma^0(z^1) < \gamma^0(x)$ and $\gamma^i(z^1) < 0$ for $i = \overline{1,K}$. Since γ^i are convex for $i \in I(x)$, these inequalities mean that $z^1 - x$ is a feasible direction at the point x and, therefore, x can be improved in (4.31).

Conversely, suppose that $\gamma^0(x) > \gamma^0(x^*)$, where x^* is an optimal solution to (4.31). The regularity condition (4.32) and the convexity of the problem under consideration imply that an arbitrary ε-neighborhood of x^* contains a point $x_\varepsilon \in S(a(\cdot))$ such that $\gamma^i(x_\varepsilon) < 0$ for $i = \overline{1,K}$. Let us take an x_ε sufficiently close to x^*, so that $\gamma^0(x) > \gamma^)(x_\varepsilon)$ (this can always be achieved, because γ^0 is continuous). Then $\xi(x_\varepsilon) < \xi(x)$ in problem (4.34), i.e., x is not optimal in this problem.

Finally, note that condition (4.33) is necessary and sufficient for the optimality of x in the discrete minimax problem (4.34), because the function ξ and the set $S(a(\cdot))$ are convex [3]. This observation completes the proof of Lemma 4.5.

Let us reformulate the optimality criterion given above in terms of the original optimal control problem (4.29), (4.30). For this purpose, take solutions $\psi^i(t)$ $(i = \overline{0,K})$ to the dual problems

$$\dot\psi(t) = -\psi(t)D(t), \quad \psi(T) = -\nabla^T\gamma^i(x(T)) \tag{4.35}$$

and consider the functions

$$H^i(\psi,\, x,\, u,\, t) = \psi^i(t)(D(t)x + b(u,t)),$$

$$\Delta H^i(v,t) = H^i(\psi(t),\, x(t),\, v,\, t) - H^i(\psi(t),\, x(t),\, u(t),\, t).$$

Applying the Cauchy formulas to the $x(t)$ from system (4.29) and considering a pair of feasible processes $x(t)$, $u(t)$ and $z(t)$, $v(t)$, we can write (in the appropriate notation)

$$\nabla^T\gamma^i(x(T))(z(T) - x(T)) = -\int_0^T \psi(\tau)[b(v(\tau),\, \tau) - b(u(\tau),\, \tau)]\, d\tau \; = \; = \; - \int_0^T \Delta H^i(v(\tau),\, \tau)\, d\tau.$$

Substituting this relation in (4.33), we obtain

$$\min\left\{ \int_0^T \max_{i\in I_a} \left(-\Delta H^i(v(\tau),\, \tau)\right) d\tau \mid v(\tau) \in U(a(\tau),\, \tau) \right\} = 0,$$

where $I_a = I(x(T))$ and $x(t)$ satisfies system (4.29) for $a(\cdot) \in \mho$.

Following the traditional scheme for deriving necessary conditions (see, e.g., [1]), we can show that (4.33) is equivalent to the following assertion.

Theorem 4.4. *Suppose that, in (4.29), (4.30), a parameter value $a(\cdot) \in \mho$ is fixed and the regularity conditions (4.32) hold. Then a feasible process $x(t)$, $u(t)$ is optimal in problem (4.29), (4.30) if and only if*

$$\max\left\{ \min_{i\in I_a} \Delta H^i(v,t) \mid v \in U(a(t),\, t) \right\} = 0,$$

almost everywhere on $[0,T]$.

This result is a formulation of the Pontryagin maximum principle; it allow us to reduce identifying a process optimal on $[0,T]$ to analyzing local finite-dimensional optimization problems at moments of time $t \in [0,T]$. The purpose of our further constructions is to aggregate these finite-dimensional problems and form an aggregated analogue of the original optimal control problem from the obtained aggregated local problems.

In what follows, we denote the optimal process and the optimal value in problem (4.29), (4.30) that correspond to the parameter $a(\cdot)$ by $x_a(t)$, $u_a(t)$ and $\varphi(a(\cdot))$, respectively.

Combining conditions (4.33) for the optimality of a process $x(t)$, $u(t)$ with the technique for consistently aggregating finite-dimensional parametric optimization problems, we can aggregate controls in the original problem (4.29), (4.30).

Suppose that the original optimal control problem is solved for the parameter value $a(\cdot) = y(\cdot)$ and, in addition to the optimal process $x_y(\cdot)$, $u_y(\cdot)$, solutions $\psi^i(t)$ $(i = \overline{0, K})$ to the dual systems (4.35) are known. Then, in accordance with the scheme considered above, we can state the following problems at separate moments of time $t \in [0, T]$:

$$\begin{aligned} &\mathcal{S}^t(v) = \min_{i \in I_y} \Delta H^i(v, t) \to \max, \\ &v \in U(a(t),\, t); \end{aligned} \tag{4.36}$$

we denote their optimal values by

$$\Delta\varphi^t(a(t)) = \max\{\mathcal{S}^t(v) \mid v \in U(a(t),\, t)\}.$$

Let us apply the apparatus of consistent aggregation and replace each problem from family (4.36) by the aggregated parametric problem

$$\begin{aligned} &\mathcal{S}^t_y(p(t)) \to \max, \\ &p(t) \in U_y(a(t),\, t) \in \mathbf{R}^{m^t_y}. \end{aligned} \tag{4.37}$$

with elements

$$\mathcal{S}^t_y(p(t)) = \mathcal{S}^t(u_y(t) + P(t)p(t)),$$

$$U_y(a(t),\, t) \subset \{p(t) \mid u_y(t) + P(t)p(t) \in U(a(t),\, t)\}$$

and optimum function $\Delta\varphi^t_y(a(t))$, where $P(t)$ is an $m \times m^t_y$ aggregating matrix $(m^t_y < m)$, $p(t)$ is the vector of aggregated controls, and $a(t)$ is the vector of parameters, all considered at time $t \in [0, T]$.

Substituting

$$U(t) = U_y(t) + P(t)p(t) \tag{4.38}$$

into (4.29), (4.30) and replacing the sets $U(a(t),t)$ with $U_y(a(t),t)$, we obtain the aggregated parametric optimal control problem

$$
\begin{aligned}
&J_y^0(p(\cdot)) = \gamma^0(x(T)) \to \min,\\
&\dot{x}(t) = D(t)x(t) + b_y(p(t),\, t), \quad x(0) = x^0,\\
&p(t) \in U_y(a(t),\, t), \quad t \in [0,T],\\
&J_y^i(p(\cdot)) = \gamma^i(x(T)) \le 0, \quad i = \overline{1,K},
\end{aligned}
\tag{4.39}
$$

where

$$b_y(p(t),\, t) = b(u_y(t) + P(t)p(t),\, t),$$

$$J_y^i(p(\cdot)) = J^i(u_y(\cdot) + P(\cdot)p(\cdot)), \quad i = \overline{0,K}.$$

The number of state variables in (4.39) remains the same, but the dimension of the control vector is lower than that in (4.29), (4.30).

Let $p_a(t)$ be the optimal control in (4.39) corresponding to the parameter value $a(\cdot)$, and let $\varphi_y(a(\cdot))$ be the optimal value of (4.39) attained at this control as a function of $a(\cdot)$. To achieve the main goal of our research, we must establish conditions for the aggregation of problem (4.29), (4.30) into (4.39) to be consistent at the base point $y(\cdot)$. The key point of this problem is determining the influence of changes in optimal values of problems from families (4.36) and (4.37) on the values of the corresponding functions φ and φ_y.

First, we prove the following auxiliary assertion, which establish a relation between the directions of optimum value variations in the original optimal control problem and those in the finite-dimensional parametric problems (4.36).

Lemma 4.6. *Suppose that $a(\cdot)$ is such that $\Delta\varphi^t(a(t)) \le \Delta\varphi^t(y(t))$ almost everywhere on $[0,T]$. Then $\varphi(a(\cdot)) \ge \varphi(y(\cdot))$.*

Proof. Consider the sets

$$U(t) = U(a(t),\, t) \cup U(y(t),\, t), \quad t \in [0,T].$$

By assumption,

$$\max\{\mathcal{S}^t(v) \mid v \in U(t)\} \le \max\{\Delta\varphi^t(a(t)),\, \Delta\varphi^t(y(t))\} = \Delta\varphi^t(y(t))$$

almost everywhere on $[0,T]$, and $\Delta\varphi^t(y(t)) = 0$. Therefore, according to Theorem 4.4, the process $x_y(t)$, $u_y(t)$ with $t \in [0,T]$ is optimal for problem (4.29), (4.30) where the set of feasible controls is $U(t)$.

This observation and the inclusion $U(a(t),t) \subset U(t)$ imply that, for any controls $v(\cdot)$ satisfying the relation $v(t) \in U(a(t),t)$ on $[0,T]$, we have $J^0(v(\cdot)) \geq J^0(u_y(\cdot)) = \varphi(y(\cdot))$. Therefore,

$$\varphi(a(\cdot)) = \min\{J^0(v(\cdot)) \mid v(t) \in U(a(t),\, t)\} \geq \varphi(y(\cdot)),$$

which proves the lemma.

Lemma 4.6 allows us to establish the main result. We assume that the set $\mho$ of feasible parameter values has the property

$$a(t),\, y(t) \in \mho^t \Longrightarrow (E_q - \alpha(t))y(t) + \alpha(t)a(t) \in \mho^t, \quad t \in [0,T], \tag{4.40}$$

where $\alpha(t) = \operatorname{diag}\{\alpha_i(t)\}$, $0 \leq \alpha_i(t) \leq 1$ for $i = \overline{1,q}$ and $t \in [0,T]$, E_q is the identity matrix of order q, and $\mho^t$ is the section of $\mho$ at time t.

Theorem 4.5. *Suppose that problem (4.29), (4.30) with $a(\cdot) = y(\cdot)$ satisfies the regularity condition (4.32), the set $\mho$ of feasible parameter values has property (4.40), and family (4.36) is aggregated in such a way that each problem (4.37) is consistent with the corresponding problem of family (4.36). Then the aggregation of the original parametric optimal control problem (4.29), (4.30) into (4.39) is consistent with respect to parameter variations at the point $y(\cdot)$.*

Proof. Let us verify the aggregation consistency conditions (1.3).

According to Remark 1.1, the special features of aggregation that are ensured by intersecting the set of feasible controls with the manifold specified by (4.28) give $\varphi(a(\cdot)) \leq \varphi_y(a(\cdot))$ for the pair of problems (4.29), (4.30) and (4.39). This directly implies the fulfillment of the first condition in (1.3); in particular, we have $\Delta\varphi_y^t(y(t)) = \Delta\varphi^t(y(t))$ for $t \in [0,T]$.

Let us verify the second condition in (1.3). For this purpose, suppose that $\Gamma(y(\cdot) \neq \varnothing$, i.e., that there exists an $a(\cdot) \in \mho$ for which $\varphi(a(\cdot)) < \varphi(y(\cdot))$. Then Lemma 4.6 implies that the subset

$$M = \{t \in [0,T] \mid \Delta\varphi^t(a(t)) > \Delta\varphi^t(y(t))\}.$$

of $[0,T]$ has positive measure.

Since the aggregation of problems (4.36) into (4.37) is consistent, there exist $\tilde{a}(t) \in \mho^t$ with $t \in [0, T]$ such that

$$\Delta\varphi_y^t(\tilde{a}(t)) > \Delta\varphi_y^t(y(t)) = \Delta\varphi^t(y(t)) = 0$$

on M. Next, take a set $M_\varepsilon \subseteq M$ with measure $\operatorname{mes} M_\varepsilon = \varepsilon > 0$ and construct the feasible control

$$p(\varepsilon, t) = \begin{cases} \tilde{p}(t), & t \in M_\varepsilon, \\ p_y(t), & t \in [0, T] \backslash M_\varepsilon \end{cases}$$

in the aggregated problem (4.39); here, $\tilde{p}(t)$ is the optimal solution to problem (4.37) corresponding to the parameter $\tilde{a}(t)$ at time t (thus, $\mathcal{S}_y^t(\tilde{p}(t)) = \Delta\varphi_y^t(\tilde{a}(t))$).

Now, let us estimate the increment of the functional φ_y in (4.39) under the replacement of $y(t)$ with $\tilde{a}(t)$ at $t \in M_\varepsilon$. According to [10], we have

$$J_y^i(p(\varepsilon, \cdot)) - J_y^i(p_y(\cdot)) = -\int_{M_\varepsilon} \Delta H^i(\tilde{p}(\tau), \tau)\, d\tau + o(\varepsilon)$$

for $i = \overline{0, K}$. By definition,

$$\min_{i \in I_y} \Delta H^i(\tilde{p}(\tau), \tau) = \Delta\varphi_y^\tau(\tilde{a}(\tau)) > 0, \quad \tau \in M_\varepsilon;$$

therefore, taking into account that $J_y^i(p_y(\cdot)) < 0$ for $i \notin I_y$, we can always select a sufficiently small $\varepsilon^0 > 0$ such that

$$J_y^i(p(\varepsilon^0, \cdot)) - J_y^i(p_y(\cdot)) < 0, \quad i = \overline{0, K}.$$

Setting

$$a_y(t) = \begin{cases} \tilde{a}(t), & t \in M_\varepsilon, \\ y(t), & t \in [0, T] \backslash M_\varepsilon, \end{cases}$$

we obtain

$$\varphi_y(a_y(\cdot)) \leq J_y^0(p(\varepsilon^0, \cdot)) < J_y^0(p_y(\cdot)) = \varphi_y(y(\cdot)).$$

Thus, $a_y(\cdot) \in \Gamma_A(y(\cdot))$, i.e., $\Gamma_A(y(\cdot)) \neq \varnothing$. This completes the proof of the theorem.

Theorem 4.5 reduces consistently aggregating the optimal control problem (4.39), (4.30) to developing methods for consistently aggregating the family of finite-dimensional problems (4.36). Some of such methods are described in the preceding sections.

Remark 4.3. More general problems where the right-hand sides of systems of differential equations also depend on parameters reduce to parametric optimal control problems of form (4.29), (4.30) considered above. Namely, suppose that the problem under consideration has the form

$$\begin{aligned} &J^0(u(\cdot)) = \gamma^0(x(T)) \to \min, \\ &\dot{x}(t) = D(t)x(t) + b(a(t),\, u(t),\, t), \quad x(0) = x^0, \\ &u(t) \in U(a(t),\, t), \quad t \in [0,T], \\ &J^i(u(\cdot)) = \gamma^i(x(T)) \le 0, \quad i = \overline{1,K}. \end{aligned} \tag{4.41}$$

To reduce it to form (4.29), (4.30), we augment $u(t)$ by an artificial control vector $\lambda(t)$ satisfying the relation

$$\lambda(t) - b(a(t),\, u(t),\, t) = 0, \quad t \in [0,T]$$

and add this relation to the constraints imposed on the controls. As a result, we obtain the parametric problem

$$\begin{aligned} &\bar{J}^0(u(\cdot),\, \lambda(\cdot)) = \gamma_0(x(T)) \to \min, \\ &\dot{x}(t) = D(t)x(t) + \lambda(t), \quad x(0) = x^0, \\ &u(t) \in U(a(t),\, t),\, \lambda(t) - b(a(t),\, u(t),\, t) = 0, \quad t \in [0,T], \\ &\bar{J}^i(u(\cdot),\, \lambda(\cdot)) = \gamma^i(x(T)) \le 0, \quad i = \overline{1,K}, \end{aligned}$$

which belongs to the class of problems of type (4.29), (4.30).

4.4. Weaker Consistency Conditions. Integer Linear Programming

Integer programming problems also involve correcting parameters needed to compensate the errors in simulating real systems and to take into account new requirements that arise in the course of solving the problem or after the solution is obtained.

A change in the parameters of an integer linear programming problem may cause no variations in the optimal solution. The boundaries of such changes (stability ranges) are studied in, e.g., [21]. Beyond the stability boundary, the optimal value of the objective function changes. In integer programming problems, the demand for seeking improving variations is even more pressing, because these problems are NP-complete [26].

This section develops a mathematical apparatus for reducing integer linear programming problems with variable parameters to simpler problems that admit estimating given coefficient variations and excluding those obviously unfit. We still call these simpler problems partially aggregated, although the aggregation scheme for constructing them includes one step more than the preceding schemes: in the partially aggregated problems, not only the dimension is reduced but the requirement that the variables must be integer is missing; these are linear programming problems whose dimensions are on the same orders as the sizes of variable parameter matrices.

The basic notions and approaches used to aggregate continuous problems apply to the discrete case too. Most substantially altered is the consistency condition (1.3). We introduce a new notion of weak consistency with respect to parameter variations; it only ensures the improvement of the optimal objective value of the aggregated problem under an arbitrary variation improving the optimal objective value of the original problem.

Definition 4.3. We say that an aggregation of the original parametric problem (1.1) in a problem of form (1.2) is quasi-consistent with respect to parameter variations at a point $a \in \mho$ if

$$\Gamma(y) \neq \varnothing \Longrightarrow \Gamma(y) \subseteq \Gamma_A(y). \tag{4.42}$$

The quasi-consistency condition allows the existence of situations where improving variations in the parameters of an aggregated problem do not lead to an improvement in the optimal objective value of the original problem.

Consider the parametric integer linear programming problem

$$\begin{aligned} &c_1x_1 + c_2x_2 \to \min, \\ &\begin{pmatrix} x_1 \\ x_2 \end{pmatrix} = X(A_{11} + \Delta A_{11}) = \\ &= \left\{ \begin{pmatrix} x_1 \\ x_2 \end{pmatrix} \middle| \begin{array}{ll} (A_{11} + \Delta A_{11})x_1 + A_{12}x_2 \leq b_1, & x_1 \geq 0 \text{ (integer)} \\ A_{21}x_1 + A_{22}x_2 \leq b_2, & x_2 \geq 0 \text{ (integer)} \end{array} \right\}, \end{aligned} \tag{4.43}$$

where $A_{11} + \Delta A_{11} \in \mho$ and all coefficients, parameters, and variables retain the same meaning as in the parametric linear programming problem (2.1), except they all are assumed to be integer. Accordingly, the set $\tilde{\mho}$ of parameter variations consists of integer matrices and, therefore, does not admit "mixtures" of old and new parameter values, unlike the set $\mho$ used in (2.1).

As the base problem, we take the problem of family (4.43) with $\Delta A_{11} = 0$, that is,

$$\begin{aligned} &cx \to \min, \\ &Ax \leq b, \quad x \geq 0 \text{ (integer)}, \end{aligned} \tag{4.44}$$

where

$$c = (c_1, c_2), \quad A = \begin{pmatrix} A_{11}, A_{12} \\ A_{21}, A_{22} \end{pmatrix}, \quad x = \begin{pmatrix} x_1 \\ x_2 \end{pmatrix}.$$

We assume that the feasible solution set of problem (4.44) is bounded. Otherwise, we impose an additional constraint of the form

$$\sum_{i=1}^{n} x^i \leq M,$$

where M is a sufficiently large constant, and assume this constraint to be already included in the matrix A and the vector b.

Suppose that, in problem (4.44), we have constructed cuts

$$Sx \leq q \tag{4.45}$$

with integer S and q possessing the following properties:

(a) all integer points satisfying the inequalities $Ax \leq b$ and $x \geq 0$ satisfy (4.45);

(b) the problem

$$\begin{aligned} &cx \to \min, \\ &Ax \leq b, \quad Sx \leq q, \quad x \geq 0 \end{aligned} \tag{4.46}$$

has the same solutions as the original integer problem (4.44).

The further considerations involve a general representation of the cuts (4.45) with these properties. Such a general representation is given in [25]; it is proved there that any inequalities of form (4.45) with integer coefficients having property (a) belong to a system of inequalities that can be obtained by applying the successive augmentation procedure to the system of inequalities

$$\bar{A}^{(0)}x \le \bar{b}^{(0)},$$

where $\bar{A}^{(0)} = \begin{pmatrix} -E \\ A \end{pmatrix}$ and $\bar{b}^{(0)} = \begin{pmatrix} 0 \\ b \end{pmatrix}$.

At the k-th step of the augmentation procedure ($k = 1, 2, \dots$), we augment the system obtained at the $(k-1)$-st step with the inequality

$$\lambda^{(k)}\bar{A}^{(k-1)}x \le \left\lfloor \lambda^{(k)}\bar{b}^{(k-1)} \right\rfloor + \beta^{(k)}, \tag{4.47}$$

where $\bar{A}^{(k-1)}$ and $\bar{b}^{(k-1)}$ are the coefficients of the system obtained at the $(k-1)$-st step, $\lambda^{(k)}\bar{A}^{(k-1)}$ is a row of integers, $\lfloor \tau \rfloor$ is the largest integer not exceeding τ, $\lambda^{(k)} \ge 0$, and $\beta^{(k)} \ge 0$.

Example 4.1. Consider the integer linear programming problem[5]

$$\begin{aligned} &-3x^1 - x^2 \to \min, \\ &2x^1 - 3x^2 \le 3, \\ &2x^1 + 3x^2 \le 6, \\ &x^1 \ge 0, \quad x^2 \ge 0 \text{ (integer)}. \end{aligned} \tag{4.48}$$

The application of the primal integer programming algorithm [9] yields the cuts

$$x^1 \le 1, \quad x^1 + x^2 \le 2. \tag{4.49}$$

It is easy to see that $\bar{x}^1 = 1$, $\bar{x}^2 = 1$ is an optimal integer point.

Cuts (4.49) are obtained from the inequalities of problem (4.48) as follows: we use $\lambda^{(1)} = \left(\frac{1}{4},\, \frac{1}{4},\, 0,\, 0\right)$ to construct the inequality

$$x^1 \le \left\lfloor \frac{9}{4} \right\rfloor = 2,$$

at the first step, $\lambda^{(2)} = \left(0,\, \frac{1}{3},\, 0,\, 0,\, \frac{1}{3}\right)$ to construct the inequality

$$x^1 + x^2 \le \left\lfloor \frac{8}{3} \right\rfloor = 2$$

[5]The example and the cuts reducing this problem to a linear programming problem are borrowed from [9].

(which coincides with the second cut in (4.49)) at the second step, $\lambda^{(3)} = \left(\frac{1}{3}, 0, 0, 0, \frac{1}{3}, 0\right)$ to construct

$$x^1 - x^2 \leq \left\lfloor \frac{3}{2} \right\rfloor = 1,$$

at the third step, and $\lambda^{(4)} = \left(0, 0, 0, 0, 0\, \frac{1}{2}\, \frac{1}{2}\right)$ to construct the inequality

$$x^1 \leq \left\lfloor \frac{3}{2} \right\rfloor = 1$$

(which coincides with the first cut in (4.49)) at the fourth step.

Remark 4.4. In the example given above, the cuts passing through vertices of an integer polyhedron were constructed with the use of $\beta^{(k)} = 0$. Positive $\beta^{(k)}$ are usually employed in representing sections not passing through vertices of integer polyhedra.

As a rule, when the base problem (4.44) is replaced by another problem of family (4.43) with $\Delta A_{11} \neq 0$, we have to construct new cuts of form (4.45). To a certain degree, this necessity can be obviated by representing an arbitrary variation ΔA_{11} in the form of a set of elementary variations and forming cuts for the problem with coefficients $A_{11} + \Delta A_{11}$ from the cuts corresponding to the elementary variations.

The matrix comprising the varied columns of the matrix A and corresponding to an arbitrary integer variation ΔA_{11} can be represented as

$$\begin{pmatrix} A_{11} + \Delta A_{11} \\ A_{21} \end{pmatrix} = \begin{pmatrix} A_{11} \\ A_{21} \end{pmatrix} + \begin{pmatrix} E, -E \\ 0, \quad 0 \end{pmatrix} \begin{pmatrix} \Delta A_{11}^{+} \\ -\Delta A_{11}^{-} \end{pmatrix},$$

where ΔA_{11}^{+} and ΔA_{11}^{-} are the matrices obtained from ΔA_{11} by replacing its positive and negative elements, respectively, with zeros. Setting

$$A_1 = \begin{pmatrix} A_{11} \\ A_{21} \end{pmatrix}, \quad H = \begin{pmatrix} E, -E \\ 0, \quad 0 \end{pmatrix}, \quad G = \begin{pmatrix} \Delta A_{11}^{+} \\ -\Delta A_{11}^{-} \end{pmatrix}, \tag{4.50}$$

we can represent an arbitrary variation ΔA_{11} via the matrix H of elementary variations as

$$\begin{pmatrix} \Delta A_{11} \\ 0 \end{pmatrix} = HG.$$

If the variations ΔA_{11} are subject to some constraints additional to the requirement of being integer, then the elementary variation matrix H used in this representation can have elements larger than 1 in absolute value provided that G is an integer matrix.

Using the elementary variation matrix H, we state the following extended problem for (4.44):

$$\begin{aligned} &cx + c_1 v \to \min, \\ &Ax + (A_1 + HG)v \le b, \\ &x \ge 0, \quad v \ge 0 \quad \text{(integer)}. \end{aligned} \tag{4.51}$$

Let $f^1(0)$, $\bar{x}$ and $f^1(\Delta A_{11})$, $\bar{\bar{x}}$ be the optimal objective values and optimal solutions to the base problem (4.44) and the problem of family (4.43) with variation $\Delta A_{11} \neq 0$, respectively. By $f^2(G)$, we denote the optimal objective value for problem (4.51). Since the points $x = \bar{x}$, $v = 0$ and $x_1 = 0$, $x_2 = \bar{\bar{x}}_2$, $v = \bar{\bar{x}}_1$ satisfy the constraints of problem (4.51), we have

$$f^2(G) \le \min\{f^1(0),\, f^1(\Delta A_{11})\}. \tag{4.52}$$

Let us show that the system of cuts for problem (4.51) can also be represented with the use of the elementary variation matrix. For this purpose, we successively augment the system of inequalities

$$\bar{A}^{(0)}x + \bar{A}_1^{(0)}v + \bar{H}^{(0)}Gv \le \bar{b}^{(0)}, \tag{4.53}$$

where $\bar{A}_1^{(0)} = \begin{pmatrix} 0 \\ A_1 \end{pmatrix}$ and $\bar{H}^{(0)} = \begin{pmatrix} 0 \\ H \end{pmatrix}$, by inequalities of the form

$$\begin{aligned} \lambda^{(k)}\bar{A}^{(k-1)}x + \lambda^{(k)}\bar{A}_1^{(k-1)}v + \left\lfloor \lambda^{(k)}\bar{H}^{(k-1)} \right\rfloor Gv \le \\ \le \left\lfloor \lambda^{(k)}\bar{b}^{(k-1)} \right\rfloor + \beta^{(k)}, \quad k = 1, 2, \ldots, \end{aligned} \tag{4.54}$$

where $\bar{A}_1^{(k-1)}$ and $\bar{H}_1^{(k-1)}$ are the matrices of the coefficients of v and Gv in the system obtained at the $(k-1)$-st step and the other coefficients coincide with those in inequality (4.47).

Lemma 4.7. *The feasible points of problem (4.51) satisfy inequalities (4.54).*

Proof. Consider an arbitrary point x', v' satisfying the conditions of problem (4.51). This point obeys the system of inequalities (4.53); therefore, to prove the lemma, it suffices to show that the point x', v' remains feasible after each augmentation step.

Suppose that x', v' satisfies the system of inequalities

$$\bar{A}^{(k-1)}x' + \bar{A}_1^{(k-1)}v' + \bar{H}^{(k-1)}Gv' \leq \bar{b}^{(k-1)}$$

obtained at the $(k-1)$-st step.

Consider the row vector $h^{(k)} \geq 0$ and the scalar $r^{(k)} \geq 0$ defined by

$$h^{(k)} = \lambda^{(k)}\bar{H}^{(k-1)} - \left\lfloor \lambda^{(k)}\bar{H}^{(k-1)} \right\rfloor,$$

$$r^{(k)} = \lambda^{(k)}\bar{b}^{(k-1)} - \left\lfloor \lambda^{(k)}\bar{b}^{(k-1)} \right\rfloor.$$

Since $\lambda^{(k)} \geq 0$, the multiplication of the $(k-1)$-st-step inequality by $\lambda^{(k)}$ on the left yields

$$\lambda^{(k)}\bar{A}^{(k-1)}x' + \lambda^{(k)}\bar{A}_1^{(k-1)}v' + \lambda^{(k)}\bar{H}^{(k-1)}Gv' \leq \lambda^{(k)}\bar{b}^{(k-1)}.$$

Taking into account that $h^{(k)} \geq$, $G \geq 0$, and $v' \geq 0$, we ob- tain

$$\lambda^{(k)}\bar{H}^{(k-1)}Gv' = \left([\lambda^{(k)}\bar{H}^{(k-1)}] + h^{(k)}\right)Gv' \geq \left\lfloor \lambda^{(k)}\bar{H}^{(k-1)} \right\rfloor Gv';$$

therefore,

$$\lambda^{(k)}\bar{A}^{(k-1)}x' + \lambda^{(k)}A_1^{(k-1)}v' + \left\lfloor \lambda^{(k)}\bar{H}^{(k-1)} \right\rfloor Gv' \leq \left\lfloor \lambda^{(k)}\bar{b}^{(k-1)} \right\rfloor + r^{(k)}.$$

The expression in the left-hand side of this inequality is integer, and $0 \leq r^{(k)} < 1$; hence,

$$\lambda^{(k)}\bar{A}^{(k-1)}x' + \lambda^{(k)}\bar{A}_1^{(k-1)}v' + \left\lfloor \lambda^{(k)}\bar{H}^{(k-1)} \right\rfloor Gv' \leq \left\lfloor \lambda^{(k)}\bar{b}^{(k-1)} \right\rfloor.$$

This proves the required assertion, because $\beta^{(k)} \geq 0$.

Separating out the inequalities corresponding to (4.45) from the constructed system (4.54), we obtain a system of cuts

$$Sx + (S_1 + QG)v \leq q, \tag{4.55}$$

for (4.51); it has the following properties:

(a) all integer points satisfying the inequalities

$$Ax + (A_1 + HG)v \leq b,$$

$$x \geq 0, \quad v \geq 0,$$

satisfy (4.55);

(b) the optimal value $\tilde{f}^2(G)$ of the objective function in the problem

$$\begin{aligned} &cx + c_1 v \to \min, \\ &Ax + (A_1 + HG)v \le b, \\ &Sx + (S_1 + QG)v \le q, \\ &x \ge 0, \quad v \ge 0 \end{aligned} \tag{4.56}$$

at $G = 0$ coincides with the optimal objective value of the base problem (4.44).

Problem (4.56) is a parametric linear programming problem with variable cuts; at $G = 0$,

$$\tilde{f}^2(0) = f^1(0), (4.57)$$

and at $G \neq 0$,

$$\tilde{f}^2(G) \le \min\{f^1(0),\ f^1(\Delta A_{11})\} \tag{4.58}$$

Inequality (4.58) follows from inequality (4.52) and the property (a) of cuts (4.55).

Problem (4.56) can be regarded as an aggregated version of the original parametric integer programming problem (4.43), because it refers to a different (simpler) class of mathematical programming problems, even though its dimension is a little higher. Thus, treating conditions (4.57) and (4.58) as an analogue of (4.42), we can conclude that the aggregation of problem (4.43) into (4.56) is quasi-consistent with respect to parameter variations at the point A_{11}.

An important special feature of problem (4.56) is that we can easily "tune" it to any variation ΔA_{11} by suitably specifying the matrix G with the use of (4.50), and there is no need to repeat the whole process of evaluating the coefficients of cuts. The number of columns to be varied in problem (4.56) equals n_1, and all new columns belong to the no higher than $(2m_1 + {}$ ${} + n_1)$-dimensional subspace spanned by the columns of the matrix $\begin{pmatrix} A_1, H \\ S_1, G \end{pmatrix}$.

If

$$m_1,\ n_1 \ll \min\{m,\ n\},$$

then we can reduce the dimension of the problem by means of a partial aggregation of (4.56) consistent with respect to parameter variations at the point A_{11}.

4.4.1. A Quasi-Consistency Theorem. Suppose that a point $\bar{x}$ corresponds to a nondegenerate vertex of the feasible solution polyhedron of problem (4.46)[6]. Let us show that it is possible to replace the n-vector of variables x in (4.56) with a $(2m_1 + n_1)$-vector of aggregated variables p in such a way that the variables v retain their original, nonaggregated, form and, thereby, problem (4.56) be consistent with the partially aggregated problem with respect to coefficient variations at the point A_{11}.

First, we reduce problem (4.46) to the standard form by introducing additional variables $u_1 \geq 0$ and $u_2 \geq 0$ into the first and second groups of its inequalities, respectively. Representing the matrix of constraints of the obtained problem in the form (B, N), we rewrite (4.46) as

$$\begin{aligned} &\tilde{c}_B\xi_B + \tilde{c}_N\xi_N \to \min, \\ &B\xi_B + N\xi_N = \tilde{b}, \\ &\xi_B \geq 0, \quad \xi_N \geq 0, \end{aligned} \tag{4.59}$$

where $\tilde{b} = \begin{pmatrix} b \\ q \end{pmatrix}$ and B is the basic corresponding to the optimal basic solution

$$\bar{\xi} = \begin{pmatrix} \bar{x} \\ \bar{u}_1 \\ \bar{u}_2 \end{pmatrix}, \qquad \bar{u}_1 = b - A\bar{x}, \quad \bar{u}_2 = q - S\bar{x}.$$

Similar transformations of problem (4.56) yield

$$\begin{aligned} &\tilde{c}_B\xi_B + \tilde{c}_B\xi_N + c_1 v \to \min, \\ &B\xi_B + N\xi_N + \begin{pmatrix} A_1 + HG \\ S_1 + QG \end{pmatrix} v = \tilde{b}, \\ &\xi_B \geq 0, \quad \xi_N \geq 0, \quad v \geq 0. \end{aligned} \tag{4.60}$$

As in Section 2, we introduce an affine manifold Π_y passing through the point $\bar{\xi}$:

$$\Pi_y = \left\{ \xi \;\middle|\; \xi_B = \bar{\xi}_B + B^{-1} \begin{pmatrix} A_1, H \\ S_1, Q \end{pmatrix} p, \ \xi_N = 0 \right\}.$$

[6]This assumption does not substantially narrow the range of problems under consideration, because cuts rarely lead to the degeneracy of optimal vertices, and a possible degeneracy can only be caused by the presence in the base problem (4.44) of linearly independent inequalities turning into equalities at the point $\bar{x}$.

The dimension of Π_y equals the rank of the matrix

$$\begin{pmatrix} A_1, H \\ S_1, Q \end{pmatrix},$$

which does not exceed the number of its columns, i.e., $2m_1 + n_1$.

Lemma 4.8. *If $\tilde{f}^2(G) < f^1(0)$, then there exist a $\xi^* \in \Pi_y$ and a v^* such that the point $\xi = \xi^*$, $v = v^*$ satisfies the conditions of problem (4.60) and*

$$\tilde{c}\xi^* + c_1 v^* < f^1(0).$$

Proof. Let us rewrite problem (4.60) in the form

$$\tilde{c}_B B^{-1}\tilde{b} + (\tilde{c}_N - \tilde{c}_B B^{-1} N)\xi_N + \left(c_1 - \tilde{c}_B B^{-1} \begin{pmatrix} A_1 + HG \\ S_1 + QG \end{pmatrix}\right) v \to \min,$$

$$\xi_B = B^{-1}\tilde{b} - B^{-1}N\xi_N - B^{-1} \begin{pmatrix} A_1 + HG \\ S_1 + QG \end{pmatrix} v \geq 0,$$

$$\xi_N \geq 0, \quad v \geq 0.$$

Since the basis B is optimal in problem (4.59), the characteristic differences $\tilde{c}_B B^{-1} N - \tilde{c}_N$ are nonpositive; therefore, $\tilde{f}^2(G) < f^1(C)$ implies that the row vector of estimates

$$\tilde{c}_B B^{-1} \begin{pmatrix} A_1 + HG \\ S_1 + QG \end{pmatrix} - c_1$$

has at least one positive component. Let us denote the column corresponding to this component by ζ_s and represent it in the form

$$\zeta_s = B^{-1} \begin{pmatrix} A_1 + HG \\ S_1 + QG \end{pmatrix} e_s,$$

where e_s is the m-vector whose s-th component is 1 and all other components are 0.

As ξ^* and v^*, we take the vectors

$$\xi_B^* = \bar{\xi}_B - \zeta_s v^{s*}, \quad \xi_N^* = 0, \quad v^* = e_s v^{s*},$$

$$v^{s*} = \min_{i \in J_s} \left\{ (\bar{\xi}_B)^i / (\zeta_s)^i \right\}, \quad J_s = \{ i \mid (\zeta_s)^i > 0 \}.$$

The point ξ^*, v^* satisfies the conditions of problem (4.60).

Since the vertex corresponding to the point $\xi = \bar{\xi}$, $v = 0$ in (4.60) is nondegenerate, we have $v^{s*} > 0$ and, hence,

$$\tilde{c}\xi^* + c_1 v^* = \tilde{c}_B \xi_B^* + c_1 v^* = \tilde{c}_B \bar{\xi}_B - \tilde{c}_B B^{-1} \begin{pmatrix} A_1 + HG \\ S_1 + QG \end{pmatrix} e_s v^{s*} +$$

$$+ c_1 e_s v^{s*} = f^1(0) - \left(\tilde{c}_B B^{-1} \begin{pmatrix} A_1 + HG \\ S_1 + QG \end{pmatrix} - c_1 \right) e_s v^{s*} < f^1(0).$$

Representing ζ_s in the form

$$\zeta_s = B^{-1} \begin{pmatrix} A - 1, H \\ S_1, \ Q \end{pmatrix} \begin{pmatrix} E \\ G \end{pmatrix} e_s$$

and setting

$$p^* = \begin{pmatrix} E \\ G \end{pmatrix} e_s v^{s*},$$

we obtain

$$\xi_B^* = \bar{\xi}_B - B^{-1} \begin{pmatrix} A_1, H \\ S_1, Q \end{pmatrix} p^*;$$

therefore, $\xi^* \in \Pi_y$, which proves the lemma.

Thus, Lemma 4.8 asserts that, if the optimal value $\tilde{f}^2(G)$ of the objective function of problem (4.43) is smaller than the optimal value $f^1(0)$ of the objective function of the base problem (4.44), then the affine manifold Π_y contains a point ξ^* at which the value of the objective function of problem (4.60) is also smaller than $f^1(0)$. In addition, the proof of this lemma contains an explicit form of the $(2m_1 + n_1)$-vector $p^* > 0$ corresponding to the vector ξ^* in the representation of the affine manifold Π_y.

Let us make the change of variables

$$\xi_B = \bar{\xi}_B - B^{-1} \begin{pmatrix} A_1, H \\ S_1, Q \end{pmatrix} p, \quad \xi_N = 0, \quad v = v$$

and replace $\xi_B \geq 0$ with the constraints

$$p \geq 0, \quad tp \leq T,$$

where t is the row of 1's and T is a sufficiently large scalar value, to bound the domain of the aggregated variable p. Now, we can construct a partially aggregated problem for (4.60) in the form

$$\begin{aligned}
&c_1 v - \tilde{c}_B B^{-1} \begin{pmatrix} A_1, H \\ S_1, Q \end{pmatrix} p + c\bar{x} \to \min, \\
&\begin{pmatrix} A_{11} + \Delta A_{11} \\ A_{21} \\ S_1 + QG \end{pmatrix} v - \begin{pmatrix} A_1, H \\ S_1, Q \end{pmatrix} p = 0, \\
&tp \le T, \quad v \ge 0, \quad p \ge 0.
\end{aligned} \tag{4.61}$$

In this problem, the coefficients $A_{11} + \Delta A_{11}$ under variation and the variables v corresponding to these coefficients are left intact, and the other variables are replaced with their aggregated analogues p. The number of variables in problem (4.61) does not exceed $2(m_1 + n_1)$, and the number of constraints is the same as in (4.60). In the space of variables of problem (4.60), to the feasible set of problem (4.61), there corresponds the intersection of the feasible set of (4.60) (without the constraint $\xi_B \ge 0$) with the affine manifold $\Pi_y \cup \{v\}$ where the additional conditions (corresponding to $p \ge 0$ and $tp \le T$) making it bounded are taken into account.

If

$$m_1, n_1 \ll \min\{m, n\},$$

then problem (4.61) contains much fewer variables than equations; linearly dependent equations can be removed.

The equations of problem (4.61) can be represented in the form

$$\begin{pmatrix} A_1, H \\ S_1, Q \end{pmatrix} \begin{pmatrix} E \\ G \end{pmatrix} v - \begin{pmatrix} A_1, H \\ S_1, Q \end{pmatrix} p = 0.$$

Take rows of the matrix

$$\begin{pmatrix} A_1, H \\ S_1, Q \end{pmatrix},$$

that complement its first m_1 rows to a maximal linearly independent system and denote the matrix composed from these rows by (U, V). Then the partially aggregated problem

$$\begin{aligned}
&c_1 v - \tilde{c}_B B^{-1} \begin{pmatrix} A_1, H \\ S_1, Q \end{pmatrix} p + c\bar{x} \to \min, \\
&(A_{11} + \Delta A_{11})v - (A_{11}, E, -E)p = 0, \qquad (4.62) \\
&(U + VG)v - (U, V)p = 0, \\
&tp \le T, \quad v \ge 0, \quad p \ge 0,
\end{aligned}$$

obtained by removing linearly dependent equations has the same number of feasible points as problem (4.61). The numbers of equations and variables in (4.62) do not exceed $2m_1 + n_1$ and $2(m_1 + n_1)$, respectively.

The linear programming problem (4.62) is the final form of the partially aggregated problem; the following theorem states that it is quasi-consistent with the original integer linear programming problem.

Theorem 4.6. *If $\bar{x}$ corresponds to a nondegenerate optimal vertex of the feasible solution polyhedron of problem (4.46), then the aggregation of the parametric integer linear programming problem (4.43) into problem (4.62) is quasi-consistent with respect to parameter variations at the point A_{11}.*

Proof. Problems (4.61) and (4.62) are equivalent, i.e., they have the same objective functions and feasible solutions. For this reason, we consider problem (4.61) instead of (4.62).

First, let us show that, at $\Delta A_{11} = 0$, we have

$$f^3(0) = f^1(0),$$

where $f^3(G)$ is the optimal objective value of problem (4.61). The point $v = 0$, $p = 0$ satisfies the conditions of problem (4.61); hence, $f^3(0) \le f^1(0)$. We prove the reverse inequality by contradiction. Suppose that there exists a point v^0, p^0 satisfying the conditions of problem (4.61) and such that $f_0^3(0) < f^1(0)$, where $f_0^3(0)$ is the objective value of (4.61) at v^0, p^0.

Put

$$\alpha_0 = \min_{i \in J^0} \left\{ (\bar{\xi}_B)^i \Big/ \left(B^{-1} \begin{pmatrix} A_1, & H \\ S-1, & Q \end{pmatrix} p^0 \right)^i \right\},$$

where

$$J^0 = \left\{ i \,\middle|\, \left(B^{-1} \begin{pmatrix} A_1, H \\ S_1, Q \end{pmatrix} p^0 \right)^i > 0 \right\},$$

then the point

$$\xi_B = \bar{\xi} - \alpha_0 B^{-1} \begin{pmatrix} A_1, H \\ S_1, Q \end{pmatrix} p^0, \quad \xi_N = 0, \quad v = \alpha_0 v^0$$

is feasible to the problem (4.60). Denote by $f_0^2(0)$ the objective function value of (4.60) at this point. We have

$$f_0^2(0) = \tilde{c}_B \xi_B - \tilde{c}_B \alpha_0 B^{-1} \begin{pmatrix} A_1, H \\ S_1, Q \end{pmatrix} p^0 + \alpha_0 c_1 v^0 =$$

$$= \tilde{f}_2(G) + \alpha_0 f_0^3(0) - \alpha_0 c\bar{x} = \tilde{f}_2(G) - \alpha_0[f^1(0) - f_0^3(0)] > \tilde{f}_2(G).$$

This inequality contradicts the property b) of the cutting planes (6.4.14). Hence, $f^3(0) \geq f^1(0)$.

Now let ΔA_{11} is such that $f^1(\Delta A_{11}) < f^1(0)$. Then by (4.58) we have $\widetilde{f_2}(G) \leq f^1(\Delta A_{11}) < f^1(0)$. By Lemma 4.2 there exist $p^* \geq 0$ and $v^* \geq 0$ such that the point with the components

$$\left(\xi_B^* = \bar{\xi}_B - B^{-1} \begin{pmatrix} A_1, H \\ S_1, Q \end{pmatrix} p^*, \quad \xi_N^* = 0, \quad v^* \right)$$

is feasible to the problem (4.60) and $\tilde{c}\xi^* + c_1 v^* < f^1(0)$. At the same time the point (v^*, p^*) is feasible to the problem (4.61) with the objective function value satisfying the inequality

$$f_*^3(G) = c_1 v^* - \tilde{c}_B B^{-1} \begin{pmatrix} A_1, H \\ S_1, Q \end{pmatrix} p^* + c\bar{x} < f^1(0).$$

This completes the proof of the theorem.

At this point, we end the consideration of partial aggregation of integer linear programming problems in the general case, where the algorithm used to represent cuts can only prove their existence. We know that coefficients $\lambda^{(k)}$ $(k = 1, 2, \dots)$ applied to obtain cuts $Sx \leq q$ with the required properties from the inequalities $Ax \leq b$ do exist, but we are not aware of an efficient algorithm for finding them. In the following section, we show how the matrix Q specifying the coefficients of cuts for elementary variations can be calculated with the use of a primal algorithm for solving integer linear programming problems.

4.4.2. Using the Cuts of the Primal Algorithm. The algorithms for solving integer linear programming problems can be divided into two classes, the algorithms based on constructing cutting planes, which reduce integer problems to continuous, and the algorithms based on various kinds of searches through feasible solutions.

The first cutting plane algorithms were suggested by Gomory; these were a cyclic algorithm and a fully integer algorithm [25]. Their shortcoming was that they would not give a feasible solution until an optimal solution was found. In this sense, these algorithms can be qualified as dual.

The computational burden associated with solving integer linear programming problems is usually very heavy, and the computations are often to be interrupted before the algorithm terminates. Such an interruption of a dual algorithm results in the impossibility of obtaining even an approximate solution to the problem. The primal algorithms provide feasible integer intermediate solutions.

An example of a primal algorithm is Young's algorithm [9].

Suppose that $b \geq 0$. Let us represent the base problem (4.44) in the form

$$-f^1 = c(-x),$$

$$x = -E(-x),$$

$$u = b + A(-x),$$

or

$$\begin{pmatrix} -z^1 \\ x \\ u \end{pmatrix} = D^{(0)} \begin{pmatrix} 1 \\ x \end{pmatrix},$$

where $D^{(0)} = \begin{pmatrix} 0, & c \\ 0, & -E \\ b, & A \end{pmatrix}$.

One step of the primal integer programming algorithm consists of selecting a leading column and a generating row, forming a from the generating row, and transforming the matrix $D^{(0)}$ with the use of this cut as the leading row (or of the generating row itself, if it coincides with the cut).

Suppose that the primal algorithm applied to problem (4.44) has yielded a sequence of matrices

$$\{D^{(k)}\}, \quad k = \overline{0, K}, \tag{4.63}$$

each obtained from the preceding matrix by applying an elementary transformation[7].

Let us augment the matrix $D^{(0)}$ by the columns corresponding to the elementary variations and denote the obtained matrix by $\bar{D}^{(0)}$; thus,

$$\bar{D}^{(0)} = \begin{pmatrix} 0, & c, & 0 \\ 0, & -E, & 0 \\ b, & A, & H \end{pmatrix}.$$

To the matrix $\bar{D}^{(0)}$, we apply the elementary transformations of the primal algorithm with the same leading elements as those used to construct the matrix sequence (4.63). We obtain a sequence of matrices

$$\{\bar{D}^{(k)}\}, \quad k = \overline{0, K}.$$

The matrices $\bar{D}^{(k+1)}$ and $\bar{D}^{(k)}$ are related by an elementary transformation equivalent to the right multiplication by the idempotent matrix

$$\bar{B}_{(k)}^{-1} = \begin{pmatrix} 1 & 0 & \dots & 0 & \dots & 0 \\ 0 & 1 & \dots & 0 & \dots & 0 \\ \dots & \dots & \dots & \dots\dots & & \dots \\ -\left[\dfrac{\bar{d}_{r0}^{(k)}}{\bar{d}_{rs}^{(k)}}\right] & -\left[\dfrac{\bar{d}_{r1}^{(k)}}{\bar{d}_{rs}^{(k)}}\right] & \dots & -1 & \dots & -\left[\dfrac{\bar{d}_{rw}^{(k)}}{\bar{d}_{rs}^{(k)}}\right] \\ \dots & \dots & \dots & \dots\dots & & \dots \\ 0 & 0 & \dots & 0 & \dots & 1 \end{pmatrix},$$

where $\bar{d}_{ij}^{(k)}$ $(i = \overline{0, m+n},\ j = \overline{0, w})$ are the elements of $\bar{D}^{(k)}$, $w = n + 2m_1$, and $\bar{d}_{rs}^{(k)}$ is the leading element of the transformation.

Thus,

$$\bar{D}^{(K)} = \bar{D}^{(0)} \bar{B}_{(0)}^{-1} \bar{B}_{(1)}^{-1} \dots \bar{B}_{(K-1)}^{-1} = \bar{D}^{(0)} \bar{B}^{-1},$$

and the inverse transformation

$$\bar{D}^{(0)} = \bar{D}^{(K)} \bar{B}$$

is determined by the matrix

$$\bar{B} = \bar{B}_{(K-1)}^{-1} \bar{B}_{(K-2)}^{-1} \dots \bar{B}_{(0)}^{-1}.$$

because $\bar{B}_{(k)}^{-1}$ are idempotent.

---

[7]The elementary transformations of the primal integer linear programming algorithm differ from those of the simplex method in that they use the row of the coefficients of the Gomory cut formed from the leading row rather than the leading row itself.

Denoting the vectors formed from the basic and nonbasic components of the vector $\begin{pmatrix} x \\ u \end{pmatrix}$ as ξ_B and ξ_N, respectively, and reordering the columns of $\bar{D}^{(K)}$ and the rows of $\bar{B}$, we obtain the following representation of problem (4.51) at the K-th step:

$$\begin{pmatrix} -f^2 \\ x \\ u \\ s \end{pmatrix} = \begin{pmatrix} \bar{D}^{(K)} \\ 0,\ -E,\ 0,\ 0 \end{pmatrix} \begin{pmatrix} 1 \\ -s \\ -\xi_N \\ -Gv \end{pmatrix},$$

where s are the slack variables corresponding to the Gomory cuts.

Let us return the original variables to the right-hand side:

$$\begin{pmatrix} -f^2 \\ x \\ u \\ s \end{pmatrix} = \begin{pmatrix} \bar{D}^{(K)} \\ 0,\ -E,\ 0,\ 0 \end{pmatrix} \bar{B} \begin{pmatrix} 1 \\ -x \\ -Gv \end{pmatrix} = \begin{pmatrix} \bar{D}^{(0)} \\ q,\ S,\ Q \end{pmatrix} \begin{pmatrix} 1 \\ -x \\ -Gv \end{pmatrix}; \tag{4.64}$$

we obtain the coefficients of the cuts in the extended problem.

To construct a partially aggregated problem of form (4.61), it remains to evaluate the row vector

$$-\tilde{c}_B B^{-1} \begin{pmatrix} A_1,\ H \\ S_1,\ Q \end{pmatrix}$$

of the coefficients in the objective function.

This can be done without the burdensome computation of the matrix B^{-1}. Indeed, the coefficients from

$$-\tilde{c}_B B^{-1} \begin{pmatrix} H \\ Q \end{pmatrix}$$

and those coefficients from

$$-\tilde{c}_B B^{-1} \begin{pmatrix} A_1 \\ S_1 \end{pmatrix},$$

that correspond to nonbasic variables in problem (4.59) are contained in the zero row of the matrix $\bar{D}^{(K)}$, and the remaining coefficient equal the corresponding components of the row $-\tilde{c}_B$.

The equations can be aggregated with the use of any numerical matrix inversion method, for example, the Gauss–Jordan method [29] with a compulsory choice of m_1 solving elements in the first m_1 rows of the matrix

$$\begin{pmatrix} A_1, H \\ S_1, Q \end{pmatrix}.$$

Remark 4.5. We have proved that the partially aggregated problem (4.62) is quasi-consistent with the original problem (4.43) with respect to variations in the elements of the matrix A_{11}. This means that, in processing problem (4.62), no variation ΔA_{11} that may improve the optimal objective value in the original problem (4.43) is missed. However, the efficiency of the use of this partially aggregated problem also depends on the fraction of the variations ΔA_{11} that do improve the optimal objective value in (4.62) but do not in the original problem.

Among the diverse factors affecting the fraction of such variations, we mention two. The first is related to the algorithm used for constructing cuts. Computations based on formulas (4.54) and (4.64) may yield different matrices Q for the same cuts (4.45). Moreover, the matrices Q obtained by (4.54) with different $\lambda^{(k)}$ also may be different.

The second factor is related to the specification of the elementary variation matrix. Under the assumption that the variations ΔA_{11} not only are integer but also can be represented in the form $\Delta A_{11} = HG$, where H is a matrix with elements larger than 1 in absolute value and $G \geq 0$ is an integer matrix, the cuts (4.55) may be "stronger".

We continue Example 4.1. Suppose that subject to variation in problem (4.48) is the coefficient a_{11}. At $a_{11} = 2$, the problem has the optimal solution

$$f^1(0) = -4, \quad \bar{x}^1 = 1, \quad \bar{x}^2 = 1.$$

Applying the variation $\Delta a_{11} = -1$, we obtain the solution

$$f^1(-1) = -9, \quad \bar{\bar{x}}^1 = 3, \quad \bar{\bar{x}}^2 = 0;$$

thus, the optimal value of the objective function decreases.

Let us construct a partially aggregated problem of form (4.62) for (4.48).

In [9], problem (4.48) is solved with the use of the primal integer programming algorithm. Let us apply the elementary transformations of the primal algorithm to the matrix

$$\bar{D}^{(0)} = \begin{pmatrix} 1 & -3 & -1 & 0 & 0 \\ 1 & -1 & 0 & 0 & 0 \\ 0 & 0 & -1 & 0 & 0 \\ 3 & 2 & -3 & 1 & -1 \\ 6 & 2 & 3 & 0 & 0 \end{pmatrix},$$

selecting the leading elements in the same rows and columns; as a result, we obtain

$$\bar{D}^{(4)} = \begin{pmatrix} 4 & 2 & 1 & 0 & -3 \\ 1 & 1 & 0 & 0 & -1 \\ 1 & -1 & 1 & 0 & 0 \\ 4 & -5 & 3 & 1 & 1 \\ 1 & 1 & -3 & 0 & 2 \end{pmatrix}, \quad \bar{B} = \begin{pmatrix} 1 & 0 & 0 & 0 & 0 \\ -1 & -1 & 0 & 0 & 1 \\ -2 & -1 & -1 & 0 & 1 \\ 0 & 0 & 0 & 1 & 0 \\ 0 & 0 & 0 & 0 & 1 \end{pmatrix}.$$

The cuts correspond to the negative second and third rows of $\bar{B}$:

$$q = \begin{pmatrix} 1 \\ 2 \end{pmatrix}, \quad S = \begin{pmatrix} 1 & 0 \\ 1 & 1 \end{pmatrix}, \quad Q = \begin{pmatrix} 0 & -1 \\ 0 & -1 \end{pmatrix};$$

the coefficients in the objective function are

$$-\tilde{c}_B B^{-1} \begin{pmatrix} H \\ Q \end{pmatrix} = \begin{pmatrix} 0 & 3 \end{pmatrix}, \quad -\tilde{c}_B B^{-1} \begin{pmatrix} A_1 \\ S_1 \end{pmatrix} = 3.$$

The last two rows of the matrix

$$\begin{pmatrix} A_1, & H \\ S_1, & Q \end{pmatrix} = \begin{pmatrix} 2 & 1 & -1 \\ 2 & 0 & 0 \\ 1 & 0 & -1 \\ 1 & 0 & -1 \end{pmatrix}$$

coincide, and

$$G = \begin{pmatrix} g_1 \\ g_2 \end{pmatrix} = \begin{pmatrix} \Delta a_{11}^{+} \\ -\Delta a_{11}^{-} \end{pmatrix}; \tag{4.65}$$

therefore, the partially aggregated problem has the form

$$-3v + 3p_1 - 3p_3 - 4 \to \min,$$
$$(2 + \Delta a_{11})v - 2p_1 - p_2 + p_3 = 0,$$
$$2v - 2p_1 = 0,$$
$$(1 - g_2)v - p_1 + p_3 = 0,$$
$$p_1 + p_2 + p_3 \leq 10,$$
$$v \geq 0, \quad p_j \geq 0, \quad j = 1, 2, 3,$$

where g_2 is specified according to (4.65).

Consider the three cases $\Delta a_{11} = 0$, $\Delta a_{11} = 1$, and $\Delta a_{11} = -1$.

(a) $\Delta a_{11} = 0$, $g_1 = g_2 = 0$. Solving the system of equations, we obtain $p_2 = p_3 = 0$; hence,

$$-3(v - p_1) - 4 \to \min,$$
$$v - p_1 = 0,$$
$$v \geq 0, \quad 0 \leq p_1 \leq 10.$$

The optimal value is $f^3(0) = -4$.

(b) $\Delta a_{11} = 1$, $g_1 = 1$, $g_2 = 0$. Solving the system of equations, we obtain $p_3 = 0$, $p_1 = p_2$; hence,

$$-3(v - p_1) - 4 \to \min,$$
$$v - p_1 = 0,$$
$$v \geq 0, \quad 0 \leq p_1 \leq 5.$$

The optimal value is $f^3(1) = -4$.

(c) $\Delta a_{11} = -1$, $g_1 = 0$, $g_2 = 1$. Solving the system of equations, we obtain $p_1 = p_3$, $p_2 = 0$; hence,

$$-3v - 4 \to \min,$$
$$v - p_1 = 0,$$
$$v \geq 0, \quad 0 \leq p_1 \leq 5.$$

The optimal value is $f^3(-1) = -19$.

Thus, in the case under consideration, the analysis of the variations of the coefficient a_{11} based on the partially aggregated problem has allowed us to select the precise variation Δa_{11} that reduces the optimal objective value of the original problem.

Comments and References to Chapter 3

Parametric optimization problems are broadly covered in the book by Gal T. [4] and recent monograph by Shapiro A. and Bonnas J.F. [28]. We would like to stress the difference between the consistent aggregation approach and the parametric analysis based on marginal values. The latter provides local characteristics, i.e., is valid for small changes of parameters, while the aggregation approach aims to analyze the parametric problem globally, allowing large parameters variations. By the consistent aggregation, the parametric analysis of the original large problem is, in certain sense, approximated by the analysis of the smaller parametric aggregated problem. The results presented in this chapter are based on the works by Ven V.L. and Krysov Yu.P. [33-35], Krysov Yu.P. [13-17] (see also Tsurkov V.I., Ven V.L., Krysov Yu.A. and Litvinchev I.S. [31]). Elements of the matrix theory, intensively used in this book, can be found in the survey monographs by Gantmaher F.R. [8], Lankaster K. [19], Markus M. and Mink X.A. [22], and Tewarson R.P. [29]. Optimality conditions and duality for the mathematical programming problems are studied in the books by Eremin I.I. and Astafiev I.N. [5], Karmanov V.G. [12], and Yudin D.B. and Golshtein E.G. [32], while these topics for infinite-dimensional extremal problems are presented in Alexeev V.M., Tikhomirov V.M. and Fomin S.V. [1], Ioffe A.D and Tikhomirov V.M. [11]. Integer programming theory and algorithms, used in Section 4, one can find in the monographs by Hu T.C. [9] and Nemhauser G.L and Wolsey L.A. [25].

References to Chapter 3

[1] Alexeev V.M., Tikhomirov V.M. and Fomin S.V., *Optimal´noe upravlenie* (Optimal Control), Moscow: Nauka, 1979.

[2] Chvatal V., Edmonds Polytopes and a Hierarchy of Combinatorial Problems, *Discrete Mathematics*, 1973, vol. 4, no 4, pp. 305-337.

[3] Demianov V.F. and Malozemov V.N., *Vvedenie v minimax* (Introduction to Minimax), Moscow: Nauka, 1972.

[4] Gal T., Postoptimal Analysis, Parametric Programming and Related Topics, New York: McGraw Hill, 1979.

[5] Eremin I.I. and Astafiev I.N., *Vvedenie v teoriyu lineinogo i vypuklogo programmirovaniya* (Introduction in Linear and Convex Programming Theory), Moscow: Nauka, 1976.

[6] Fiacco A.V., McCormick G.P., Nonlinear Programming: Sequential Unconstrained Minimization Techniques, Chichester, England: Wiley, 1968.

[7] Finkel′shtein Yu.Yu., *Algoritmy resheniya zadachi parametricheskogo programmirovaniya dlia slychaiya, kogda A-matritsa zavisit ot parametra* (Parametric Programming Algorithms for the Case of A-Matrix Depending on a Parameter), *Econ Mat. Metody*, 1977, vol.13, no. 2, pp. 342-347.

[8] Gantmaher F.R., The Theory of Matrices, New York: Chelsea, 1959.

[9] Hu T.C., Integer Programming and Network Flows, New York: Addison-Wesley, 1969.

[10] Iliutovich A.E. and Levitin E.S., *Metody decompositsii zadach optimal′nogo upravleniya so smeshannymi reguliarnymi ogranicheniymi i so svobodnym kontsom traektorii* (Decomposition Methods for Optimal Control Problems with Mixed Regular Constraints and without Terminal Restrictions), Preprint VNIISI, Moscow, 1987.

[11] Ioffe A.D. and Tikhomirov V.M., *Teoriya extremal ′nykh zadatch* (Theory of Extremal Problems), Moscow: Nauka, 1974.

[12] Karmanov V.G., *Matematicheskoe programmirovaniye* (Mathematical Programming), Moscow: Nauka, 1980.

[13] Krysov Yu.A., *O sovmestnom agregirovanii v nekotoryh classakh parametricheskikh optimizatsionnykh zadach v banakhovom prostranstve* (On Consistent Aggregation in Some Classes of Parametric Optimizations Problems in Banach Space), *Avtom. Telemekh.*, 1987, no.12, pp. 147-157.

[14] Krysov Yu.A., *O sovmestnosti i ystoichivosti agregirovaniya v parametricheskikh zadachakh matematicheskogo programmirovaniya* (On Consistency and Stability of Aggregation in Parametric Mathematical Programming Problems), *Zh. Vych. Mat. Mat. Fiz.*, 1988, vol. 28, no. 12, pp. 1789-1798.

[15] Krysov, Yu. A., *Approksimatsiya optimal'nogo znacheniya v parametricheskikh optimizatsionnykh protsedurakh* (Approximation of the Optimum Function in Parametric Optimization Procedures), *Avtom. Telemekh.*, 1992, no. 12, pp. 106–114.

[16] Krysov, Yu. A., *Prinyatie i coordinatsiya reshenii v nekotorykh protsedurakh parametricheskogo analiza optimizatsionnykh zadach* (Making and Coordinating Decisions in Certain Procedures of Parametric Analysis in Optimization Problems), *Izv. Akad. Nauk SSSR. Tekhn. Kibernet.,* 1990, no. 6, pp. 43–57.

[17] Krysov, Yu. A. *Optimizatsiya parametricheskikh sistem na osnove agregirovaniya upravlenii* (Optimization of Parametric Systems Based on Aggregation of Controls), *Izv. Akad. Nauk SSSR. Tekhn. Kibernet.,* 1990, no. 3, pp. 50–56.

[18] Kyparisis J., On Uniqueness of Kuhn-Tucker Multipliers in Nonlinear Programming, *Math. Progr.*, 1985, vol. 32, pp. 242-246.

[19] Lankaster K., *Matematicheskaia economica* (Mathematical Economics), Moscow: Sovietskoe Radio, 1972.

[20] Lasdon L.S., Optimization Theory for Large Systems, London: McMillan, 1970.

[21] Leontiev V.K., *Ustoichivost' v kombinatornykh zadachakh vybora* (Stability in Combinatorial Problems of Choice), *Dokl. Acad. Nauk SSSR*, 1976, vol. 228, no1, pp. 23-25.

[22] Markus M. and Mink X., A Survey of Matrix Theory and Matrix Inequalities, Prindle: Weber and Schmidt, 1964.

[23] Mordukhovich B.S., *Metody approximatsii v zadachakh optimizatsii i upravleniya* (Approximation Methods in Problems of Optimization and Control), Moscow: Nauka, 1988.

[24] Murtagh B.A., Advanced Linear Programming, New York: McGraw-Hill, 1981.

[25] Nemhauser G.L. and Wolsey L.A., Integer and Combinatorial Optimization, New York: Wiley; 1988.

[26] Papadimitriou C.H. and Stieglitz K., Combinatorial Optimization: Algorithms and Complexity, Englewood Cliffs: Prentice-Hall, 1982.

[27] Pshenichnyi B.N., *Metod linearizatsii* (Linearization Method), Moscow: Nauka, 1983.

[28] Shapiro A. and Bonnans .J.F., Perturbation Analysis of Optimization Problems, New York: Springer, 1999.

[29] Tewarson R.P., Sparce Matrices, New York: Academic Press, 1973.

[30] Tsurkov V.I., *Dinamitcheskie zadachi bol´shoi razmernosti* (Large-Scale Dynamical Problems), Moscow: Nauka, 1988.

[31] Tsurkov V.I., Ven V.L., Krysov Yu.A. and Litvinchev I.S., *Metody decompozitsii i agregirovaniya v mnogomernykh zadachakh optimizatsii* (Decomposition and Aggregation Methods in Multidimensional Optimization Problems), in *Probl. Prikl. Mat. Informat. Part 2,* 1992, pp. 65-88, Moscow: Comp. Ctr. Russ. Acad. Scis.

[32] Yudin D.B. and Gol´shtein E.G., *Lineinoe programmirovanie. Teoriya, metody i prilozjeniya* (Linear Programming: Theory, Methods and Applications), Moscow: Nauka, 1969.

[33] Ven V.L., *Chastichnoe agregirovanie v zadachakh tselochislennogo lineinogo programmirovaniya s var'iryemymi parametrami* (Partial Aggregation in Parametric Integer Linear Programming), *Econ. Mat. Metody,* 1985, vol. 21, no. 3, pp. 501-510.

[34] Ven V.L. and Krysov Yu.P., *Chastichnoe agregirovanie v zadachakh lineinogo programmirovaniya s var´iryemymi parametrami* (Partial Aggregation in Parametric Linear Programming Problems), *Izv. Akad. Nauk SSSR. Tekhn. Kibernet.,* 1983, no. 3, pp. 3-10.

[35] Ven, V. L. and Krysov, Yu. A., *Agregirovanie i approximatsiya v zadachakh parametricheskogo programmirovaniya* (Aggregation and Approximation in Problems of Parametric Programming), *Dokl. Akad. Nauk SSSR,* 1990, vol. 311, no. 6, pp. 1289–1292.

[36] Zangwill W.I., Nonlinear Programming, Englewood Cliffs: Prentice Hall, 1969.

Index

MIX
Papier aus verantwortungsvollen Quellen
Paper from responsible sources
FSC® C105338

If you have any concerns about our products,
you can contact us on
ProductSafety@springernature.com

In case Publisher is established outside the EU,
the EU authorized representative is:
Springer Nature Customer Service Center GmbH
Europaplatz 3, 69115 Heidelberg, Germany

Printed by Libri Plureos GmbH
in Hamburg, Germany